D0148252

THE APARTHEID CITY AND BEYOND

Apartheid as legislated racial separation substantially changed the South African urban scene. Race 'group areas' remodelled the cities, while the creation of 'homelands', mini-states and the 'pass laws' controlling population migration constrained urbanization itself.

In the mid-1980s the old system – having proved economically inefficient and politically divisive – was replaced by a new policy of 'orderly urbanization'. This sought to accelerate industrialization and cultural change by relaxing the constraints on urbanization imposed by state planning. The result: further political instability and a quarter of the black (or African) population housed in shanty towns.

Negotiations between the nationalist government and the African National Congress are working towards the end of the old apartheid system. Yet the negation of apartheid is only the beginning of the creation of a new society.

The vested interests and entrenched ideologies behind the existing pattern of property ownership survive the abolition of apartheid laws. Beyond race, class and ethnicity will continue to divide urban life. If the cities of South Africa are to serve all the people, the accelerating process of urbanization must be brought under control and harnessed to a new purpose.

The contributors to this volume draw on a broad range of experience and disciplines to present a variety of perspectives on urban South Africa.

GRACE LIBRARY CARLOW COLLEGE
PITTSBURGH PA 15213

THE APARTHEID CITY AND BEYOND

Urbanization and Social Change in South Africa

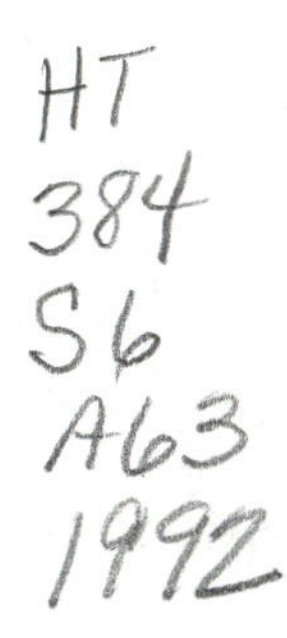

edited by

David M. Smith

Professor of Geography, Queen Mary and Westfield College, University of London

London and New York *Witwatersrand University Press*

First published 1992
by Routledge
11 New Fetter Lane, London EC4P 4EE

Simultaneously published in the USA and Canada
by Routledge
a division of Routledge, Chapman and Hall, Inc.
29 West 35th Street, New York, NY 10001

Published in the Republic of South Africa
by Witwatersrand University Press
1 Jan Smuts Avenue, Johannesburg 2001, South Africa

Copyright © David M. Smith

Typeset in 10 on 12 point Bembo by
Falcon Typographic Art Ltd, Edinburgh
Printed and bound in Great Britain by
Biddles Ltd, Guildford and King's Lynn

All rights reserved. No part of this book may be reprinted or reproduced or utilized in any form or by any electronic, mechanical, or other means, now known or hereafter invented, including photocopying and recording, or in any information storage or retrieval system, without permission in writing from the publishers.

British Library Cataloguing in Publication Data
A catalogue record for this book is
available from the British Library.

Library of Congress Cataloging in Publication Data
The Apartheid city and beyond: urbanization and social change in
South Africa / edited by David M. Smith.
p. cm.
Includes bibliographical references and index.
1. Urbanization – South Africa. 2. Urban policy – South Africa.
3. Apartheid – South Africa. 4. South Africa – Social
condition – 1961– I. Smith, David Marshall, 1936–
HT384.S6A63 1992
307.76′0968 – dc 20 91–39303 CIP

ISBN 0–415–07601–3 ISBN 0–415–07602–1 (pbk)

South African ISBN: 1–86814–207–8 (pbk)

Contents

List of figures

List of tables

List of Contributors

K. S. O. Beavon Professor of Human Geography and Head of the Department of Geography and Environmental Studies, University of the Witwatersrand.

Bruce Boaden Professor of Construction Management, Department of Construction Economics and Management, University of Cape Town.

Patrick Bond doctoral candidate, Johns Hopkins University; journalist, and consultant to the Johannesburg service organization PLANACT.

John Butler-Adam Professor and Director, Institute for Social and Economic Research, University of Durban–Westville.

K. S. Chetty Department of Community Health, University of Cape Town; Vice President for International Affairs, National Medical and Dental Association.

Gillian P. Cook visiting fellow at Southampton University; previously lectured at the University of Cape Town.

Peter Corbett lecturer in economics, University of Natal, Durban; member of Durban City Council.

Owen Crankshaw senior researcher, National Institute for Personnel Research, Human Sciences Research Council.

R. J. Davies Professor of Geography, Department of Environmental and Geographical Sciences, University of Cape Town.

David Dewar Director, School of Architecture and Planning, and Urban Problems Research Unit, University of Cape Town; principal in planning consulting firms.

Linda J. Grant associated with Department of Geography, University of Natal, Pietermaritzburg; MA in geography, University of Natal, Durban.

Tim Hart a Director, National Institute for Personnel Research, Human Sciences Research Council; honorary lecturer, Department of Geography and Environmental Studies, University of the Witwatersrand.

Phillip S. Hattingh Professor and Head of Department of Geography, University of Pretoria.

Gavin Heron assistant researcher, National Institute for Personnel Research, Human Sciences Research Council.

André C. Horn lecturer in geography, University of Pretoria.

Meshack M. Khosa completing doctorate at School of Geography, Oxford University; MA in geography, University of the Witwatersrand.

Malcolm Lupton post graduate research in Department of Geography and Environmental Studies, University of the Witwatersrand.

Brij Maharaj senior lecturer in geography, University of Zululand, Umlazi.
Jeff McCarthy Professor and Head of Department of Geography, University of Natal, Pietermaritzburg; previously Senior Manager, Urban Policy at the Urban Foundation.
Susan Parnell lecturer in geography, University of the Witwatersrand.
Udeshtra Pillay doctoral candidate in geography, University of Minnesota; MA in geography, University of Natal, Durban.
G. H. Pirie lecturer in geography, University of the Witwatersrand.
Claudia Reintges lecturer in geography, Rhodes University.
Jennifer Robinson lecturer in geography, University of Natal, Durban.
C. M. Rogerson Associate Professor and Reader in Human Geography, University of the Witwatersrand.
Dianne Scott lecturer in geography, University of Natal, Durban.
David M. Smith Professor of Geography, Queen Mary and Westfield College, University of London.
Dhiru V. Soni lecturer in geography, University of Durban–Westville.
Jonathan Steinberg student of sociology and African literature, University of the Witwatersrand.
Rob Taylor General Manager, Informal Settlements Division of the Urban Foundation in Natal.
Keyan Tomaselli Professor and Director, Centre for Cultural and Media Studies, University of Natal, Durban.
Ruth Tomaselli part-time lecturer, Centre for Media and Cultural Studies, University of Natal, Durban; MA in development studies, University of the Witwatersrand.
Jan L. Vermaak town and regional planner, Department of Public Works and Land Affairs.
Paul van Zyl student of sociology and law, University of the Witwatersrand.

Preface

This book is designed to replace an earlier volume entitled *Living under Apartheid: Aspects of Urbanization and Social Change in South Africa*, published by George Allen & Unwin in 1982. The intervening years have seen significant and at times dramatic change in South Africa, culminating in the release of Nelson Mandela in 1990 and the subsequent engagement of the National Party government and the African National Congress in negotiation towards a new constitution. The erosion of aspects of racial discrimination which began in the early 1980s has continued, and attention is now focused on the formation of a democratic and non-racial 'post-apartheid' society.

But despite these developments, much remains to be done. In particular, the accelerating process of urbanization has to be brought under control, and harnessed to a new social purpose. South African life will continue to become increasingly urban, offering scope for material advancement for some but perhaps merely the relocation of dire poverty for others. Turning cities which were substantially remodelled under half a century of apartheid into places of real opportunity for the mass of the people, rather than of privilege for a minority, is one of the major challenges facing the new South Africa.

This volume provides a series of original contributions on a variety of topics related to urbanization and social change in South Africa. Most are concerned with the impact of apartheid, as it effected housing, community life, settlement forms, and the servicing of the cities. Far from merely dwelling on a best-forgotten past, these studies help to explain how the cities of South Africa came to be as they are: the locus of people's present lives and a major constraint on new urban forms. But we also try to look ahead, to a post-apartheid city, with hope tempered by understanding that the struggle for more egalitarian cities has only just begun.

The authors whose research is assembled here come from a variety of backgrounds, generations and institutions (and indeed racial classifications, as along as apartheid's odious race-group fetishism survives). Although many practise geography in South African universities, some affiliate with other disciplines or professions requiring or sensitive to a spatial perspective. They demonstrate a diversity of theoretical orientation and methodological practice, which gives the collection a healthy eclecticism. They also reflect different prescriptions for the post-apartheid order. What brings the authors together here (in addition to the intrinsic quality of their work) is the conviction, shared with the editor, that careful analysis of the past and present, brought to bear on the problems of the future, is a necessary (if not sufficient) condition for the creation of cities to serve all the people.

Grateful acknowledgements are due to Roger Jones at what was then Unwin Hyman for initial encouragement of this project, to Susan West for seeing it through and to the authors for their contributions and cooperative responses to editorial reaction. The compilation and editing of the collection was greatly assisted by a visit to South Africa in 1989, supported by the Students' Visiting Lecturers Trust Fund of the University of Natal and by the Hayter Fund of the University of London. A further visit in 1990, supported by the Human Sciences Research Council through the good offices of the University of Cape Town, facilitated discussions with most of the contributors at the draft stage. The opportunity provided by various individuals and institutions to combine work on this book with other research, wider academic interaction and the occasional relaxation was very much appreciated. I am also grateful for the encouragement of the London Office of the African National Congress, with whom consultation took place before visiting South Africa.

Special thanks are due to Keith and Pat Beavon, Ron and Shirley Davies and Denis and Betty Fair, for their limitless hospitality. To drop, almost literally, out of the sky to a welcoming 'brai' or bed has made visiting South Africa so much more than an academic experience. And to have had such distinguished South African geographers as friends as well as professional colleagues, over so many years, has been a great privilege, and also a frequent reminder of how much some of us still have to learn. Admiration as well as acknowledgement also goes to those who my own advancing years tempt me to term the younger generation, prominently represented in the pages of this volume. Their commitment to a new society is building bridges between those hitherto separated by the idiocy of academic apartheid, helping to bury at least that part of the past in the process of forging a new and truly progressive South African geography and urban studies.

Finally, to the one who went while I was away, and to those who miss him.

DAVID M. SMITH
Loughton, Essex
February 1991

Introduction

DAVID M. SMITH

> These restless broken streets where definitions fail – the houses the outhouses of white suburbs, two-windows-one-door, multiplied in institutional rows; the hovels with tin lean-tos sheltering huger old American cars blowzy with gadgets; the fancy suburban burglar bars on mean windows of tiny cabins; the roaming children, wolverine dogs, hobbled donkeys, fat naked babies, vagabond chickens and drunks weaving, old men staring, authoritative women shouting, boys in rags, tarts in finery, the smell of offal cooking, the neat patches of mealies between shebeen yards stinking of beer and urine, the litter of twice-discarded possessions, first thrown out by the white man and then picked over by the black – is this conglomerate urban or rural? No electricity in the houses, a telephone an almost impossible luxury: is this a suburb or a strange kind of junk yard? The enormous backyard of the whole white city, where categories and functions lose their ordination and logic . . .
> . . . a 'place'; a position whose contradictions those who impose them don't see, and from which will come a resolution they haven't provided for.
>
> Nadine Gordimer, *Burger's Daughter* (1979)

This commentary on a black 'township' seems no less apposite as an evocation of settlement in South Africa today than it did a decade ago. It captures something of both the life and the landscape of apartheid, reflecting the prevailing confusion as to the very nature of the urban condition in this strange society. It also hints at the central significance of urbanization under apartheid: that those places imposed by the white government on the black majority have taken on a life of their own, rebounding on the system to its discomfort and ultimate demise. Very simply, urbanization under apartheid, no matter how carefully the state contrived to control it, has undermined apartheid itself, bringing South African society and its cities to the brink of significant if still uncertain change.

Since the doctrine of apartheid as legislated racial separation was introduced following the National Party's assumption of power in 1948, it has been inextricably bound up with urbanization. At the national scale, the creation of so-called 'homelands' or 'bantustans' for the African majority of the population, was largely an attempt to constrain urbanization, with the notorious 'pass laws' controlling entry to the cities to levels consistent with demands for labour. Once in the cities, Africans were expected to be no more than temporary sojourners, there only as long as required by the white economy. Ideally they would be single migrant workers, but those who qualified for permanent and perhaps family settlement were still expected to look to the homelands for their 'political rights'. And residential segregation was obligatory, along with conformity to the day-to-day indignity of 'petty apartheid' under legislation enabling the provision of separate amenities for different race groups.

That rigid constraints on urbanization were inconsistent with economic efficiency as well as with personal liberty soon became clear. A sophisticated workforce, of the kind required by the manufacturing and service industries steadily displacing mining from its earlier pre-eminence, could not be expected to emerge from a disenfranchised and insecure proletariat who were supposed to identify with an often unfamiliar mini-state many miles away. A large and evidently permanently settled urban African population existed well before the Nationalists were finally forced to come to terms with reality, in a new policy of 'orderly urbanization' set out in the 1986 White Paper on Urbanization. Now, the urbanization of the African population was to be turned from problem to solution, with the processes of industrialization and cultural change expected to transform a discontented and threatening people into more compliant members of a mass-consumption society.

The outcome has hardly been orderly, however. Coinciding with the privatization impulse elsewhere (in Britain under the Thatcher government, for example), the state largely abandoned its earlier role as direct provider of housing for urban Africans, manifest in construction of the familiar townships. Very simply, it sought the benefits of accelerated urbanization but without bearing all the enormous costs. A consequence has been the spread of spontaneous or 'shack' settlements around the major metropolitan areas, to the extent of accommodating an estimated 7 million or a quarter of the African population today. And strict segregation broke down in parts of some cities, as black people seeking the advantages of inner-city residence were able to evade the restrictions of racial 'group areas' legislation. Thus the archetype apartheid city as elaborated by Davies (1981; see also Lemon 1987: 220–21; Fig. 19.1 in this volume), with its racially exclusive as well as class-divided wedges of formal urban development, has undergone substantial change during the past decade (Simon 1990; Lemon 1991).

This book addresses various aspects of the creation and transformation of the apartheid city, in the general context of the urbanization process in a changing society. The purpose of this Introduction is briefly to set the scene, providing such basic information as is required for readers largely unfamiliar with urban South Africa to engage the chapters that follow.

Up-to-date background on apartheid more generally can be found in Smith (1990).

The people of South Africa

Not even basic population data for South Africa can be provided without qualification and explanation of its peculiarities. The latest (1989) mid-year estimate of the total population of the Republic of South Africa (RSA) is 30.2 million. However, this excludes about 6.5 million people resident in four of the homelands which have the official status of 'independent republics' (see Fig. 1), and are therefore no longer considered the responsibility of the South African government. The remaining six 'self-governing territories' are included as part of the Republic, however, although they are not directly administered from Pretoria. How long this distinction, and indeed the

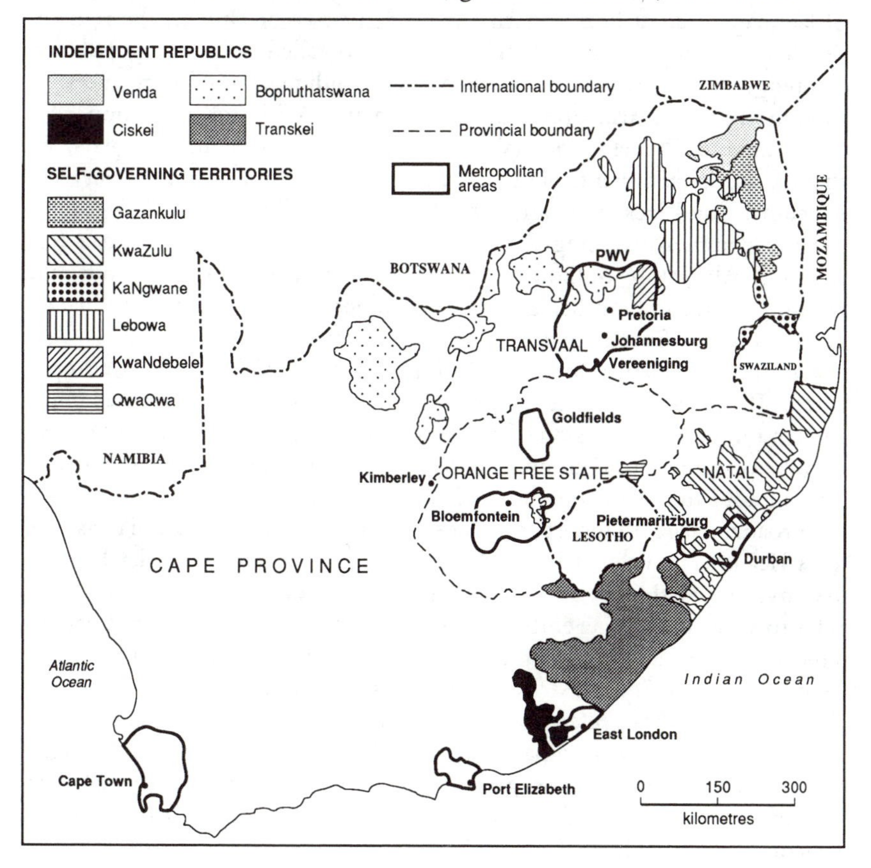

Figure 1 South Africa's homelands and metropolitan areas.

homelands themselves, will survive the current process of political change remains to be seen; reincorporation within the RSA of those territories currently considered independent seems increasingly likely.

The next, crucial complication is the racial classification of the population, itself central to the implementation of apartheid. The African or negroid population of the RSA, officially termed 'Black', was estimated to be 21.1 million in mid-1989, almost exactly three-quarters of the total, and to this can be added virtually all the population of the independent homelands. People classified as 'white' numbered almost 5.0 million, or 16.5 per cent (falling to 13.6 per cent if the independent homelands are included). The population classified as 'coloured' is 3.2 million (10.5 per cent of the RSA total). The final official category is the 'Asians', often referred to as Indians, comprising 940,000 or 3.1 per cent of the Republic's total.

Each of the four official population groups is subject to internal differentiation. About 55 per cent of the whites are Afrikaners by the criterion of language used at home (Afrikaans, a derivative of Dutch), the remainder being largely of British decent. The so-called coloured population is predominantly of mixed European and indigenous African ancestry, though there are also about 200,000 Malays. With the exception of a few thousand Chinese, the Asians are descended from immigrants from the Indian subcontinent; about 70 per cent are Hindus and 20 per cent Muslims. The Black population is officially divided into ten tribal groups, the most numerous being the Zulu (about 6.4 million) and Xhosa (6.2 million). Almost 16 million of the Blacks live in the ten homelands, the remaining 11.5 million being in what is sometimes described, with blatant error, as 'white South Africa'.

Racial nomenclature in South Africa is controversial as well as complicated. 'Black' is sometimes used to refer to all those not classified as white (which can be an expression of political solidarity), as well as to the African population previously known officially as Bantu and earlier as Natives. Throughout this book the term 'black' with lower-case 'b' is adopted for the broader usage, to incorporate those classified coloured and Asian as well as Africans. The term 'Black' capitalised is confined to the official usage, synonymous with African. All such terms, including 'white', are taken to refer to social constructs embedded in the apartheid system, and not natural subdivisions of humankind in South Africa or anywhere else. Similarly, to describe part of South Africa or its cities as 'white' refers to an official designation, without reference to the legitimacy or even factual accuracy of such racially exclusive occupancy. And, to complete this necessary apologia, such terms as 'homeland' or 'group area' are used throughout this book as part of what is to be understood, and in no way to dignify the contorted lexicon of apartheid.

Enumeration of the urban population of South Africa depends crucially on the definition of urban. Official census figures for 1985 class 89.6 per cent of the white population of the RSA as urban (about 4.1 million), 77.8 per cent of the coloureds (2.2 million), 93.4 per cent of the Asians (767,000) and 39.6 per cent of the Blacks or Africans (6.0 million), to give a total urban population approaching 13.7 million. At about 58 per cent of the Republic's total population, this is roughly comparable with the proportion living in urban areas in such countries as Hungary, Poland and Tunisia. However, these findings depend crucially on the validity of the figure for African urbanization. Some authorities consider that this could actually be as high as 60 per cent, allowing for under-counting and defining as urban the growing shack accretions around South Africa's formal cities and similar so-called 'closer settlements' in the homelands. In any event, the present level of urbanization in South Africa is well below what is typical for advanced industrial nations, such as those of western Europe, a fact that can be attributed almost entirely to the relatively low figure for Africans irrespective of how 'urban' is defined.

The cities

In the conventional terms of urban geography, South Africa's cities conform to a fairly simple hierarchy. At the top is the Pretoria–Witwatersrand–Vereeniging metropolitan region (usually abbreviated to PWV), covering the country's economic heartland in the southern Transvaal (see Fig. 1). Within this loose but increasingly integrated conurbation, the Witwatersrand centred on Johannesburg accounted for somewhat over 4.0 million people in 1985, greater Pretoria for about 2.5 million, and the southern section comprising Vereeniging and its neighbours over 600,000 (most of the population figures here are derived from the Urban Foundation, 1990). In the PWV as a whole, there were some 635,000 informal dwellings accommodating over 2.5 million people in 1989 – rather more than in the formal Black townships of which Soweto adjoining the city of Johannesburg is by far the largest.

The next level in the hierarchy is occupied by the Durban metropolis with about 2.6 million people in 1985, and by metropolitan Cape Town with 2.25 million. The inclusion of rapidly growing shack settlements across the border in the KwaZulu homeland would make Durban's figure much higher, however; recent estimates put the shack population in the wider metropolitan region as 1.7 to 2.0 million, perhaps half the total number living there. There are also areas of rapid growth on the eastern edge of the Cape Town metropolis, the large informal component making precise population counts almost as hard as for Durban.

The third level of cities comprises Port Elizabeth (816,000 in 1985), Bloemfontein (525,000), Pietermaritzburg (425,000), the Orange Free State Goldfields (395,000) and East London (380,000). Port Elizabeth in particular has experienced spectacular recent growth, much of it informal, resembling Durban on a smaller scale. Also in this size category is the informal settlement of Botshabelo in a detached bit of the Bophuthatswana homeland to the east of Bloemfontein; some estimates put the population of Botshabelo at close to 0.5 million, not much short of that of Bloemfontein itself. The urban hierarchy is completed by a fourth rung of smaller cities and towns.

The administrative structure of South Africa's cities reveals a complexity befitting the divided society of apartheid. White areas function as local government authorities similar to those in Britain and the United States, for example. Coloured and Asian areas have some degree of local autonomy, with their own elected representatives, though turnout at elections is so poor as to give this form of local government little popular support. Black local authorities with some responsibility for the townships have less autonomy, and even less legitimacy. The informal settlements combine some elements of formal government with other means of administration ranging from democratic local control to brutal coercion.

Urban local government represents one of three tiers of administration in South Africa, those above being provincial (the Cape, Natal, Transvaal and Orange Free State) and national. Since the introduction of a new constitution in 1984, with the whites-only parliament replaced by three separate houses (the House of Assembly for whites, House of Representatives for coloureds and House of Delegates for Asians), there has been an explicit distinction in government between what are regarded as the 'own affairs' of one race group and the 'general affairs' held to affect all groups. Local authorities deal essentially with the 'own affairs' of the race group in question, but Regional Services Councils set up in the latter part of the 1980s are able to exercise some functions over entire metropolitan areas. The conspicuous omission in the so-called 'tricameral' national parliament is, of course, the Black/African population, resolution of which is central to negotiations over the country's political future. The virtual coincidence of race, residential area and local government within the cities is a major obstacle to non-racial representation and administration more generally.

The future

While it may be tempting to explain the demise of apartheid in terms of such forces as internal struggle and external sanctions, the contradictions built into the system itself must bear major responsibility. The attempt to take advantage of (cheap) Black labour without conceding the franchise and other commonly accepted rights of citizenship foundered on the fact that labour, unlike other commodities, has a human embodiment that cannot for long be denied. Exploitation was transparent, and moral indignation if not revolt inevitable. But more practically, the vision of a white heartland into which African workers were drawn from their peripheral reserves on a purely temporary basis was patently unsustainable. By the time the government was ready to concede and in part plan for a large and permanent African presence in the cities, it had been overtaken by events. An economy never seriously tempted by incentives to decentralize to the homelands or 'border areas' needed a local and settled workforce, and this coincided with the wishes of the workers themselves. A largely spontaneous reorganization of South Africa's spatial structure centred on the major metropolitan nodes was already under way well before the government proclaimed its policy of 'orderly urbanization', although there had been strong hints of a change in official thinking on regional development strategy at the beginning of the 1980s (for discussion, see Tomlinson and Addleson 1987; Tomlinson 1990). The emerging 'deconcentrated' urban regional structure in effect transfers part of the old rural labour reserve to the growing peri-urban shack settlements on the outer metropolitan periphery.

But it is not only a changing spatial form with which the authorities have to come to terms. The pace of urbanization along with the anticipated size of the cities of the future is generating a crisis, in the sense of serious doubts as to the capacity of the state to manage the process, even without the apartheid distractions of keeping Black people in their place. At the heart of the problem is the 'time bomb' of Black population growth, fuelled by a widening gap between continuing high birth rates and falling death rates. This situation is characteristic of societies during the period of 'demographic transition', which is supposed to end in falling birth rates leading to the roughly zero population growth experienced by advanced industrial societies. Current estimates suggest that it could take South Africa 30 years to reach this population balance.

Herein lies possibly the most serious problem facing the new South Africa. The more rapid the process of economic development and associated socio-cultural change usually referred to as 'modernization', including voluntary reduction in family size, the shorter the demographic transition and the sooner population will be held to levels which the economy can sustain with some semblance of decent living standards. But this very process entails accelerated urbanization, initially at the currently high levels of natural population growth, and under conditions of primitive accommodation and even worse services for most of the people involved.

Research undertaken by the Urban Foundation (1990) provides the most authoritative numerical predictions. The population of South Africa as originally constituted (including the independent homelands), enumerated at 33.1 million in 1985, is expected to reach 47.6 million by the year

2000 and 59.7 million by 2010 – not much short of twice what it is today. Whereas the white population will have risen from 4.9 to 5.8 million between 1985 and 2010, the coloureds from 3.0 to 4.2 million and the Asians from 0.9 to 1.2 million, the number of Blacks/Africans will have virtually doubled – from 24.5 to 48.5 million. South Africa's population is thus destined to become increasingly Black, and youthful, and insofar as its growth continues to outstrip that of the economy, it will be increasingly poor.

Classical apartheid at the national scale sought to externalize African population growth, surplus to what the economy could absorb, by confining it to the peripheral homeland reserves. This arrangement has now collapsed, with respect to the power of the state to enforce it as well as to the capacity of the homelands to support continuing population increase. Poor people are making their own cities, not necessarily in conditions of their own choosing but increasingly defying the ability of the state to mould them to its own order. The inevitable consequence is a massive shift of (African) population from rural to metropolitan areas.

Following the Urban Foundation (1990) again, it is predicted that the urban African population of 13 million (53 per cent of their total) in 1985 will have risen to over 33 million (69 per cent) by the year 2010. The metropolitan areas, including their extensions across the increasingly irrelevant homeland borders, contained 8.7 million Africans in 1985; the figure for 2010 is expected to be 23.6 million. Over the same quarter of a century the African population of other urban areas in South Africa, excluding the homelands, is predicted to rise from 1.6 to 3.3 million, with other homeland urban areas registering an increase from 2.7 to 6.3 million.

The impact on individual metropolitan areas will be spectacular. The PWV will have a population of 12.3 million by the turn of the century and 16.5 million by 2010 – similar to New York and São Paulo today (Urban Foundation 1990). Durban will have some 4.4 million people by the year 2000, and 6 million by 2010. Cape Town will grow to 3.3 million by 2000 and 4 million by 2010. In the next tier of the urban hierarchy, Port Elizabeth is predicted to reach 1.9 million by 2010, and the other cities should double in size. In all cases, Africans will substantially increase their share of the total metropolitan or city population. Most of them will be poor, many incapable of affording anything at all for shelter other than what they construct for themselves.

Against these predictions, the legacy of urban apartheid, in the form of residential segregation, buffer zones between races, peripheralization of the black population and long dislocation between residence and workplace, may appear largely irrelevant. But the kind of cities constructed or reconstructed under apartheid themselves constrain the capacity of any government, present or future, to respond to the rising tide of urbanization and especially to the needs of the poor. Indeed, the very practice of urban planning and state housing provision in South Africa is itself so tainted by apartheid as to require considerable effort to regain popular confidence among those whose actual experience is of community destruction and forced relocation. In the mean time, people will find their own solutions to the daunting challenges of day-to-day living, seeking this generation's survival while perhaps unconsciously compounding the problems facing those to come.

The book and its content

The content of this book has been chosen to exemplify and illuminate various issues concerning the past, present and future of the South African city. To be comprehensive is impossible: the intention is to combine a reasonable breadth of treatment with the inclusion of topics of particular interest. The structure and ordering of content requires only brief explanation here. The book is divided into five parts, or groups of chapters on related issues. Part One comprises an historical overview of urbanization in South Africa, and a review of the changing context of urban and regional government. Part Two brings together pieces on various aspects of housing and community, as imposed, struggled over and reconstructed under apartheid. Part Three provides case studies of informal settlement. Part Four looks at the servicing of cities, including the informal economy, travel, tourism and health care. Part Five turns to the post-apartheid city, considering some of its challenges (or problems) and precedents from which the prospect for change might be judged. There is a brief editorial conclusion.

While each chapter can be read on its own as a contribution to understanding the South African city, together and in sequence they should help to underline the continuity and contradictions built into urban life, and above all its spatiality. Apartheid required the imposition of its own spatial order on human settlement, the pre-existing structure of which could be modified, with considerable expense and human suffering, but never made entirely subservient to the purpose of the state. New spatial forms became the locus of struggle, as black people sought control over their immediate environment even if denied broader political participation. The informal process of settlement, with people explicitly denying their allotted place, contributed significantly to the erosion of the apartheid urban order. Now the apartheid city, and the challenges to it created within the constraints of apartheid society, impose their own problems and limitations on the formation of a post-apartheid city. Apartheid may have been overcome, but the struggle for liberation from its legacy has barely started.

References

Davies, R. J. 1981. The spatial formation of the South African city. *GeoJournal*, supplementary issue 2: 59–72.
Lemon, A. 1987. *Apartheid in Transition*. Aldershot: Gower.
Lemon, A. (ed.) 1991. *Homes Apart: South Africa's Segregated Cities*. London: Paul Chapman.
Simon, D. 1990. Crisis and change in South Africa: implications for the apartheid city. *Transactions of the Institute of British Geographers* 14: 198–206.
Smith, D.M. 1990. *Apartheid in South Africa. UpDate* series. Cambridge: Cambridge University Press.
Tomlinson, R. 1990. *Urbanization in Post-Apartheid South Africa*. London: Unwin Hyman.
Tomlinson, R. and Addleson, M. (eds) 1987. *Regional Restructuring under Apartheid: Urban and Regional Policies in Contemporary South Africa*. Johannesburg: Ravan.
Urban Foundation 1990. *Policies for a New Urban Future: 1. Population Trends*. Johannesburg: The Urban Foundation.

PART ONE

Background

The first two chapters provide broad overviews, to act as background for those in subsequent parts of the book.

In the first essay (Chapter 1), Mabin sets urbanization in South Africa in its historical context. He links population movement to the towns and cities to a long-standing practice on the part of African people to seek employment away from their traditional tribal reserves, often oscillating between rural and urban residence. Added to this have been various state and private strategies of dispossession, which have forced Africans off their land. Individual households have often deliberately separated, to maintain both a rural and an urban base for securing the means of subsistence, inextricably binding the one to the other. As new forms of urbanization arise from the impossibility of maintaining a rigid separation between town or city and country, core and periphery, most obviously in the expansion of peri-urban informal settlements, the meaning of 'urban' in South Africa continues to challenge conventional interpretations.

McCarthy (Chapter 2) addresses the issue of urban and regional government, central to the control of urbanization. He argues that these levels of administration tend to have been neglected in the literature, in favour of an interpretation of apartheid which stresses the role of the central state. His emphasis on sub-national processes echoes Mabin's argument concerning the role of local popular resistance to central government. The contradictions built into classical apartheid have required new spatial structures of control, some of which will carry over into the post-apartheid state in the management of metropolitan regions.

1 *Dispossession, exploitation and struggle: an historical overview of South African urbanization*

ALAN MABIN

The tendency for urban scholars to dismiss South African urbanization as an aberration has a strong following. To most casual observers, apartheid shaped the country's peculiar forms of urbanism. Its uniqueness arises from the result of the mapping of white political power onto the country. This standard view contains considerable dangers. Politically, the result is to emphasize ideology and the state (at the expense of economics and daily life) as the primary spheres of struggle against the oppressive order. Intellectually, the consequences include an aversion to probing the real material conditions and social character of urbanization.

This chapter seeks to place material issues at the core of a view of South African urbanization[1] over more than a century. It makes no claim to being definitive; but it does hope to provide a coherent account which can inform understanding of the dynamics of contemporary urban processes in the country, and thus, debates over their future.

The origins of urban South Africa

Legal slavery ended in the Cape Colony with the British imperial emancipation of 1834. Until that time such towns as existed in southern Africa were few and tiny; the largest concentrated, non-rural settlements probably consisted of the enormous residential 'villages' of Tswana chiefdoms and perhaps the large capitals of the Zulu kingdom. But both were devoid of the commercial and financial institutions which grew rapidly in colonial ports such as Cape Town and Port Elizabeth and country towns like Graaff-Reinet and Beaufort West from the 1830s onwards (Mabin 1984). Such institutions were replicated in Boer centres like Potchefstroom from

[1]For the purposes of the chapter, 'urbanization' as a process is understood primarily as population movement towards densely populated and mainly non-agricultural settlements.

the 1840s (Christopher 1976). Rapidly expanding exports of staple products – wine to begin with, then wool – fuelled the growth of trading centres. White expansion into more remote reaches of the interior brought conflict with established polities. For the African communities already settled in these areas and subjected to colonial and Boer expansion, the results generally meant anything from declining independence of the chiefdoms to forced labour for white settlers. The pressures of land loss, military exigency and a growing commercialization of exchange relationships rendered both individuals and whole communities susceptible to involvement in the growing wage-labour economy of the towns by the 1850s.

Urban–rural migration on any scale is often taken to have begun in South Africa with conscious attempts on the part of white colonial and Boer republican authorities to extract labour for farms and mines late in the nineteenth century. However, recent historiography shows that rural people in South Africa have engaged in substantial migration to non-rural activities and places for well over a century (Delius 1980; Harries 1980). A generation or more before colonial authorities achieved direct military, political and economic control over the Pedi, Zulu, Mpondo, Ndebele and Venda, Africans began in growing numbers to join others who found themselves pressured to seek wage work on docks, in railway works, at warehouses and in the small manufacturing enterprises of the towns.

From the 1850s onwards, a number of economic changes wrought a revolution in the urban pattern. An influx of foreign investment occurred, a massive expansion of economic activity began, and a new export – diamonds – rapidly grew to the status of the leading staple, surpassing wool by the late 1870s. By then not only the town most closely associated with diamonds, Kimberley, but also the ports, transport points in between, and agricultural–commercial centres supplying produce to Kimberley had begun to change both in size and character more rapidly than before. To the opportunity for pressured rural communities to tap into a small urban economy which the towns had provided prior to the 1870s was added a new phenomenon: aggressive recruiting for mining, construction and other urban activities (Jeeves 1985).

Thus, in rural-to-urban migration up to 1880, the period which migrants spent at urban destinations varied greatly, ranging from very short to lifetime terms. This variety has persisted to the present day, and encouraged Simkins (1983) to deny the usefulness of the simplistic permanent-versus-temporary distinction. Equally, household or family participation in such migration has also varied, involving parties from individuals to whole extended families (Murray 1987a). The vital point is that entire households have frequently not migrated as a whole, and while a base has been maintained by some members in rural (more recently simply non-formally-urban) areas, other household members have moved to town for longer or shorter periods. This simple fact meant that the reproduction of the workforce did not take place completely within urban (including mining) environments, making urban areas to some extent dependent on the reproductive functions of rural areas. In summary, from mid nineteenth century onwards a part of the African population always lived in essentially urban households which by varied means provided for their own reproduction in that environment. Many urban households, however, combined resources from both urban and rural activities (Martin and Beittel 1987).

The corporate economy and urbanization

If the changes wrought in South African society prior to 1880 were substantial, they seem dwarfed by the revolution which private companies initiated during the decade of the 1880s. Hastened by speculative collapse and severe depression, diamond mining companies centralized and merged so rapidly that De Beers Consolidated Mines monopolized the industry by 1889. Corporate endeavour moved on to open the gold-fields of the Transvaal: first at Barberton and then on the Witwatersrand in the latter half of the decade. The scale and pace of foreign investment, of technological change, of infrastructural development and of urban growth went far beyond anything previously experienced.

Furthermore, this economic expansion took place under governments with rapidly increasing capacity to rule their territories effectively. Not a single part of rural South Africa reached the turn of the present century with a substantial body of people able to escape the pressures of incorporation into a rapidly growing capitalist economy. Most were deprived of independent control of what they saw as their land, and most rural households henceforth found it difficult to avoid participation in the urban economy through selling the labour of one or more of their members in the towns or mines.

Nevertheless, and importantly, most South African households remained based in rural areas, in actual occupation if not legal possession of some piece of land. This particular combination of powerlessness and possession of land strengthened a circular system of migration which gained support and eventually enforcement from large companies and the state.

Rural dispossession and urbanization

The South African War of 1899–1902, resulting in the defeat of the Boer republics and their annexation to Britain, opened the path to constructing a still more effective state. The government of the Union of South Africa, with its racist constitution providing for an almost exclusively white vote, took control of its million-square-kilometre territory in 1910. Among its explicit intentions was to give effect to the recommendations of the inter-colonial Native Affairs Commission report of 1904. In coordinating 'native policy', the corner-stone was to be a land policy. The policy arrived at, although not fully legislated (let alone implemented) until the late 1930s, had the long-term effect of further entrenching the circular migration system.

Prior to union, all four colonies which came together in 1910 had some division of land between Africans and other inhabitants. Areas of greatly varied size, mostly small, had been retained or set aside as 'reserves', within which only African people could live. The intention of the 1913 Land Act was in part to continue a process of adding land to the reserves which the Native Affairs Commission had begun. For most of the period since 1913, outside (as well as at some places inside) the reserves, much of the African population occupied land as tenants or squatters. On farms outside the reserves, some labour went to production on the farm of residence, sometimes allocated by the household itself under sharecropping or rental arrangements. The subjection of labour to control by the landowner or manager was the focus of intense struggle throughout rural South Africa, particularly in the 1920s (Bradford 1987).

Whatever the pattern of resistance, however, 'farmers' assisted by the state increasingly determined part of the labour allocation of rural households through labour tenancy or wage relationships (Van der Horst 1943). Decreasing ability to cultivate crops and run cattle on most non-reserve farms of residence encouraged households to attempt to export labour to other markets, with the result that the numbers of migrant workers originating from rural areas other than the reserves grew rapidly. According to estimates by Nattrass (1981), even in 1970 the number of migrant workers with homes in white-owned rural areas working in the non-agricultural parts of the economy exceeded 400,000 – and prior to that date, the numbers may have been even larger. The idea that the *reserves* supported the remainder of the economy through assumption of reproduction functions would, if not already dubious, receive a serious blow from the admission of the fact that rural, non-reserve households supplied a large proportion of migrant labour to urban areas throughout the twentieth century, and especially from the 1930s to the 1960s.

Evictions from private farms and a variety of measures adopted in the reserves, including the replanning of agricultural communities known as 'betterment', had the effect of creating a large landless population by the time the National Party government of D.F. Malan came to power in 1948. In many reserve areas, 'miserable', 'bleak and bare' settlements of the landless began to develop (Walker 1948), from which, inevitably, most households had to send members to participate in the urban economy. During the 1950s the pace of rural eviction began to increase, and it accelerated greatly in the 1960s and 1970s, until literally millions of people had directly experienced eviction from land on which, in most cases, their family histories were much longer than those of the titular owners (Platzky and Walker 1985).

It should not be assumed that these evictions affected Africans alone. Many thousands of coloured and Indian households also experienced eviction as did some whites. The effect on the urban population was, of course, profound. By the start of the Second World War two-thirds of Indians and whites were living in urban areas, half of coloureds, but only a fifth of Africans (Cilliers and Groenewald 1982).

From the 1930s, informal settlement on the fringes of the cities and many towns began to become common. In 1938, the central state's Ministry of Health initiated an enquiry into 'areas which are becoming urbanised' but which fell outside local authority boundaries. Its main report was completed only in 1941, by which time the exigencies of the war economy precluded much action being taken. Despite the amount of deliberation which the Smuts (United Party) government gave to urban issues in a spirit of reconstruction at the end of the war, the state took few positive actions, while overcrowding and informal residence developed apace. In some respects apartheid was a (racist) response to previous failure to develop coherent urbanization policy.

The apartheid era

Inadequate urbanization policy threatened the system of municipally-controlled passes which had been instituted from 1923 onwards under the Natives (Urban Areas) Act. Yet little housing of any kind was constructed during the war years, so that overcrowding in existing areas, especially in the African 'locations', reached extreme levels by 1945. The result was a series of land invasions and the development of other forms of informal urbanization (Bonner 1990), a situation which the National Party promised to attend to in its manifesto of 1948. Once in government it did so through a series of measures which strengthened the pass system and the police force, while at the same time it adopted policies which channelled the expanding landless population into both non-agricultural settlements in the reserves and into urban townships. In many cases, of course, the same households divided themselves between one or more rural bases and some form of access to shelter in the urban townships.

Especially after 1960, both the possibility and the utility of remaining on farms outside the reserves declined more precipitously, though with considerable spatial variation. Though eviction and relocation to the reserves were central, the reasons for this massive relative and absolute population shift were, and continue to be, by no means simple. Most accounts have stressed an ideologically-based role of the apartheid state (Baldwin 1975; Platzky and Walker 1985; Unterhalter 1987). But changes in the character of agricultural production yielded pressures towards evictions of tenants and other resident labour. Thus the 1960s became the decade of massive but not necessarily state-sponsored removals of labour tenants and squatters – 'removals of a quiet kind' (Donald 1984) which continue today.

In the bantustans, increasing numbers of households had little or no involvement in agriculture under their own aegis and growing dependence on wages from members finding work in towns or industries. The obvious question, then, is why these rural but largely non-agricultural households established and maintained bases in the reserves and did not move completely to towns. The answer may be more elusive than a purely state-centred analysis would suggest and has changed substantially over time.

In the first instance, many ex-farm households (or some of their members) probably did move more or less directly to the towns, though more so in the 1950s than the 1960s. In so doing households followed individuals who had already migrated to seek work. Considerable housing construction in new townships from Daveyton and Soweto to Guguletu and Zwide made it possible for such people to find shelter, even though that new housing also had to absorb many forcibly relocated (for example, from the old locations) under the increasingly strict urban segregation practices of the era. But in the townships, two further aspects of the new regime of apartheid gradually made life more difficult. The passage of control over urbanward movement out of the hands of local authorities and into the hands of the central state was one; the erection of the labour bureau system was the other (Hindson 1985).

By the 1960s, then, a system existed which, at least for a time, 'provided for the legitimate labour requirements of employers' (Posel 1989) while allocating massive forced migration off the farms in a cold-blooded manner to closer settlements of various kinds in the reserves. In 1960 the closer settlements were more or less non-existent: by 1980 they contained, according to Simkins's (1983) probably low estimate, 3.7 million people. One effect of this massive population growth was to strain beyond any capability the meagre resources of the new (but rapidly growing) bantustan administrations of the era: the results included desperate conditions such as extreme infant mortality rates. These peculiar features of the landscape prompted the view prevalent in the literature that specific and conscious state actions underpinned by an ideology called apartheid 'contained' African urbanization – or, in later views, 'displaced' that urbanization (Fair and Schmidt 1974; Letsoalo 1983; Murray 1987b).

During the 1960s this system maintained its stability partly through growth of the characteristic 'townships' of urban South Africa – where local authorities built much of the housing, or, just across bantustan boundaries, the Bantu Trust as well as neighbouring white local authorities did the same. Little collective resistance to the system crystallized: low rents ruled, political organization remained repressed; residents commuted oppressively long but practically manageable distances to work on heavily subsidized buses and trains and, at a relatively slow pace, township housing grew overcrowded through subletting rooms. After 1969, most new housing construction in areas open to Africans took place in bantustans. But those new houses could not compensate for the loss of new construction in the urban townships, thereby generating a rapid increase in subletting with the unsurprising result of extreme overcrowding.

Internal problems of apartheid urbanization

In the 1970s a number of changes disrupted the apparent stability. Thus the massive strike actions of 1973–4 over low wages might be interpreted as addressing the mounting problems of adequate reproduction for urban dwellers, 'migrant' or not, without a substantial rural (re)productive base. Increasing bantustan populations delivered large numbers to the job queues at the rural labour bureaux, while the rate of job creation and labour requisitions slowed. With a great increase in domestic (and corresponding decline in foreign) recruitment of mine labour, many ex-farm residents found themselves forced to join the hard core of migrant labour in the mines. As labour demand stagnated in the later 1970s, rural labour bureaux ceased to have any substantial recruiting function at all, to the point where 'for many blacks in the rural area there is no labour market' (Greenberg and Giliomee 1985). For sheer survival, supposedly 'rural' households, huge numbers of which had no prospect of supporting themselves solely through rural activities, had to find access to urban economies.

The state's refusal to build houses in sufficient numbers to meet needs and its exclusion of 'illegals' from official tenantry in formal townships forced people to build for themselves; the poverty of the great majority meant that the results are often massively inadequate. The overcrowding of township houses and the growth of shack populations in back yards and on open spaces in and around formal townships demonstrated some of the results.

In short, booms in the recruitment of domestic migrant workers for the mines and construction of houses in the often remote bantustans could not shore up a crumbling regime of population management. That regime had produced numerous bureaucratic problems and material difficulties for urban as well as rural people. It had also coexisted with the development of the quite unintended consequence of massive 'informal' population concentrations. Recognizing that its policies were in disarray, the government appointed numerous commissions of enquiry; but the pace of urban change eclipsed recommendations such as those of the Riekert Commission (Hindson 1985).

Struggling to survive and gain greater access to the accumulations of wealth represented by the cities, African people all over South Africa, by individual and collective actions, began to remake the nature of urbanism in the country. That struggle assumed obvious and intense forms first in the Western Cape, where African households with little or no base in the bantustans sought most vigorously to create urban space for themselves as intact units – and met with both victories and defeats. Modderdam and Unibel both disappeared under state bulldozers in 1977–78. But Crossroads survived, grew, and developed a defiant and uncontrolled culture which challenged the bases of an earlier urban regime. It did so at exactly the time at which state officials had to face both their inability to impose full control on the urban population, and the new, unapproved, unintended concentration of population in unserviced areas.

Townships within bantustan boundaries already fringed towns and cities closer to bantustans. For example, 25 to 30 km north-west of Pretoria, Ga-Rankuwa and Mabopane lay just inside Bophuthatswana. Unlike Cape Town, where church land provided the nucleus of eviction-free squatting, privately-owned small land holdings on the Bophuthatswana side of the townships offered sites on which to live at low rentals. In this area, the Winterveld, population rapidly grew to some hundreds of thousands. Around Durban, similar development took place not only inside bantustans; just as in the Western Cape, privately-held and church-owned non-reserve land became more densely settled by Africans, as the examples of Inanda and St Wendolins show. Even around the Witwatersrand, squatting developed rapidly in the late 1970s and early 1980s. Many of those attempting to create an urban life in these strictly illegal and unapproved circumstances faced defeat at the hands of the state as well as private landowners; eviction and relocation has been common. Persistence has been rewarded for some who have struggled to create new communities, though many have found themselves accepting relocation to distant 'approved squatting' sites. Even further away, just within daily reach of Pretoria and even the Witwatersrand, KwaNdebele grew from almost no population to enormous size from the later 1970s (Murray 1987b). In some parts of the country, a large proportion of the 'rural slum' population lives at such distances from the metropolitan centres that the implication of peripherality has to be extended greatly.

South Africa's informal settlements vary greatly in their setting, population size, density, social stratification, levels of wealth and poverty and social organization, political division and conflict. People come to live in such settlements for a variety of reasons. Fundamental to their motives is usually the question of finding places to live. Thus, ex-farm residents do not simply live where they were dumped by private or public evictors. Some went from village to village. Those who have had experience of forced removal, such as eviction from farms, have frequently tried to find agricultural land on which to settle; but after several attempts, there is a strong tendency to abandon the search and to accept the relative security available in resettlement areas or on residential sites allocated through tribal authorities (Mabin 1989). It is the need to find a place to live under severely constrained circumstances which has led to the growth of a new form of urbanism. Those constraints are experienced by most of the residents of the non-formal settlements more as material than state-authority constraints; but the factors giving rise to the new forms of urbanism – of 'urbanization' – in South Africa have roots in a complex history of state policy, of household organization of labour and of struggles to create and sustain communities.

Not content to wait for the millennium, the inhabitants of the bantustans and evictees from the farms built new informal 'urban' environments which gave them as much access to the benefits of an urban life as they could achieve. Places to live, some security, access to varying levels of participation in the real urban economy, lower costs of living than encountered either in formal urban environments or in remote bantustan districts: these achievements have redrawn the map of population distribution, and greatly affect the political landscape. For a (small) class of informal settlement entrepreneurs (in one view) or exploiters (in another) they provide the base for substantial accumulation of wealth. But the majority of their residents have not yet been able to challenge the central controls over their lives – propertylessness and state power. The prospects for them to do so seem bleak at present. One factor militating against the people's hopes lies in the violent conflicts which have tragically characterized so many informal settlements since the mid 1980s.

Renewed apartheid? Land invasion and state land allocation

In some townships, community organizations have responded to the pressing demand for relief of oppressive material conditions by fostering invasions of open land. Civic associations in places as diverse as Mangaung (Bloemfontein) and Wattville (Benoni, east Witwatersrand) have planned and executed land invasions in which members of the township communities concerned have taken over land adjacent to the townships, and erected settlements. Through a variety of tactics they have encouraged authorities such as local white town councils, development agencies such as the Urban Foundation and branches of the state such as provincial administrations to negotiate on their security, and even more significantly, on the provision of basic services to these new urban communities (Mabin and Klein 1991).

But these movements are not without their problems. Amongst other things, they tend to reinforce the broad apartheid geography of the cities rather than to fundamentally challenge it. By establishing themselves next to the large townships created in the 1950s, the land invaders reinforce the pattern, created under Prime Minister Verwoerd in the 1950s, of peripheral, segregated African residential areas.

State planners have also engaged vigorously in the allocation of large tracts of land for legalized informal residence since the mid-1980s. Using the provisions of Section 6A of the amended Prevention of Illegal Squatting Act, minimally serviced areas have been opened to settlement in the hinterlands of the established townships. Orange Farm in the area between Soweto and Sebokeng, Motherwell across the Zwartkops river from the sprawling Port Elizabeth townships and parts of Khayelitsha on the Cape Flats provide well-known illustrations. Apartheid in this sense of the broad allocation of segregated, remote land to black urban residents is very much alive, though it is continuously challenged by squatters who occupy land far from the approved townships. Through the actions of squatters, support groups and even officials in places such as Hout Bay and Noordhoek in the Cape, Midrand in the Transvaal and a few instances in Natal, there are prospects that the apartheid land allocation pattern may at last begin to break down.

In these new African communities, the overwhelming majority of residents would appear to come from existing townships rather than directly from rural areas. It is common cause that most population growth in South African urban communities is supplied internally. Indeed, the peak era of African migration from rural areas to the cities, which probably began in the late 1970s, seems to have subsided by the time influx control was abolished in 1986.

Conclusion

A survey of the history of South African urbanization, such as that outlined in this chapter, must provide a reminder of the extent to which the processes involved consist of dispossession and exploitation. Indeed, rural dispossession lies behind almost every form of urbanization. In South Africa, where dispossession is recent and even continues in obvious and sometimes bloody ways, discussion of urbanization and appropriate policy is increasingly conducted within opulent venues and between glossy covers, as though such tactics would deprive the urban poor of their collective memory of these active processes of underdevelopment. Of course such is not the case, though the struggles which the 'urbanizers' have waged against dispossession and exploitation have a terrible tendency, in the almost unliveable urban environments of late apartheid, to mutate into internecine conflict in which the poor find themselves pitted against each other – with renewed consequences of dispossession and exploitation.

The struggles of residents in the informal settlements defy any simple categorization. Increasingly, the same point might be made about the entire spectrum of 'struggles for the city' (Mabin 1989). If those struggles are to be adequately related to the national struggle for development, several emphases of the existing literature require elaboration and in some ways modification.

In the new urban regime, struggles within informal settlements and, increasingly, formal townships as well, often involve no direct confrontation with the state. Material circumstances in the urban sphere – themselves the culmination of decades of massive urbanward movement, forty years of apartheid and a century of low wages – provide an adequate environment for considerable struggle along other cleavages. The new urban regime poses profound problems for those engaged in struggles for a humane, democratic and environmentally liveable urban future. Simkins (1983) stressed that new developments in South African urbanization would depend to a high degree on struggles over the de facto ability (and right) to live, work or enjoy facilities in particular places; on struggles over access to urban life, rather than merely on policies from above. This chapter has placed much emphasis on such struggles. But economic conditions and state policies will inevitably continue to shape South African urbanization too.

References

Baldwin, A. 1975. Forced removals and separate development. *Journal of Southern African Studies* 1: 215–27.

Bradford, H. 1987. *A taste of Freedom: the ICU in Rural South Africa 1924–1930.* New Haven: Yale University Press.

Bonner, P. 1990. The politics of black squatter movements on the Rand 1944–1952. *Radical History Review* 46/7: 89–115.

Christopher, A.J. 1976. *Southern Africa.* Folkestone: Dawson.

Cilliers, S.P., and Groenewald, C.J. 1982. *Urban Growth in South Africa 1936–2000: a deomographic overview.* Occasional Paper 6, Research Unit for the Sociology of Development, University of Stellenbosch.

Delius, P. 1980. Migrant labour and the Pedi. In S. Marks and A. Atmore (eds), *Economy and Society in Preindustrial South Africa*, 293–312. London: Longman.

Donald, I. 1984. Removals of a quiet kind: removals from Indian, coloured and white owned land in Natal. Carnegie Conference paper 75. Southern African Labour and Development Research Unit (SALDRU), University of Cape Town.

Fair, T.J.D., and Schmidt, C. 1974. Contained urbanization: a case study. *South African Geographical Journal* 56: 155–66.

Graaff, J. 1987. The current state of urbanization in the South African homelands. *Development Southern Africa* 4: 46–66.

Greenberg, S., and Giliomee, H. 1985. Managing influx control from the rural end: the black homelands and the underbelly of privilege. In H. Giliomee and L. Schlemmer (eds.), *Up Against the Fences*, pp. 68–84. Cape Town: Oxford University Press.

Harries, P. 1980. Kinship, ideology and the origins of migrant labour. Paper presented at Class Formation, Culture and Consciousness conference, Centre of International and Area Studies, University of London.

Hindson, D. 1985. *Pass Controls and the Urban African Proletariat*. Johannesburg: Ravan.

Jeeves, A. 1985. *Migrant Labour in South Africa's Mining Economy: the Struggle for the Gold Mines Labour Supply 1890–1920*. Johannesburg: Witwatersrand University Press.

Letsoalo, E. 1983. Displaced urbanization: the settlement system of Lebowa. *Development Studies Southern Africa* 5: 371–87.

Mabin, A. 1984. The making of colonial capitalism: intensification and expansion in the economic geography of the Cape Colony, South Africa, 1854–99. PhD thesis, Simon Fraser University.

Mabin, A. 1987. The land clearances at Pilgrims Rest. *Journal of Southern African Studies* 13: 400–16.

Mabin, A. 1989. Struggle for the city: urbanization and political strategies of the South African state. *Social Dynamics* 15: 1–28.

Mabin, A. 1990. Limits of urban transition models in understanding South African urbanization. *Development Southern Africa* 7: 311–22.

Mabin, A., and Klein, G. 1991. Victories in the struggle? Land invasions and the right to the city in South Africa. *Transformation* (forthcoming).

Martin, W., and Beittel, M. 1987. The hidden abode of reproduction: conceptualizing households in southern Africa. *Development and Change* 18: 215–34.

Murray, C. 1987a. Class, gender and the household: the development cycle in southern Africa. *Development and Change* 18: 235–49.

Murray, C. 1987b. Displaced urbanization: South Africa's rural slums. *African Affairs* 86: 311–29.

Nattrass, J. 1981. *The South African Economy*. Cape Town: Oxford University Press.

Platzky, L., and Walker, C. 1985. *The Surplus People: Forced Removals in South Africa*. Johannesburg: Ravan.

Posel, D. 1989. 'Providing for the legitimate labour requirements of employers': secondary industry, commerce and the state in South Africa during the 1950s and 1960s. In A. Mabin (ed.), *Organization and Economic Change: Southern African Studies, Vol. 5*, pp. 199–220. Johannesburg: Ravan.

Seekings, J., Graaff, J. and Joubert, P. 1990. *Survey of Residential and Migration Histories of Residents of the Shack Areas of Khayelitsha*. Occasional Paper 15, Research Unit for the Sociology of Development, University of Stellenbosch.

Simkins, C. 1983. *Four Essays on the Past, Present and Possible Future Distribution of the Black Population of South Africa*. Cape Town: Southern Africa Labour and Development Research Unit, University of Cape Town.

Urban Foundation 1990. *Population Trends*. Policies for a New Urban Future Series, No. 1. Johannesburg: The Urban Foundation.

Unterhalter, E. 1987. *Forced Removal: the Division, Segregation and Control of the People of South Africa*. London: International Defence and Aid Fund.

Van der Horst, S. 1943. *Native Labour in South Africa*. London: Oxford University Press.

Walker, O. 1948. *Kaffirs are Lively: being some Backstage Impressions of the South African Democracy*. London: Gollancz.

2 *Local and regional government: from rigidity to crisis to flux*

JEFF McCARTHY

Introduction

The world media has historically adopted an interest in specific manifestations of the political crisis in South Africa: ugly spectacles of state repression and mass revolt; white opulence versus black poverty; instances of racial discrimination and inter-racial violence; and, above all, the apparent dominance of key 'national actors' within a political drama about contrasting visions of the desirable form of national state. More recently, the focus of media attention has shifted to the so-called 'negotiation process' in South Africa: a process in which the same key national actors are apparently engaged in pre-negotiation posturing prior to settling their differences concerning the future form and character of the post-apartheid national state.

Whilst there is much that is plausible in this imagery, it remains restricted in its value. This is because, firstly, it fails to identify the properties of broader social processes that make the tactics of supposedly 'independent actors' variously sensible or incoherent. Secondly, the almost exclusive emphasis upon debates and negotiations on the nature of the national state is deceptive. In particular, this emphasis tends to obscure from view a set of politico-economic processes operating at the local and regional scales which have not only influenced the broader directions of 'popular' and 'power bloc' politics in the past, but are also now influencing debates and negotiations on the future form and character of the new national state itself. In order to understand emerging debates on a possible national federal constitution for example, these local and regional processes need to be appraised.

Against such a background, the present chapter reviews developments in the field of local and regional politics in South Africa during the 1980s, and situates these within the context of the more generalized political conflicts of that time. Following on from this, the more recent trends towards restructuring of local and regional government, and their relationships to change at the political centre, are analysed. The chapter is therefore structured in four subsequent sections. First, the historical context for the development of so-called Verwoerdian apartheid – the political geography for which South Africa became infamous – is described. This is followed by an account of what might be termed the 'internal contradictions' of the apartheid scheme. The implications of these contradictions within the so-called popular versus power bloc political conflicts of the early to mid 1980s provide the context for the third section, where neo-apartheid adaptations to local and regional state structures are analysed. These occurred in the mid to late 1980s, during a period of increasing repression of popular politics. Finally, the chapter concludes by examining current trends in the restructuring of local and regional government, and their relationship to negotiations at the political centre, during a period of reduced repression and political realignment between popular and power bloc forces themselves.

Historical context: the forging of Verwoerdian apartheid

Until approximately 1975, most works on South African politics assumed the near omnipotence of the central state in the forging of the political geography of apartheid. Developments at the level of local and regional government – with the partial exception of the establishment of some so-called homeland governments – received scarcely a footnote (see Carter 1962; Adam 1971). Moreover, whilst the politics of popular resistance featured slightly more prominently than the local state, this politics was largely identified as a reaction to central state initiatives, rather than as having internal dynamics of its own.

The assumptions of central state power in such studies were not altogether inconsistent with the realities of power at the time. Organized popular resistance had been rendered less effective through central state repression by the early 1960s (Lodge 1983). In addition, some recalcitrant local authorities which initially opposed certain local implications of the central impositions of apartheid had, by that time, also been effectively undermined by the governing National Party (Western 1981). It should be noted, nevertheless, that more recent historical research is beginning to call the central state power argument into question. The continuity of popular resistance to the implementation of apartheid, and local government complicity in the exercise – even in more liberal urban areas – now appear to be rather more significant than was initially recognized (Lodge 1983; McCarthy and Smit 1988; Bonner *et al.* 1989).

The debate on the origins of, and resistance to, apartheid notwithstanding, it is clear that by the 1950s, those in control of central government had modified and adapted a variety of local systems of segregation, and codified them into a singular national policy framework. It was Prime Minister Verwoerd who was to lend a grand geopolitical 'vision' to this policy framework. This framework, in turn, was founded upon two major politico-geographic concepts: (a) a 'group areas' concept applicable to trading, residence and related local political rights in 'white' South Africa, and (b) a 'homelands' concept applicable to the separate political and economic 'development' of Blacks outside of 'white' South Africa.

Superimposed upon these systems of segregation were the local and regional political structures of the apartheid state. 'White' South Africa had effectively been defined in terms of the provisions of 1910 Constitution, and the 1913 and 1936 Land Acts. It comprised some three-quarters of the country, and included within it the best agricultural areas, virtually the entire urban hierarchy, and its associated infrastructure. Prior to the 1950s informal discrimination, the Land Acts (1913 and 1936) and local government by-laws had forged a high level of urban and rural segregation within 'white' South Africa.

The Group Areas Act of 1950, however, required the strict segregation, within discrete areas, of the four groups recognized in terms of the Population Registration Act of 1950 (White, Coloured, Asian, Black). Within towns and cities, separate residential and commercial districts were demarcated, and ownership and occupation of property was restricted to members of the race group to which the district had been assigned.

Black South Africans were completely denied ownership rights in 'white' South Africa, and were required to live in 'townships' which were owned and administered by local agencies of the central state. The administration of these townships was closely correlated with the application of the so-called 'pass laws'. In terms of the latter, 'labour bureaux' in the urban areas allocated permits to Black people seeking permanent residence in the cities. The quantity of permits was effectively determined by employers' demands for labour in different places, and at different periods. These same labour bureaux coordinated their planning with township administration offices, which in turn allocated housing and collected rents (Mabin and Parnell 1983; Hindson 1987).

Outside of the townships, whites, coloured and Asian residents' needs were administered by white-elected City Councils. These Councils, in turn, were 'advised' by Local Affairs Committees drawn from coloured and Asian group areas. The resulting imbalance in access to decision-making power led to the allocation of most salutary public goods to white group areas, and most noxious facilities either near to, or within, coloured, Asian or Black areas (McCarthy and Smit 1984). Blacks had no representation on City Councils and their local political rights were restricted to advisory roles in respect of township administration. The Urban Bantu Councils established for the townships, however, had little de facto or de jure control over planning, services or rental levels within the townships, which were determined by white administrators.

The model, in essence, was one of enforced Black patronage with white administrators, insofar as the latter required obedience from 'representatives' of the former in return for 'favours' with regard to resource allocation.

For the rest, Verwoerd's government anticipated that Black communities would 'develop' both politically and economically, within the geographically remote homelands, thereby possibly distracting from national political ambitions. Blacks were permitted no formal access to the central government of 'white' South Africa, although coloureds and Asians were permitted to make some 'advisory' input through government-established Advisory Councils.

Contradictions of the Verwoerdian scheme

By the mid 1970s, the political geography of Verwoerdian apartheid began to manifest a number of contradictions, and these had become magnified by the 1980s. Such contradictions largely conditioned the neo-apartheid reforms to local and regional government in the mid 1980s.

The contradictions of Verwoerdian apartheid were initially registered in the mounting economic costs of maintaining a racial–geographic fragmentation of settlement and local and regional political control. For example, expensive programmes of rural and industrial job creation in the homelands – apparently designed to make them economically and politically viable entities – failed to keep pace with the rate of rural impoverishment (Daniel 1981; Wellings and Black 1986). Despite annual subsidies of hundreds of millions of Rand, and despite millions of pass-law arrests, Black people continued to migrate to the cities in search of a better life (Giliomee and Schlemmer 1985).

The South African government, apart from persisting with pass-law prosecutions of migrants and demolishing informal settlements near cities, attempted to constrain Black migration by providing almost no new formal township housing for Blacks since 1970 (Sutcliffe 1986). The migrants improvised, however, and large-scale informal settlements grew up especially within the homelands themselves where development control was slight, and borders were within commuting range of the cities (e.g. the Winterveld in Bophuthatswana north of Pretoria, Inanda in KwaZulu north-west of Durban).

This pattern of settlement, in turn, exacerbated a growing commuter problem, in which increasing numbers of Black workers travelled long distances to work in the cities on parastatal bus services. These transport services were subsidized by central government, and the rate of increase in subsidy demands was rising geometrically. Moreover, public transport was becoming politicized, with bus boycotts becoming increasingly common whenever fare increases were effected (McCarthy and Swilling 1985).

The mounting costs of homeland development and commuting were just two examples of the growing costs to the state of territorial apartheid. In a context of the declining international competitiveness of South African industry and the slowing down of the economy in general during the late 1970s and early 1980s, there were mounting pressures to both constrain and restructure state expenditures. The state therefore embarked upon a programme of privatization of its assets, and restructured taxes in such a fashion as to discourage capital flight. In respect of the latter, for example, corporate taxes became a decreasing proportion of the total, and the general sales tax gradually escalated from nil to 12 per cent by 1985.

With respect to privatization, some of the first areas tackled by the central government were those that would not alienate its own (white voter) constituency. Black housing and commuter transport, for example, each became logical targets, and in 1983 major initiatives were taken to sell all the government's Black housing stock to individuals, and to privatize bus companies and promote the kombi-taxi alternative (Mabin and Parnell, 1983; McCarthy and Swilling, 1985; see also Ch. 14 and 15 in this volume).

By extending the tax burden increasingly to the broader population, and privatizing the consumption necessities of the poor, the state inadvertently deepened a groundswell of popular opposition to apartheid rule. This opposition had been largely triggered by 1983 'reforms' at central government level, when the government attempted to co-opt coloureds and Indians into central government, but excluded Black representation at the centre, through the establishment of the so-called tricameral parliament (Morris and Padayachee 1988). A crisis of political legitimacy was, therefore, inflamed by the state's responses to broader economic and fiscal crises, which in turn were partly rooted in the inefficiencies of apartheid geography.

Neo-apartheid political geography: the search for new spatial coordinates of control

As Cobbett *et al.* (1985) have noted, by the mid 1980s the South African government had been forced to search for new geographical units of planning and administration in response to the contradictions of the Verwoerdian vision, popular insurrection within the country, and international pressures for change. By the mid 1980s, Black permanence in 'white' cities for example, had been conceded by government as part of the rationale for the privatization of Black urban township housing. Influx control had been repealed in response to its *de facto* failure, thereby weakening some of the rationale for the separate 'nationhoods' of the homelands. Moreover, the formal administration of Black townships was in chaos since, although there had been some upgrading of the autonomy and legal standing of Black representation on township councils, they lacked the revenues necessary to maintain or extend basic services (Heymans and Totemeyer 1987).

Whilst there were periodic insurrections in the homelands, resistance to apartheid rule was centred principally in the cities and metropolitan areas during the mid 1980s. It was in the urban areas, therefore, that the initial efforts were made by central government in respect of the restructuring of local and regional government. Specifically, the Regional Services Council Act of 1985 provided for the creation of a new tier of local government, the intentions of which were to overcome the costly duplication of service provision implied by racial fragmentation of local government, and to provide services to the fiscally-deficient Black local authority areas. Regional Services Councils (RSCs) now operate at the metropolitan scale, and councillors are nominated from the various racially-specific locales within the jurisdiction of an RSC area. That is to say, councillors are nominated from the Black Local Authorities, Asian or Coloured Local Affairs Committees, or White Councils within the area of concern to the RSC. New levies on the payrolls and turnovers of firms provide the funding for RSCs, and the voting powers of councillors are determined according to the service consumption levels of councillors' constituencies.

To the extent the RSCs allowed, for the first time in South Africa, for joint decision-making by whites, coloureds, Asians and Blacks in a common forum, and to the extent that they allow for some redistribution, they marked a slightly different point of departure for government in conceding racial–political interdependence. However, the racial basis of the primary local authorities from which Regional Services Councillors derive; the fact that such councillors are nominated by government and not elected; and the bias of voting powers towards representatives from wealthy (white) areas has endangered the broader legitimacy of RSCs from the outset (Watson and Todes 1986).

The RSCs went a short distance towards ameliorating apartheid contradictions in the urban areas, but the question of the homeland/South African divide, at a broader regional level, remained unresolved by 1985. Indeed, attempts to implement RSCs in Natal ran into such intense conflict with the KwaZulu homeland leadership (see Ch. 16 in this volume) that a separate initiative on the restructuring of regional government was spawned. Whilst this so-called 'Kwa-Natal Indaba' was originally conceived by Natal and KwaZulu ruling groups as an alternative, more wide-ranging project to promote a common administration for the whole of Natal and KwaZulu, central government and major Black opposition groupings remained aloof from the exercise. Nevertheless, as Glaser (1986) has observed, the Indaba took as its point of departure the broader issues of functional and economic interdependence at the regional level between Natal and KwaZulu, and central government personnel appeared to take an interest in the experiment as a possible 'laboratory' for the future evolution of regional political reforms.

There were, in addition to Kwa-Natal Indaba, a number of other important initiatives in respect of the possible restructuring of local and regional government during the mid 1980s. One of the more influential was a report commissioned by the Associated Chambers of Commerce for South Africa (ASSOCOM). This report (Lombard and du Pisane 1985) appears to have initially been conceived as organized commerce's (somewhat critical) response to RSCs, but its authors took a broader view of local and regional government in a possible post-apartheid South Africa. They argued for a future federal constitution with maximum possible devolution of powers to the most local of levels:

> Should regional authorities on the basis of one man one vote be established in any set of regions which are bigger than individual municipal areas, a single group would obviously dominate political processes in most of such regions. . . . It follows that . . . should territorially-based federal states be established in South Africa . . . local authorities with extensive powers covering as many culturally sensitive government functions as possible will be most important . . . (Lombard and du Pisane 1985).

What is striking in all these efforts at either implementing, or conceptualizing, new systems of local and regional government however, is that they failed to constructively engage with the protagonists of the major forces of population opposition – for example, the United Democratic Front (UDF), the African National Congress (ANC), Inkatha or the Azanian Peoples Organization (AZAPO). Indeed, part of the rationale of such initiatives (with the possible exception of the Indaba) seems to have been to divide and frustrate the objectives of popular opposition.

Popular opposition to apartheid rule during the 1980s

The repression of popular opposition to apartheid rule during the 1960s and 1970s had driven the two major opposition groupings – the ANC and PAC (Pan African Congress) – into exile. The ANC was the largest and most effective, and the perception arose amongst power bloc figures in South Africa during the 1980s that the groundswells of popular opposition at that time were orchestrated from abroad. However, as Slabbert (1986) has pointed out, there was little evidence for this theory.

The ANC's enforced external presence, in fact, determined its major strategic initiatives: efforts at delegitimizing the South African state in international forums, and the organization of relatively low-key guerrilla incursions into South Africa. The intensity of guerrilla incursions did coincide with groundswells of popular protest activity inside South Africa, but the latter appear to have been motivated largely by the objective, politico-economic conditions that developed within South Africa itself, rather than vice versa.

It is widely acknowledged, for example, that one of the most significant developments within the ambit of popular opposition politics was the exceptionally rapid growth of the progressive trade-union movement during the late 1970s and early 1980s (Sutcliffe and Wellings 1985). The leading union federation of the early 1980s – the Federation of South African Trade Unions (FOSATU) – initially concentrated heavily on shop-floor organization, and avoided issues beyond the factory gates. Indeed, by as late as 1983 FOSATU and other union officials described both political and community issues as 'partisan and divisive', and they preferred to emphasize the need to build working-class solidarity at the point of production (Foster 1982; Lewis 1983).

By 1983, however, the state's racially-based tricameral reform initiatives at the political centre, together with its privatization of collective consumption goods and promotion of ineffective structures at the local level, had engendered considerable community-based mobilization. National stay-aways from work for one, two or more days became an important tactic. As a consequence, unions such as FOSATU found that their strictly shop-floor approach became difficult to sustain. Soon the class politics of the workplace became more integrated with the broader popular politics of anti-apartheid opposition, and mobilisation against rising rents in Black townships, rising bus transport costs, poor education facilities, and so on.

Several trade unionists interpreted this as a shift in working-class politics towards issues of the 'social wage' (Erwin 1985). In one study of factory organizer and shop steward attitudes, Webster (1984) found that such issues were becoming increasingly central to the politics of the progressive labour movement:

> The frontier of control (is being pushed) forward beyond production to the question of the reproduction of the labour force. . . . The demand for what has been called the *social wage* is now being put on the table and in the process of a new form of unionism is in the making that seeks to challenge poverty in South Africa. (Italics in original)

Of course, issues of reproduction and the social wage had been the stock-in-trade of Black civic associations for some time. In the view of one federation of local civics, the Durban Housing Action Committee:

> The allocation of national and local resources is precisely the object of community struggles for better housing and lower rents. Our communities remain to be convinced that 'there is simply not enough money to provide decent housing for all'. . . . any reduction in the average quantity and quality of housing for the average worker is tantamount to a reduction in wages. (Durban Housing Action Committee 1982)

Similar positions had been adopted by other civic associations such as the Cape Areas Housing Action Committee and the Soweto Peoples Delegation. Organizations such as these have included in their activities the co-ordination of so-called 'rent boycotts' in Black residential areas. These boycotts originated during the early 1980s when housing rentals escalated, partly as a consequence of state attempts to privatize black housing, and partly as a result of rising services charges ('rates') levied by the financially constrained Black Local Authorities (BLAs) (Reintges, 1990). Indeed, councillors participating in BLAs were often perceived as being responsible for such increases, and they became unpopular 'collaborators' in the eyes of many – a view that corresponded with perceptions of many rural people of the rulers of most homelands.

A set of socialist material claims therefore became increasingly associated with critiques of apartheid-based local and regional government structures. The formation of the United Democratic Front (UDF) in 1983 first signalled the growth of an internal coalition of forces dedicated to the establishment of a strong unitary state pursuing non-racial and social-democratic policies – a position similar to that of the then exiled ANC. The UDF coalition of civic associations, smaller trade unions and related groups organized programmes of protest politics against the state's reform moves, including the tricameral parliament and the RSCs. Its rallying cry was 'Apartheid Divides, UDF Unites'. Soon, the UDF became the prime target of state repression.

At much the same time the establishment of the new union 'superfederation', the Congress of South African Trade Unions (COSATU), indicated a more political role for the organized urban industrial working classes in alliance with the UDF and ANC. For example, one of the resolutions of the first COSATU congress was 'to reject as a total fraud the proposed new federal constitution; to reaffirm our belief in a unitary state based upon one person one vote; to work towards the destruction of all barriers and divisions so that we are united irrespective of language, race or creed' (*South African Labour Bulletin* (1986) 11:45).

This approach was, of course, diametrically opposed to the directions of state restructuring then being encouraged by certain business groups such as ASSOCOM. Such differences should not be entirely surprising. Apart from the obvious differences of class interest, the two tendencies were being informed by different historical experiences. As Adam and Moodley (1986:217) observe:

> Because of the state strategy of fragmentation and, of course, the prospect of a reversal of power by the numerical racial majority, the disfranchised have always demanded a centralist rather than a federal state . . . Genuine democracy, in the eyes of radical South African democrats, requires strong centralist intervention to bring regional disparities in line with a political formula that guarantees greater equality.

It is generally supposed that the reversal of power to which Adam and Moodley (1986) refer would be used to extend or redistribute the social wage, one of the key demands of popular political opposition: that is, more and better hospitals, residential services, schools, water and electricity supply and so on, would be extended to the majority via their elected representatives who, it was hoped, might control the organs of central state power. Historically, it has been this emphasis upon the prospects for centrally-imposed redistribution which have been resisted by power-bloc groupings, and which have also become associated with a conflict over federalist approaches to the restructuring of central–regional–local government relationships.

The era of negotiation politics

The 1980s were characterized by an effective stalemate between the proto-revolutionary forces of popular opposition and the resilience of the power bloc and its forces of repression. It is generally agreed, however, that F.W. de Klerk's opening speech to Parliament on 2 February 1990 marked a watershed in South African politics. He announced the unbanning of the ANC and other popular opposition organizations, and invited them to join in a negotiation process to determine a future national constitution.

The reaction of the most important popular opposition groupings, including the ANC, UDF and COSATU, has been broadly positive. As a result, the contours of popular/power-bloc political alignments have become potentially available for redefinition. Whilst the ANC continues its support for social-democratic centralism, de Klerk and the National Party stress the merits of capitalism, and the need for devolution of state powers. Nevertheless, there is also now common ground between these two major parties in their mutual commitment to the politics of negotiation and emphasis upon non-racialism (the latter is of course a new-found emphasis within National Party).

It is difficult to predict the outcome of the era of negotiation politics in terms of restructuring of local and regional government. Current indications are that the National Party has already accepted the principle of local negotiations leading to non-racial local government, although it is possible that its concurrent emphasis upon 'local option' in this process may lead to differences with the ANC. Local option would, for example, permit more conservative local authorities to resist change, and it is not clear how this could coexist with a non-racial constitution implemented at the centre.

Debates on a new intermediate (or regional) tier of government in the future are less advanced than those on local government restructuring. The deep divisions and inequalities between homelands and 'white' South Africa will obviously require redress in the process of regional restructuring. The

1982 Good Hope Conference on regional policy unveiled a set of nine 'planning regions' which transcended the RSA/homeland boundaries and which have subsequently provided the basis for some cross-border cooperation around regional development planning. There has been speculation that these planning regions could form the basis for an intermediate tier of government in a future regional – federal political order (Cobbett *et al.* 1985). What is not yet clear is how such a restructuring could occur. The Kwa-Natal Indaba offered one model in a single planning region, but the Indaba approach unfortunately effectively excluded popular political participation apart from Inkatha supporters. This, in turn, has led to negative perceptions of the model on the part of the ANC/UDF/COSATU grouping.

Conclusion

As South Africa is increasingly forced, by political and economic realities, towards a post-apartheid era the configuration of local and regional government powers is in flux. In Johannesburg, for example, 1990 saw the formation of a Metropolitan Chamber, including representation from the Soweto Peoples Delegation (a popular civic association) and the Johannesburg City Council, and sanctioned by the Transvaal Provincial Administration. The Chamber is seen as an interim step until non-racial democratic local government is established in the greater Johannesburg area. Elsewhere, similar initiatives are afoot. The future contours of regional government are more shadowy, but it seems reasonable to expect similar changes on this front soon. Given the historical legacy of apartheid, it can be confidently predicted that the transition will be protracted and painful, but the outcome will be of considerable importance to the future of urban and regional development in South Africa, and South African society generally.

References

Adam, H. 1971. *Modernizing Racial Domination*. Berkeley: University of California Press.

Adam, H. and Moodley, K. 1986. *South Africa without Apartheid*. Berkeley: University of California Press.

Bonner, P., Hofmeyer, I., James, D. and Lodge, T. 1989. *Holding their Ground: Class, Locality and Culture in 19th and 20th Century South Africa*. Johannesburg: Ravan.

Carter, G. M. 1962. *The Politics of Inequality: South Africa since 1948*. New York: Praeger.

Cobbett, W., Glaser, D., Hindson, D. and Swilling, M. 1985. Regionalism, federalism and the reconstruction of the South African state. *South African Labour Bulletin* 10:87–116.

Daniel, J.B. 1981. Agricultural development in the Ciskei. *South African Geographical Journal* 63:3–23.
Durban Housing Action Committee 1982. *Discussion Paper on Indian Housing Study: Metropolitan Durban*. Durban: Community Research Unit.
Erwin, A. 1985. The question of unity in the struggle. *South African Labour Bulletin* 11:51–70.
Foster, J. 1982. The workers struggle: where does FOSATU stand? *South African Labour Bulletin* 7:67–86.
Giliomee, H. and Schlemmer, L. 1985. *Up against the Fences: Poverty, Passes and Privilege in South Africa*. Cape Town: David Philip.
Glaser, D. 1986. Behind the Indaba: the making of the KwaNatal option. *Transformation* 2: 9–30.
Heymans, C. and Totemeyer, G. 1987. *Government by the People*. Cape Town: Juta.
Hindson, D. 1987. *Pass Laws and the Urban African Proletariat*. Johannesburg: Ravan.
Lewis, D. 1983. General Workers Union and the UDF. *Work in Progress* 29: 11–18.
Lodge, T. 1983. *Black Politics in South Africa since 1945*. Johannesburg: Ravan.
Lombard, J. and du Pisane, J. A. 1985. *Towards the Removal of Racial Discrimination in the Political Economy of the Republic of South Africa*. Johannesburg: Associated Chambers of Commerce of South Africa.
Mabin, A. S. and Parnell, S. 1983. Remodification and working class homeownership: new directions for South African cities? *South African Geographical Journal* 65: 148–66.
McCarthy, J. J. and Smit, D. P. 1984. *South African City: Theory in Analysis and Planning*. Cape Town: Juta.
McCarthy, J. J. and Smit, D. P. 1988. *The History of the Urban Land User Process, Urban Planning and Urban Social Movements in South Africa*. Pretoria: Human Sciences Research Council.
McCarthy, J. J. and Swilling, M. 1985. South Africa's emerging politics of bus transportation. *Political Geography Quarterly* 4: 235–49.
Morris, M. and Padayachee, V. 1988. State reform policy in South Africa. *Transformation* 7: 1–26.
Reintges, C. 1990. Urban movements in South African black townships. *International Journal of Urban and Regional Research* 14: 109–34.
Slabbert, F. van Zyl 1986. Speech. *Hansard* (Debates of the House of Assembly) No.1, Col. 419.
Sutcliffe, M. O. 1986. The crisis in South Africa: material conditions and reformist response. *Geoforum* 17: 141–59.
Sutcliffe, M. O. and Wellings, P. 1985. Worker militance in South Africa: a socio-spatial analysis of trade union activism in the manufacturing sector. *Society and Space* 3: 357–79.
Watson, V. and Todes, A. 1986. Local government reform, urban crisis and development in South Africa. *Geoforum* 17: 251–66.
Webster, E. 1984. Poverty and trade unions. In *Proceedings of the conference on development in South Africa*. Cape Town: Southern African Labour and Development Research Unit, University of Cape Town.
Wellings, P. and Black, A. 1986. Industrial decentralization under apartheid: the relocation of industry to the South African periphery. *World Development* 14: 1–38.
Western, J. 1981. *Outcast Cape Town*. Minneapolis: University of Minnesota Press.

PART TWO

Housing and community under apartheid

This part of the book brings together essays on various aspects of the provision of housing and the creation (or destruction) of local communities under apartheid. It begins with an attempt to identify the main phases of capitalist development and associated state activity relevant to the housing question (Chapter 3). In this context, Soni examines and exemplifies black struggles centred on housing, in the face of the state's pursuit of strategies to accommodate labour where capital required it but at minimal cost. This is followed by a more detailed examination of state intervention in housing provision in the 1980s (Chapter 4). Parnell shows that the state remains closely involved in housing, despite attempts to abandon direct responsibility for the provision of shelter, and that the outcome tends to advantage whites over blacks while encouraging the emergence of a privileged property-owning class among the black population.

In the first of four case studies (Chapter 5), Lupton returns to the theme of struggle involving labour, capital and the state, initiated by Soni, to show the outcome of conflicting meanings and definitions of the built environment in the coloured townships of Johannesburg. The payment of rent and the physical form of the environment provided specific grounds for popular protest. Next (Chapter 6), Maharaj details the impact of segregation in Durban, revealing the role of local government in the form of the City Council and how it influenced and was influenced by the central state. The Group Areas Act provided Durban's whites with a means of fulfilling long-standing aspirations to relocate and separate the Indian population, though on some important matters central government had the last word. Scott provides further insight into the treatment of Indians in Durban, with a study of the development and subsequent destruction of Clairwood (Chapter 7). She shows how non-racial legislation concerning housing standards, health and planning was used to break up this cohesive community standing in the way of industrial expansion. In the final essay in Part Two (Chapter 8), the focus shifts to the impact of the policy of so-called orderly urbanization, introduced in the 1980s, on a Black/African township adjoining the 'white' city of East London. Reintges reveals some of the contradictions in the policy of privatization of housing, generating advantages for the few but only the prospect of shacks for the rest, in a process where the capacity of the authorities to manage urbanization is rapidly being overtaken by the actual practice of people desperate for shelter.

3 The apartheid state and Black housing struggles

J71
R31
South Africa

DHIRU V. SONI

Introduction

Recent Marxian interpretations have argued that land-use arrangements in a capitalist city reflect the conditions of capital accumulation and the class struggles that derive therefrom. The conflictual situation is exaggerated in the apartheid city, where the struggle against deprivation due to racial segregation and the continual battle to maintain an acceptable quality of life explains much of the dissension and in black residential areas. It is within this context that the question of housing and its inextricable relationship to capital and the state takes on a distinctive dynamic in South Africa.

The genesis of urban social theory

In the last two decades the world has experienced a dramatic increase in urban protest. Castells originally suggested that urban social movements could usher in fundamental social changes (Castells 1977: 271–2; 432–3), but his subsequent writings indicate that this task must be left to political parties (Castells 1983). More recent comparative research on urban social movements (e.g. Smith and Tardanico 1987) seems to indicate that, in order to advance fundamental social change, urban social movements must transcend parochial issues and mobilize as an integral part of a broader counter-hegemonic struggle within a particular capitalist social formation. The legacy of this contentious debate has now shifted to South Africa (Sutcliffe and Wellings 1985; Swilling 1986; Hendler 1987; McCarthy 1987; Bond 1990).

Critical urban social theory accepts the fundamental importance of the organization of production in creating and structuring all social processes. However, this does not mean ignoring the issues of consumption and reproduction. On the contrary, they are all inextricable parts of the organic whole. Even Marx, in dealing with the fundamental linkages of the capitalist mode of production, expressed difficulty in singling out the complexity of links that make up the system (Marx [1953] 1973:44). He saw the issues of production and consumption not as mutually exclusive elements but as givens in a presupposing relationship within the whole. If we take this proposition to its logical conclusion, then we should consider the historical specificity of the capitalist state and its social formation in the context of contradictions, tensions and cleavages which define the conjunctural possibilities of contestatory and transformative action both within the state on the one hand, and in relationship to its social formation on the other.

It is with this background in mind that an attempt will be made to analyse the question of housing and social movements in South Africa. This will be attempted empirically by investigating state intervention and urban and housing development as part of wider processes of capital accumulation and class relations, and theoretically by linking local housing provision to broader conceptual notions of political economy. If we accept that urban processes are intrinsically linked to the capitalist imperative of accumulation, then it is necessary to understand the specifics of each phase of capitalism. It is also necessary to understand how the dialectical progression of capitalist accumulation during these phases not only informs state action and class struggles, but also how it is informed by it. Accordingly, in order to understand the specific moment of the state in the housing struggle it is necessary to understand the sucessive phases of capital accumulation.

1948: the birth of the apartheid state

After gaining political power in 1948 the Nationalists were faced with the task of reconstruction and recreating the migrant labour system. The state now sought not only to secure a cheap and controlled labour supply but also to allocate it to the sectors that required it. The policy of separate development which was characterized by the intensive segregation of races, the more effective control of Blacks in urban areas, and the introduction of policies aimed at reducing the financial burden of Blacks in white areas on the national and local state (Morris 1981: 42).

During this period the Group Areas Act was enshrined in the statute books, providing for stricter control of race separation in the city and also for compulsory removal of existing residents. Housing financial policies introduced during this period were aimed at reducing the fiscal burden on central and local states. The policy also was geared towards 'economic homes' so that monies loaned could be recouped via rental or purchase schemes. For the first time an attempt was made to coerce all sectors of capital to contribute to the financing of Black housing. In a speech to the Union Senate in 1952 Dr H.F. Verwoerd, then Minister of Native Affairs, stressed the need for employers to contribute to a fund to pay for the provision of housing and services in Black areas (Morris 1981: 47). A

Native Services Levy Act was passed and employers had to contribute to the fund on a monthly basis. In 1954 another policy was introduced by the Department of Native Affairs specifying that all township housing was to be allocated according to ethnic group. The prime intention was to reinforce the tribal barriers that were prevalent in the rural areas to prevent working-class organization and mobilization in urban areas.

It was estimated that at the beginning of 1950 the Black housing shortage was in the region of 250,000 units (South African Institute of Race Relations, 1950–1). The Black squatter population in urban and peri-urban communities had reached a formidable 200,000 (Minister of Native Affairs, Senate 23 April 1951). In 1955 after having realized that the backlog in housing had increased substantially and that squatter settlements were being established despite the harsh measures imposed, the idea of site and service schemes was introduced by the Department of Native Affairs. In order to stimulate this programme, the government advocated the principle of home ownership.

With the Group Areas Act as a legal foundation, the removal of Africans, Indians and coloureds was intensified. At the same time a substantial housing programme was initiated, especially stimulated by growing public and private sector awareness of critical housing shortages.

In Durban, for example, Umlazi and Kwa Mashu townships were established in 1950 and 1953 respectively. In the existing settlement of Cato Manor, wholesale removal of Africans had commenced towards the latter half of the 1950s. Residents were settled either in Kwa Mashu or Umlazi. This action did not go without protest. Edwards (1983) notes that with the situation being so fluid in Cato Manor, one popular organization became influential, the Natal African Tenants and Peasants Association, which had strong links to both the South African Indian Congress and the Communist Party of South Africa. The struggle was over the demolition of housing provided for Blacks by themselves. Demonstrations were also held against the sale of kaffir beer and municipal beer halls were boycotted. The unrest culminated in early 1960 when nine policemen were killed during a routine liquor raid in Cato Manor. Clearance of shacks by this time was almost completed and it is estimated that 82,826 persons were resettled in Kwa Mashu and Umlazi (Morris 1981: 55).

Thus, during this period, the main goal was the eventual removal of Africans to either townships or 'homeland' areas. This was generally achieved through the removal of Black freehold rights, segregation of races, control of movement and the steady reduction of the economic burden of Blacks on the central and local states. Nowhere within these programmes of removals and resettlement were Blacks ever consulted.

This period was very significant vis-à-vis the contradictions that arose between labour (Black and white) and between capital (mining, commercial and industrial) and the state. As the state was encouraging a Black industrial labour force, it simultaneously threatened the privileged white counterpart.

As Wolpe (1988: 66) states: 'The National Party came to power on a policy aimed at suppressing the emergent black opposition which threatened the reproduction of white domination, that is, threatened the conditions which would enable the regime to meet, inter alia, the demands of white farmers and protect the interests of the white working class.' Draconian measures were thus introduced by the state to implement and give effect to its policy of separate development. However, as the state's policies were intensified, mass mobilization increased and the state in turn began to institute legislation and build its security apparatus to deal with the struggle. Against the backdrop of these actions and counteractions, profound structural changes were taking shape both within the state and in the broader political terrain. Herein, we can see how the state is being informed by the struggle, and in turn determines the nature of the struggle. Historically, ever since the union in 1910, the power of the central state has increased and this has had serious repercussions not only for capital but also for the local state. Capital, however, benefited in many ways by working with the central state, primarily because the state provided the necessary social and physical infrastructure for its expansion.

The period 1960 to 1973

The 1960s was a turbulent period for the state. Towards the end of 1959 the ANC had decided upon a major campaign and the Anti Pass Day was scheduled for 31 March 1960. Africans would leave their passes at home and march to the nearest police station in defiance. The crisis for the state erupted when police opened fire on a group of between 5000 and 7000 demonstrators, killing 69 people and wounding some 200. The event sparked strikes and demonstrations throughout the country. Although the state wavered for a short while, it gained its normal repressive composure and responded by banning all public meetings. On 30 March 1960 a state of emergency was declared and many Black leaders were arrested.

Sharpeville provided the state an occasion to tighten up on its existing laws regarding Blacks on the one hand and extra-parliamentary forces on the other. It passed the Unlawful Organizations Act and in the process banned the ANC (African National Congress) and the PAC (Pan African Congress). Thus together with the state of emergency it began to restructure the political arena in its own image.

The Bantu Laws Amendment Act of 1963 tightened up the Bantu Areas Act of 1952 and the state began to impose serious punishments if these laws were infringed in urban areas. In effect, the amendment gave the state complete authority vis-à-vis the influx of Africans to urban areas. The state began a concerted drive to establish the homelands as a panacea for the 'African problem'. In order to encourage the development of housing schemes within the homelands, leasehold tenure was suspended in 1968 and Africans could only rent houses in white areas and they were encouraged to build their own homes in the homelands (Ratcliffe 1980: 9). One of the effects of the homelands policy was that, through the spatial division of the working class, mass organization was constrained.

In 1973 the control of most Black areas was handed over from the respective councils and the non-European Affairs Departments to Boards. The Boards were staffed by all-white civil servants who were supposedly experts in the field of administration.

The state's policy was to decentralize the Black working classes, and as a result, housing priority was given to homeland development. Contradictions between the local states, capital and the central state arose out of the provision, or rather the lack of provision, for housing in urban areas. Large scale removals under the Group Areas Act became the norm. The removals not only affected the housing arena but also economic activity such as the commercial sector.

The inevitable urbanization process resulting from poverty in rural areas and job opportunities in the urban areas meant that migration was accelerated into those areas where there were relatively few controls. Large uncontrolled settlements mushroomed in the periphery of most urban areas of South Africa.

The period 1973 to 1980

The political repression of the state laid down in the previous period had to be readjusted due largely to widespread rioting which reflected the Black man's resentment with the way he was treated in all spheres of life. The effects of streamlining administrative and financial policies had serious repercussions for housing – the crisis was exacerbated with critical shortages.

The 'stability' created by harsh legislation set the scene for international economic confidence in the country. The emergence of the industrial sector and the concomitant need for labour with greater skill and supervisory capacity to operate machinery required a greater degree of education for workers and stability in the living environments. Simultaneously a large part of the white labour force moved out of blue-collar jobs into administrative and technical jobs and this in turn created a void which was readily filled by Blacks. In short, there was a shift in the pattern of labour utilization and concomitantly more and more Blacks were required within urban areas to service industrial development and a growing tertiary sector.

June 1976 was a watershed in the history of South Africa's Black disenchantment with their living environment. Black rioting spread from Soweto throughout townships in South Africa. Immediately after the Soweto protest, the state appointed the Cillie Commission, and its findings were far-reaching. For example, it acknowledged that the Group Areas Act had contributed in no insignificant way to the rioting. Housing was recognized as a major source of grievance: one of the major bones of contention was the fact that Blacks could not own their homes in urban areas. Amongst the other housing issues which paved the way for resentment were: a great shortage, overcrowding, lack of funding, too-high rents and the serious lack of services and facilities in most Black townships (Morris 1981).

Despite these findings, the state came down very heavily on Black activists and in 1977 most Black Consciousness movements were outlawed. Although many of these movements, such as the South African Students Organization (SASO) were educational-based, they were almost fully comprised of children who had grown up in deprived Black townships.

During the latter part of the 1970s monopoly capital was faced with a major conflictual situation. It required a large reservoir of Black labour to fulfil its production needs. It also required cheap labour and one way of attaining this was to cut down on social costs. This was primarily achieved through the failure on the part of capital to provide further housing (in the form of hostels or compounds) for its Black labour force. However, the unrests of the 1960s and 1970s seriously affected the production capability and confidence in the economy as a whole. In fact in 1976, after the Soweto struggle, the Transvaal Chamber of Commerce advocated the need for a stable urban Black middle class and called for, *inter alia*, greater expenditure on housing and other amenities. Faced with this contradictory relationship in the drive for accumulation, capital, for the first time, resolved to participate in general provision of housing. Thus, in 1976 the Urban Foundation was formed in the hope of improving the quality of life of the Black townships and contributions to the value of R25 million were pledged.

During this period the state intensified, through legislation, the removal of squatter settlements. As a result further shortages were experienced. With reference to Durban, housing conditions had deteriorated further and Black townships were reported to be critically overcrowded. Despite the harsh restrictions, squatter settlements spread rapidly on the peripheral areas of the city.

The state responded to the needs of capital via the Riekert and Wiehahn Commissions. The Riekert Commission recommended the improvement of certain conditions for Blacks who qualify for urban rights. This further reinforced the differences between 'urban' dwellers and 'contract' workers. Firm legislation came into being to deal with the streamlined influx control during this period and a large share of policing the system was shifted to employers. With the demand for labour increasing, the state was forced to reconsider its stance on the impermanence of the Black urban dwellers. In 1975 the state made a concession by reintroducing leasehold rights on a 30-year basis, though it was again stressed that the higher political aspirations of the Blacks would have to be exercised in the homelands. The 30-year leasehold right, however, did not succeed in promoting home ownership, largely because building societies did not consider that it provided adequate security.

Strict policies aimed at furthering the ideology of separate development were thus modified, due partly to the widespread dissent on the part of Black labour and partly to a fundamental shift in capital's needs for a stable labour force. The overall aim, though, of capital and the state was to extend political and ideological control though granting material concessions to the Black proletariat – a new means of reproduction. The Urban Foundation manifests capital's attempt to maintain the existing process of capital accumulation (Ratcliffe 1980: 11). This point was made very clear by Mr Justice Steyn, its executive director, when he said: 'If people want the free enterprise system to continue, they are going to have to get off their butts, and make contributions towards housing' (*The Star*, 11 April 1980).

Moreover, it seemed that certain sectors in capital were becoming less interested in preserving the infrastructure of control. Rather they were concerned with forming or encouraging skilled labour and also 'institutionalizing industrial relations in the hope of evading costly and disruptive "wildcat" strikes' (Wolpe 1988: 72). There was thus a restructuring of the axes of development between capital and labour on the one hand, and capital and the state on the other. Despite the state's 'reform' initiatives with regard to urbanization and housing, the housing situation remained critical. The Soweto uprising and other townships' struggles bear adequate testimony to this. Capital, likewise, was very slow with its own initiatives. The central state's introduction of the 99-year leasehold plan did not stimulate capital's response to the housing crises and this was due largely to the poor investment returns, contributing to the contention that capital is reluctant to contribute to 'social' or public housing, primarily because of the risk.

What we have seen in this particular period is state reform with regard to housing and Black urban rights due largely to the demands of the urban struggles and capital's needs for a stable labour force. The major legislative reform procedures were the Community Council Act of 1977, the 99-year leasehold legislation in 1978, the Bantu Laws Amendment Act of 1978 and new regulations governing conditions of residence in townships. These attempts on the part of the state were to 'humanize' apartheid and separate development. The previous stance on urban Black rights had finally shifted in the direction of recognizing that the urbanization process of Blacks could not be constrained any more.

The present conjuncture

The apartheid vision of separate development for separate groups was crucially blurred by major constitutional changes in the early 1980s. The major thrust of legislation and policy arising from these fundamental changes included the recognition that the urbanization of Africans was inevitable and in some respects desirable. Thus state strategy shifted from trying to control all aspects of urbanization to intervening in very specific areas. The emphasis in the early 1980s seemed to be geared towards management. The state, it appeared, had finally conceded that urbanization of Blacks was here to stay and that its role was to direct and manage that process. Since 1984 the Departments of Housing, for example in the (Indian) House of Delegates and (coloured) House of Representatives, have emerged as important actors in the demand for land and housing for their respective constituencies. Moreover, government initiatives since 1983 have allowed private enterprise and reconstructed local and regional governments to influence urbanization.

Before the 1980s almost all housing provided for Blacks was on a rental basis from the state. There has since been a shift towards market-oriented forms of housing. Privatization has meant that the state has begun to withdraw from the direct provision of services and to some extent, from financing of housing. This has meant that for the first time private developers, financial institutions, individuals and employers have become increasingly involved in Black housing. However, the private sector only helped fund housing that provided adequate security in case of foreclosure. About 40 per cent of the money spent on housing in 1987 went to houses costing more than R40,000 and only 15 per cent was invested in housing costing less than R15,000 (*Sunday Tribune*, 4 June 1989). In effect, this meant that 40 per cent of the money was spent on meeting the needs of 10 per cent of the urban Black population while only 15 per cent of it went towards meeting the needs of 57 per cent of these people. Where private developers (housing capital) control the production of housing, people face great problems in gaining access to houses and land.

In 1983 serious dislocations were witnessed in the structure of the central state. The Cabinet was given much more power than hitherto. This was achieved in various ways and amongst them was weakening the power base of parliament. The state president was allocated supreme power in the appointment of ministers' councils for each of the Houses in the tricameral parliament and the appointment of the President's Council. Over and above this, the state president was given sole power in grievance of 'Black Affairs'. As a result, the state apparatus was penetrated by the military, security forces and the police. Prior to this, although these forms had a key role in the state, this was performed under the supervision of an administration which was largely civilian in its composition and structure.

In this way, the security forces had found their way into policy-making and administrative apparatuses of the state. The 1980s also evidenced the reconstruction of the extra-parliamentary forces. Despite the repressive structure of the central state, the 1980s heralded the formation of the United Democratic Front (UDF). In 1983, the UDF brought together some 700 community and labour-based organizations, representing some 2 million people in a common front based on the Freedom Charter (Wolpe 1988: 78). Thus, in terms of the Black urban struggles, for the first time there was a wedding of production-based and consumption-based struggles. As Johnson Mpukumpa, National Chairman of the General Workers' Union, remarked:

> As a progressive trade union, we feel we must be involved in community activities. To have a good relationship and work together with the communities, will also help us to have more strength. The reason I say so is that the community is the workers and the workers are also the community. I see no difference. It would be a strange thing if I say I'm opposed to community associations, whilst I'm a worker from the community. The job of the trade union in the factory is to look after members in the factory but the members in the factory are from the community. We should, therefore, have a good relationship with community associations, since we are workers from the community. (Cited in McCarthy and Smit 1984: 179)

The establishment of these extra-parliamentary forces provided a new political terrain which had virtually been obliterated in the 1960s. It also set the scene for the development of social movements which were never before witnessed in this country.

The state did not take lightly the threat presented to it by popular Black struggles. The establishment of National Security Management System (NSMS) shows how seriously they had considered this threat. The NSMS had empowered 9 Joint Management Centres (JMCs), 82 sub-JMCs and 320 mini-Joint Management Centres. All JMCs and their sub-committees were chaired by military and police officers. The effect was to give a security slant to all administrative decisions and to subordinate all interests to security. The idea of joint action on this scale was an embodiment of the 'total strategy' philosophy – a reaction to the perceived 'total onslaught' by the revolutionary forces, which was conceived by military strategists (*Daily News*, 3 November 1989). The total strategy approach dictated that the revolution should be fought by a combination of material upgrading of Black living conditions, and stern security action. It was in effect a local application of the 'hearts and minds' strategy of winning revolutionary wars which the military had applied in Namibia. This was the state's response to the growing unrest of the mid-1980s – a time when the most intense rioting and unrest was witnessed in South Africa's history.

Nevertheless, despite these new developments in the state apparatus, unrest and disobedience continued both on the shop floor and in the living environment. In particular, during the mid 1980s 'people's power' through urban struggles became popular. Street committees were established in Black townships. As the President of the Border Region of the UDF put it:

> Within the ensuing vacuum we have seen the remarkable . . . emergence countrywide of rudimentary organs of people's power. In particular, democratic street committees, elected house by house, street by street, have developed . . . in a large number of townships and villages. Suddenly, many of the most severe problems that had plagued our ghettoes have been resolved, like weekend violence, gangsterism, rape. A strange paradox this, as the authorities have collapsed, as black policemen have packed up and fled from the townships, so, strange to say, we have seen a dramatic drop in the level of violent crime. This has been the experience countrywide. It is limited to the development of basic democratic control over the streets by the people themselves. These have become our own liberated zones, not in the remote countryside but in the backyard of an industrial society. (Cited in Wolpe 1988: 102)

In short, the previous dichotomy that existed between production and reproduction terrains was now welded into a common arena within which the Black urban struggle was waged.

The housing backlog in the country during this period reached critical proportions. It was estimated that South Africa will have to build almost 5 million homes in the next decade at a rate of 400,000 a year if housing needs are to be met (*Daily News*, 29 August 1989). During 1986 four contingent Acts were passed based on the repeal of influx control. The Identification Act made provision for the scrapping of the reference book for Africans. The abolition of the Influx Control Act removed all racial directions on movement of Blacks in urban areas, and the Black Communities Development Act made provision for the designation of development areas and the acquisition and vesting of land therein for the establishment of townships by development boards. Finally the Restoration of the South African Citizenship Act restored citizenship to three categories of Africans – those born in republic prior to TBVC (Transkei, Bophathatswana, Venda, Ciskei) independence, those who are citizens of TBVC homelands by birth and who have been permanently resident in South Africa for at least five years, and, those who are citizens of homelands but wish to apply for naturalization after living in South Africa for five years. These measures gave new a dimension not only to the process of urbanization but also to housing in South Africa. Whereas in the previous periods, especially prior to 1948, the urban process was characterized by low urbanization due to influx control, the present era is noted for its high migration rates to the urban areas. As a result, demand for housing has increased substantially and informal settlements have increased enormously. Overcrowding and a lack of proper amenities contribute signally to a poor quality of life. However in 1989 the Prevention of Illegal Squatting Act has made the process of squatting a severe breach of the law. Since the state no longer controls all urban housing and development, it is expected in that in the 1990s life chances will become more difficult for the black proletariat living on the peripheries of most cities.

Social movements in Durban

The crisis in housing in Durban exemplifies the context within much urban social movements have emerged. Mobilization over rent issues began to crystalize from 1980 in almost all the townships of Durban. In the Indian and coloured townships of Chatsworth and Phoenix and Wentworth and Newlands East, respectively, rent-based protests gave rise to a new social movement – the Durban Housing Action Committee (DHAC). DHAC is a federation of several civic associations and initially opposed the Durban City Council over rental issues or the Department of Community Development over removals under the Group Areas Act. It is a housing protest movement whose prime motivating force is enshrined in the Freedom Charter, which views housing as much more than a physical artifact and related to many other issues, both production and consumption based. The movement has since its inception aligned itself to the UDF.

The Joint Action Rent Committee (JORAC) was formed in 1983 primarily to promote rent-based issues in African townships. The organization came to the fore after an activist, Harrison Dube, was assassinated (Reintges 1986: 90). The protest over state rentals was highly successful, in that 75 per cent of tenants did not pay (*Daily News*, 31 August 1983). JORAC thus became another social movement which emerged to demonstrate against the state's undemocratic decision-making and the imposition of higher rentals.

In Kwa Mashu and Umlazi the most significant protest movement was the Congress of South African Students (COSAS). That it had contacts with trade unions was evident when the students organized a mass stay-away in these townships to support them in their protest over education-based struggles. These were largely uncoordinated protests which led to looting and burning of shops and houses. Indian store-owners were also attacked, so that the struggle in the living environments was fractionalized.

Another social movement has also arisen in the Durban area – the Durban

Central Residents' Association (DCRA). The DCRA was formed in 1984 to oppose the eviction of illegal black residents in the supposedly white 'Warwick Avenue Triangle' and Albert Park area in Durban. Since then the Association has taken up the issue of rentals and living conditions of people in Durban and surrounding areas.

In addition to the above-mentioned social movements, various other civic organizations have coalesced under the banner of UDF. During the parliamentary elections in 1989 political unrest heightened, and once again, in Durban and elsewhere, grass-roots organizations militated against Indians and Coloureds being represented in the tricameral parliament. Amongst their major grievances were inadequate housing facilities, unemployment and inadequate amenities. From 1987 the local state adopted a new policy with reference to squatters in the Durban region. Accordingly, policy required that all squatters' settlements within and outside the city be identified and accepted as permanent communities and their infrastructure should be upgraded (Institute of Race Relations *Survey*, 1987–8).

With the passing of the Prevention of Illegal Squatting Act, however, the onus of removing squatters from 'white' urban areas shifted to private landowners. The situation led to a number of internecine struggles between African squatters and Indian landowners. The situation exacerbated itself until the Natal Democratic Lawyers Association, a progressive organisation with strong ties to the UDF, intervened on behalf of the squatters and defended their claims in the Supreme Court. Whilst a short respite was granted, the majority of the squatters are still without proper shelter.

The situation in respect of the struggle has shifted somewhat, especially after the historic moment of February 1990 when the ANC and various other extra-parliamentary forces were unbanned. The protracted black struggle for a decent living environment and the inherent contradictions have intensified. As the possibility of a post-apartheid South Africa looms on the horizon, various internecine power struggles have also escalated.

Synthesis and conclusion

In the context of attempting to understand the housing question, this chapter has abstracted a few historical moments of urban struggles that have generated as a result of the blatant discriminatory policies of the apartheid capitalist state. More specifically, though, what ought to be clear from this presentation is not so much the historical development of social movements, but rather, their importance for emancipatory action. Our conceptual understanding of the housing question is not simply a matter of production or consumption. The housing question in South

Africa, especially for blacks, pervades their very existence: who they are, what they are, and where they stay. The historical moments of the struggle for survival that were alluded to previously have exemplified the nature of the dialectical relationships between the state, the mode of production and society.

Thus the question of housing needs to be contextualised in all its manifestations. Consequently, it is imperative that we frame our analysis to seek the dialectical relationships between state and capital interventions on the structure of housing opportunities and the fragmentation of class interests in housing. Moreover, the principal consideration should encompass the role of knowledge in the arena of emancipatory action. As Sayer (1984: 230) avers, 'what is learning for, if not to change peoples' understanding of their world and themselves?'

Housing, therefore, 'becomes an indicator and a potent symbol of the shifting power relations between classes and within different sectors of capital. The historical and immediate place of locality and its population in the capitalist schema are visible in the manner in which reproduction and production link.' (Klausner 1986: 38). The study of the dialectical relationship between production and consumption and their specificities become, as a result, imperative in understanding the organic whole.

References

Bond, P. 1990. Deracialized urbanization: for the benefit of the middle class? *New African*, Urbanization Supplement, Durban.

Castells, M. 1977. *The Urban Question*. London: Edward Arnold.

Castells, M. 1983. *The City and the Grassroots*. London: Edward Arnold.

Edwards, I. 1983. Living on the 'smell of an oilrag': African life in Cato Manor in the late 1940s. Paper presented to a workshop on 'African life in Durban in the Twentieth Century', University of Natal, Durban, October 1983.

Hendler, P. 1987. Capital accumulation and conurbation: rethinking the social geography of the African townships. *South African Geographical Journal* 61: 60–85.

Klausner, D. 1986. Beyond separate spheres: linking production with social reproduction and consumption. *Environment and Planning, D: Society and Space* 4: 29–40.

Marx, K. [1953] 1973. *Marx's Grundisse*. Trans. D. McLellan. London: Paladin.

McCarthy, J. 1987. Paul Hendler's rethinking of the social geography of black townships: a brief reply. *South African Geographical Journal* 69: 86–88.

McCarthy, J. and Smit, D. 1984. *The South African City: Theory in Analysis and Planning*. Johannesburg: Juta.

Morris, P. 1981. *A History of Black Housing in South Africa*. Johannesburg: South Africa Foundation.

Ratcliffe, S. 1980. A political economy of housing. In Development Studies Group and Southern African Research Service (eds), *Debate on Housing*; pp. 1–14. Johannesburg: University of Witswatersrand.

Reintges, C. 1986. Rents and urban political geography, the case of Lamontville. Master's thesis, University of Natal, Durban.

Sayer, A. 1984. *Method in Social Science: a Realist Approach.* London: Hutchinson.
Smith, M. P. and Tardanico, R. 1987. Urban theory reconsidered: production, reproduction and collective action. In M. P. Smith and J. R. Feagin (eds), *The Capitalist City – Global Restructuring and Community Politics*, pp. 87–110. New York: Basil Blackwell.
Southern African Institute of Race Relations, 1950–1. *A Survey of Race Relations in South Africa.* Johannesburg: South African Institute of Race Relations.
Sutcliffe, M. and Wellings, P. 1985. Worker militancy in South Africa: a sociospatial analysis of trade union activism in the manufacturing sector. *Environment and Planning, D: Society and Space* 3: 357–79.
Swilling, M. 1986. The United Democratic Front and township revolt. In W. Cobbett and R. Cohen (eds), *Popular Struggles in South Africa*, pp. 90–113. New Jersey: Africa World Press.
Wolpe, H. 1988. *Race, Class and the Apartheid State.* Paris: Unesco Press.

4 *State intervention in housing provision in the 1980s*

SUSAN PARNELL

Despite the South African government's official position that it is no longer responsible for *providing* shelter, other than in exceptional cases to the indigent (South Africa 1983: para. 8.4), state *involvement* in housing persists and has even expanded in the last decade. The rate of government building of housing remains constant and the amount of state money allocated to shelter has increased through state-subsidized home ownership and informal housing. However, the administration of housing finance in the declining years of apartheid continues to favour whites over blacks, and the rich over the poor. As a result the most notable features of the 1980s urban landscape are vast informal shack settlements, the development of an elite African housing market, the persistent oversupply of housing for whites relative to the gross shortage of African shelter, and the involvement of the tricameral parliament in the supply of housing for coloureds and Indians. The housing question in South Africa is more politically charged now than it has been at any time since the Second World War.

Conventionally, the restructuring of apartheid housing policy that followed the 1978 Riekert and the 1982 Viljoen Commissions is understood to centre on the state's abdication from its once ubiquitous role as township developer and landlord. Academic focus is generally on the implications of private-sector intervention in black housing supply (Mabin and Parnell 1983; Wilkinson 1983; Bond 1990a). Perhaps the strongest argument to emerge from the focus on capital's expanding construction programme is that corporate and union involvement in providing shelter has made housing a critical issue for negotiating a resolution to the crippling urban political crisis (Hendler 1988, 1989). The recent proliferation of squatting and site and service schemes widens the housing debate to embrace informal accommodation and such issues as the impact of the abolition of influx controls on urbanization (Hindson 1987; Watson 1987; Mabin 1989; Crankshaw and Hart 1990). Fresh awareness of the spread of shack settlements in metropolitan areas places the spotlight on the role of the state in providing affordable shelter. By posing problems such as the cost of a serviced site and the necessary state subsidy required to provide adequate infrastructure, planning dilemmas of the new South Africa are already being confronted. Central to discussion on state housing policy is the extent to which current spending should be modified (see African National Congress 1990), an area that requires considerably greater background knowledge than exists to date.

The government's current position on housing provision makes it difficult to ascertain the precise amount or priority of state involvement in housing. In an attempt to depoliticize the government's role, particularly in the black townships, the extent of state participation in housing provision has diversified away from state-managed public housing that is the hallmark of apartheid cities. Officials now publicly encourage individual and corporate responsibility for shelter. The argument is that home ownership provides a 'highly desirable social objective', whilst for the private sector the privatization of construction 'presents a substantial business opportunity' (*Housing in Southern Africa*, October 1989, p. 59). Since 1983, virtually every policy statement, parliamentary debate and departmental report stresses the increased private sector responsibility for housing supply and the diminishing construction function of government itself. In reality, however, the state's commitment to housing has increased in the 1980s, most notably through home-ownership subsidies and infrastructural investments (Hendler and Parnell 1987).

State house building in the 1980s

Despite the widely publicized shift in policy, privatization has had only minimal impact on government's traditional function in supplying low-income public housing. A survey of the state's construction record over the past decade reveals no significant overall pattern of change. The number of units erected annually remains relatively constant at around the 25,000 unit mark. The escalating shortage of housing units in urban areas (Fig. 4.1) suggests that the government has never adequately addressed the problem of providing sufficient housing. The failure to supply adequate shelter for the black urban population is not purely a post-privatization phenomenon.

Notwithstanding the state's ideological commitment to black home ownership and the reform of apartheid a huge section of the urban population, notably Africans, depend on welfare support or sub-economic housing for their position in the formal housing market. The 1983 announcement that public housing was to be privatized was accompanied by the warning that no new state housing was to be erected, other than for those welfare cases earning less than R150 per month (the breadline is estimated at approximately R600 for a family of six (South African Institute of Race Relations 1989)). In fact, the overall rate of public housing construction in the 1980s has fluctuated, but has not altered dramatically. There are,

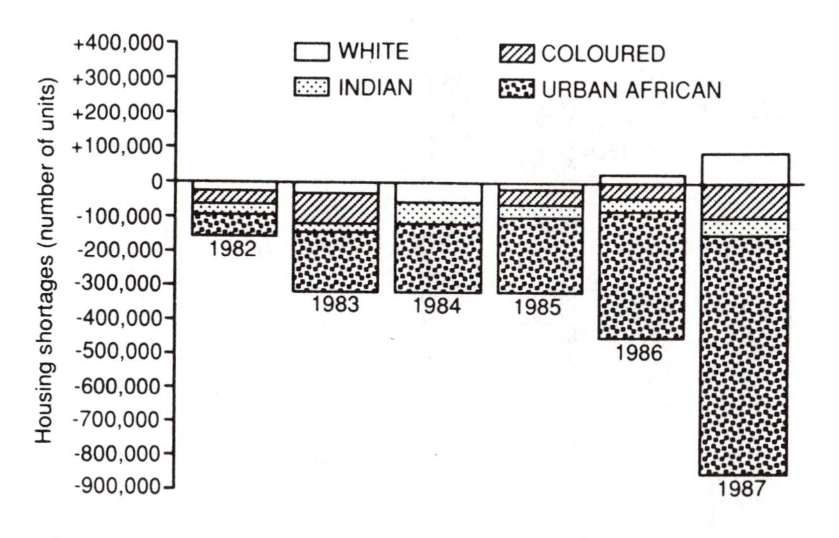

Figure 4.1 Official housing shortage by race, 1982–7.
Source: South African Institute of Race Relations.

however, important shifts in the pattern of supply across the race groups. There is a notable decline in the number of units built for whites. Supply of housing to Africans has remained relatively constant, but construction for coloured and Indian occupation has increased steadily throughout the decade.

The percentage of the budget (Fig. 4.2) allocated to white public housing is surprisingly low, given the apartheid state's legendary support of the white working class. However, an estimated 75 per cent of the white population are housed in individually owned homes (*Hansard A* 1983: col. 776), and only about 14 per cent of people in South Africa are classified white. The number of whites dependent on state assistance is therefore relatively low, and is largely restricted to the aged. Nevertheless, it is true that the poorest whites, approximately 8 per cent of the white population, who hitherto enjoyed state support on a scale unrivalled even by welfare states such as Holland (*Hansard A* 1984: col. 5982), can no longer count on the same degree of assistance in securing affordable shelter.

Coloureds, who comprise about 9 per cent of the total population, and Indians, who are only about 3 per cent of the total population, also received a disproportionate allocation of both funds and state-managed units in the 1980s (Figs 4.2 and 4.3). The bias toward coloured and Indian housing provision has increased since the introduction of tricameralism. Having assumed control of their own budgets in 1985 visible areas of spending such as housing and education were prioritized by both the Indian House of Delegates (*Hansard D* 1987: col. 2084 ff.) and the coloured House of Representatives (*Hansard R* 1987: col. 386). For example, in the 1988/9

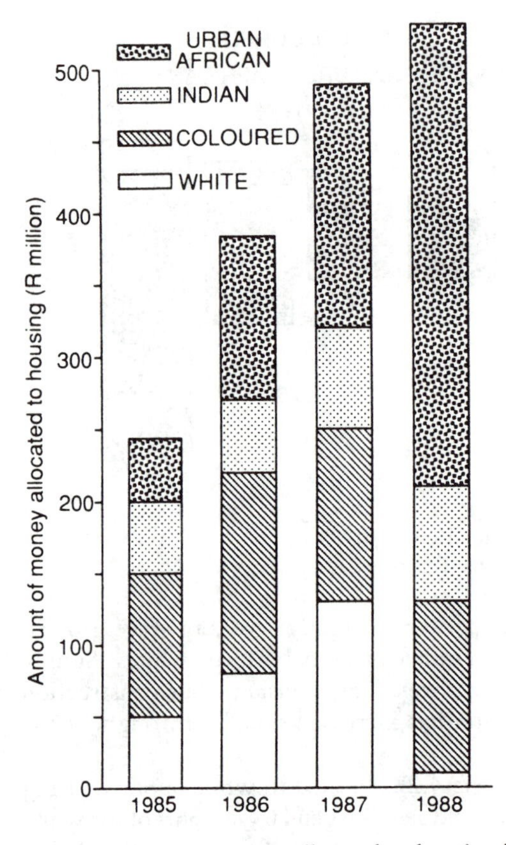

Figure 4.2 State money allocated to housing by race, 1985–8.
Source: South African Institute of Race Relations.

budget the House of Delegates housing allocation was increased by 400 per cent on the previous year (*Housing in Southern Africa*, August 1988). The bulk of the increased allocation for coloured and Indian housing has been for home ownership rather than welfare assistance and in this respect the trend corresponds with the allocation to Africans, but not to assistance for white housing (Table 4.1).

The impact of state involvement in African housing changes dramatically if money rather than the number of units constructed is considered. Although the per capita expenditure on Africans remains low, the 1980s have seen huge increases in the proportion of the housing budget allocated to African housing, especially in the latter part of the decade (Fig.4.2). As the constant rate of house production demonstrates, little of the extra money has been spent on house building. Instead there has been a concerted effort to improve services and acquire additional township land to allow private-sector housing projects to go ahead (South Africa

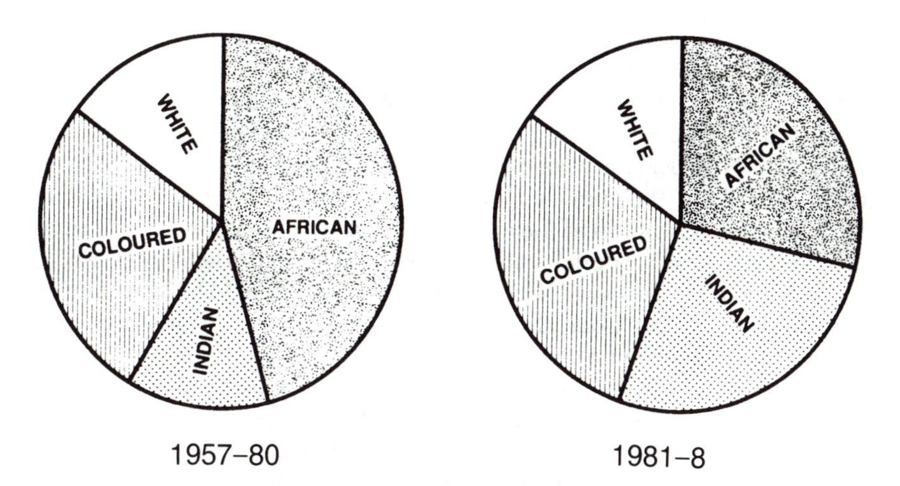

Figure 4.3 Proportion of housing units allocated to each race.
Source: South African Institute of Race Relations.

1986). The resultant tariff hikes levied on African township residents, whose residential areas are now forced to be financially self-sufficient, sparked extensive township violence and successful rent boycotts in several parts of the country (Chaskalson *et al.* 1987).

Table 4.1 Summary of National Housing Committee expenditure proportions by race group and subsidy type, 1988.

Type of subsidy	Whites %	Coloureds %	Indians %	Africans %
Family housing (FTHB subsidy)	15	81	91	73
Serviced sites and materials	0	13	8	26
Welfare housing	84	3	1	1
90 % loans	1	3	0	0
TOTAL	100	100	100	100

Source: Goede 1990: 7.

The township unrest of 1984–5 was the catalyst for the restructuring of privatization practices. The policy of black home ownership that was floated as early as 1976 and was fully operational by the early 1980s, was a self-conscious attempt to stabilize the urban population by creating a black middle class through home ownership and other social and economic reforms. Mysteriously, the possibility of similarly buying off the black working class appears to have been overlooked. In the drive to reduce government spending on sub-economic shelter the full costs of African township development were passed on to tenants, while those who could afford home ownership were subsidized. Traditional methods of reducing township rentals, the municipal beer monopolies and the support of white local authorities who drew income from black residents, were revoked in the late 1970s when Administration Boards were made wholly responsible for financing African areas (Bekker and Humphries 1985). The introduction of Regional Services Councils has marginally offset the greatly increased costs incurred in the improvement and extension of African areas, but the majority cannot and will not accept higher levies for housing.

The 1985 Development and Housing Act affirmed the government's commitment to home ownership, but also introduced important concessions to welfare candidates unable to afford conventional shelter. State acceptance of responsibility for assisting this section of the population has thus far involved endorsing site and service schemes, rather than extending the practice of rent subsidization that was common in African townships in the 1950s and 1960s. Concessions to the expansion of serviced sites for Africans in urban areas represents a significant departure from previous policies of orderly urbanization (Hindson 1987). It is also a pragmatic response to the fact that the majority of the black urban population are not potential home owners because they cannot afford the costs of formal housing. In response to the affordability crisis and prevailing township violence, the government is now rumoured to be giving favourable consideration to the Urban Foundation's proposed capital subsidy scheme (*Update*, July 1989). The suggestion is to give a once-off grant to finance infrastructural investments, thereby lowering the cost of core housing for those with incomes less than R1,000 per month. Cosser (1990) estimates that such a subsidy would apply to 80 per cent of Africans, 55 per cent of coloureds, 33 per cent of Indians and 10 per cent of all whites. Should the scheme be implemented, it would require a fundamental restructuring of resources to facilitate the necessary five-fold increase in housing spending. Clearly current state attitudes to subsidizing working class housing are in state of flux. Growing concessions to popular demands for affordable land and housing have in no way lessened the resolve to create a home-owning black middle class.

State home ownership incentives

The housing crisis in South African cities is primarily a legacy of apartheid policies. Although as part of the reform of apartheid, Africans are now accepted as permanent urban dwellers, the available facilities are severely deficient. After 1968 the apartheid state confined the construction of African family housing to the bantustans. Over the next 10 years of 'grand apartheid', spending on African housing in 'white South Africa' was cut by a factor of seven (Hendler 1989). Funding for African township development began to increase only after the 1977 advent of Administration Boards, a trend that persists to this day. The shortage of housing in metropolitan areas has, however, not abated. An estimated 40 per cent of Soweto houses now have backyard shacks. Many squatters are employed urbanites who are unable to find formal accommodation (Crankshaw and Hart 1990). The state has not therefore simply abdicated a position of successful supplier of shelter that it once fulfilled in favour of privatized housing delivery. Rather, in the face of an unmanageable shortfall in black housing, the private sector *and individuals themselves* have been called upon to assume their 'rightful place' in supplying shelter and redressing apartheid's failures.

In theory the private sector supported the introduction of black home ownership, particularly insofar as it was seen to enhance racial equality (Mather and Parnell 1990). However, low cost margins discouraged large-scale cost-effective housing development and poor wages were not accepted as sufficient security for financial institutions to grant assistance to buy property. In 1981 less than a tenth of all black housing developments were private-sector initiated. Reluctance to participate in the deregulation of low-income housing delivery is averted by financial incentives, generally mortgage subsidies, that promote the new ideological hegemony of 'home ownership for all' (*Housing Southern Africa*, October 1987). The construction industry, finance capital and some segments of the urban population are the primary beneficiaries of the new 'privatized' dispensation which is really just state intervention to increase the proportion of the population that can afford to buy a house.

In 1983, tenants of 500,000 public housing units were given the option of purchasing their homes at greatly reduced prices. Ironically, although the state sought to divest itself of its existing housing assets, the sale of public housing has necessitated extensive high-profile interventions to persuade tenants to purchase their houses. In addition to the costs of publicizing and administering the 1983 'Great Sale', purchase discounts have been periodically increased to encourage tenants to buy the stock. The success of the selling scheme has, nevertheless, been limited. In September 1989, only 39 per cent of coloured, 77 per cent of Indian, 37 per cent of white and 34 per cent of urban African occupied government stock had been transferred to individual ownership (*Update*, July–December 1989). The most obvious explanation for the poor response is that tenants cannot afford to purchase. Escalating mortgage rates (from 12 per cent to over 20 per cent in the post 1983 sale period) have put home ownership beyond the reach of the average family. Thus the government continues to be the single most important landlord in the country and in many areas the majority of legal residents are still state tenants.

The extent of indirect government involvement in housing in South Africa is vast. An estimated 70 per cent of all mortgages are subsidized by the state (Bond 1990b). Civil servants (36 per cent of the economically active white population work for the state (*Hansard A* 1982: col. 2745)) qualify for mortgage assistance, unless they are married women in which case this employment benefit is withheld. The practice of assisting employees with housing costs has spread to the private sector where 63 per cent of companies, including the mines, subsidize accommodation costs (Hendler 1989). Because these subsidies are applicable to both new and old housing the majority of established income earners are not encouraged to acquire new properties where building costs are higher. In an attempt to boost the total housing stock and to facilitate a differentiated township housing market, thereby giving a large percentage of the black population a stake in the future of urban areas, generous state assistance is made available to prospective home owners.

There are different types of state home ownership assistance offered to first-time buyers (see Bond 1990b and Goede 1990 for comprehensive details). Without exception applicants for government subsidies are subject to an income restriction, and the total cost of the new house is limited. By far the most popular method of financing housing is the first-time home owner's subsidy (Table 4.1), a scheme that has had significant impact on who is able to afford a house. In 1990 17 per cent of the total housing budget was set aside for this form of assistance (Cosser 1990). By mid year, however, all the funds were depleted (*The Star*, 24 July 1990). The first-time owner's scheme was first introduced as a mechanism for assisting whites at the lower end of the housing market who were struggling to obtain their own properties. Of particular concern were young people who, due to escalating building costs and declining standards of living associated with the economic slump of the early 1980s, could no longer afford to buy a house (see *Hansard A* 1981: col. 696, 1222). At the same time low-income whites who lived in rented accommodation were being squeezed out of affordable housing. The Rent Control Act was amended in 1980 and only buildings erected before 1949 remained protected by the Act. Also of significance was the introduction of sectional title on multiple unit residences (*Hansard A* 1980: col. 6288). As property speculators bought up the cheaper inner-city flats many whites found themselves homeless. Thus, despite the glut of over 83,000 units in white group areas, 26,000 working-class whites were reportedly without shelter in 1987 (*Housing Southern Africa*, February 1988).

From its inception the first-time home buyer's scheme was open to all races, but it was effectively only used by whites. Delays, measured in years rather than months, in proclaiming coloured and Indian group areas prevent individuals from taking transfer of their sites and thus prevent financial institutions from granting mortgages. Individuals therefore live in houses that they will one day buy, but that they do not actually own; they therefore do not qualify for the first-time home owner's subsidy (*Hansard D* 1985: col. 2716). Although Africans were never excluded from the first-time owner's subsidy they were only able to participate once they were granted ownership rights (99-year leasehold was accepted in 1976 and freehold in 1985). As in the case of coloureds and Indians, red tape in African township development reduced the number of applicants who could take advantage of the subsidy. At least some of the delays in surveying and registering individual title appear to have been overcome and African participation in the scheme has escalated beyond expectation (South Africa 1989).

The rapid increase in African demand for home ownership assistance and declining white participation has meant that the first-time subsidy has effectively become a mechanism for creating a black middle class. However, despite overt state support of reform and the creation of a racially integrated home-owning urban population, discriminatory housing assistance persists. Africans wishing to draw on the scheme must earn under R2,000 per month while whites may earn R3,500. Since 1987 whites have been allowed to use the subsidy on existing stock, blacks are forced to build from scratch. In all other respects the regulations are the same for all applicants. Yet the distribution of the first-time home owner's subsidies has further class and race implications. The manner in which the subsidy currently operates means that those who earn more enjoy greater state assistance. In addition to the differential wage criteria, income determines the total grant. An applicant earning R500 per month would only qualify for a R7,000 loan, whereas a R35,000 loan would be allowed for a family with a monthly income of R2,000. The monthly state subsidy on the R35,000 loan is R193, three times greater than the R66 paid out on loans of R7,000 (Table 4.2).

The confluence of economic and racial discrimination in South Africa means that whites, and to a lesser extent Indians, have higher incomes than Africans and coloureds (Table 4.3). Because of wage discrepancies, the first-time home owner's subsidy benefits people who already have greater economic power. Although the proportion of whites who gain from the subsidy has declined in the past five years (South Africa 1989), the 1988 figures for the first-time scheme show that the average white applicant received a subsidy of R3,047, in contrast to the average R1,116 subsidy given to African buyers. More importantly, the 1989 statistics reveal that the average white subsidy increased by 36 per cent, but that African applicants received only 30 per cent higher than the previous period (Table 4.3). The discrepancy between the amount individual white and African families can expect to receive from the home ownership incentives is increasing.

Table 4.2 Housing subsidy allocations according to income.

Monthly household income, R.	Proportion of applicants	Amount lent to applicant, R.	Number of loans granted	Average monthly subsidy, R.	Total state subsidy, R.
500–599	17.99	11 930	7 196	66	5723 438
600–699	10.21	14 090	4 086	78	3837 970
700–799	14.84	16 280	5 934	90	6423 864
800–899	12.26	18 430	4 904	102	6025 995
900–999	7.89	20 600	3 156	114	4333 891
1000–1099	10.85	22 770	4 339	127	6587 331
1100–1199	5.07	24 930	2 030	139	3371 244
1200–1999	20.89	34 690	8 355	193	19323 143
	100.00%	20 863	40 000		55626 876

Adapted from: *Housing in Southern Africa*, February 1990, p. 12.

Conclusion

Notwithstanding the tremendous changes in urban policy and practice that have been wrought in South Africa in the past decade, allocation of housing assistance remains unequal. The so-called privatization of housing has not left normal market forces to dictate access to formal housing. Instead, the state's proactive role in fostering a home-owning class has seen greater government participation in housing delivery than was the case when the government accepted total responsibility for African township development and white welfare housing.

The injustices of racial segregation and discrimination forged by the apartheid regime have not disappeared with the 1983 proposals to reform state involvement in financing low-income housing. Widespread unrest, especially in the mid-decade, heightened state awareness of African people's inability to afford formal housing, and achieved some measure of support for subsidizing site and service provision. State support of home ownership, abortively launched in 1983, is finally being widely implemented. The 'Great Sale' has been less successful than schemes such as the first-time home owner's subsidy in fostering an urban elite. However, the differential benefits granted to buyers with higher purchasing power exacerbates wealth discrepancies within the black community and fails to fully redress racial inequalities in state housing support.

Table 4.3 First-time home owners' interest subsidy scheme.

	1989 average income	Cumulative provision to December 1988		Cumulative provision to December 1989	
		Number of units	Per unit subsidy	Number of units	Per unit subsidy
White	3 297	15 696	3 047	18 618	4 820
Coloured	1 059	6 714	1 774	10 532	3 141
Indian	1 604	5 667	2 266	7 452	3 259
Urban	521	6 877	1 116	16 241	1 598

Adapted from: Cosser 1990: 66; Goede 1990: 6.

Acknowledgement

I wish to thank Alan Mabin for unpublished source material.

References

African National Congress 1990. *Discussion Document on Economic Policy*. Johannesburg: African National Congress Department of Economic Policy.

Bekker, S. and Humphries, R. 1985. *From Control to Confusion*. Pietermaritzburg: Shuter and Shooter.

Bond, P. 1990a. State housing subsidies. Unpublished mimeo. Johannesburg: PLANACT.

Bond, P. 1990b. Township housing and South Africa's 'financial explosion': theory and practice of finance capital in Alexandra. *Urban Forum* 2: in press.

Chaskalson, M., Jochelson, K., and Seekings, J. 1987. Rent boycotts and the urban political economy. In G. Moss and I. Oberry (eds), *South African Review 4*, pp. 53–74. Johannesburg: Ravan.

Cosser, E. (ed.) 1990. *Homes for All*. Johannesburg: CREID.

Crankshaw, O. and Hart, T. 1990. The roots of homelessness: causes of squatting in the Vlakfontein settlement south of Johannesburg. *South African Geographical Journal* 73: 65–70.

Goede, P. 1990. Subsidies and other housing assistance. Paper delivered to National Association of Home Builders Conference, Pretoria, May.

Hansard, A. (House of Assembly)

Hansard, D. (House of Delegates)

Hansard, R. (House of Representatives)

Hendler, P. 1988. *Urban Policy and Housing*. Johannesburg: South African Institute of Race Relations.

Hendler, P. 1989. *Politics on the Home Front*. Johannesburg: South African Institute of Race Relations.

Hendler, P. and Parnell, S. 1987. Land and finance under the new housing dispensation. In G. Moss and I. Oberry (eds), *South African Review 4*, 423–31. Johannesburg: Ravan.

Hindson, D. 1987. *Pass Controls and the Urban African Proletariat*. Johannesburg: Ravan.

Mabin, A. 1989. Struggle for the city: urbanization and political strategies of the South African state. *Social Dynamics* 15: 1–28.

Mabin, A. and Parnell, S. 1983. Recommodification and working-class home ownership: new directions for South African cities? *South African Geographical Journal* 65: 148–66.

Mather, C. and Parnell, S. 1990. Upgrading 'matchboxes': urban renewal in Soweto, 1976–1986. In D. Drakakis-Smith (ed.), *Economic Growth and Urbanization in Developing Areas*, pp. 238–50. London: Routledge.

South Africa 1983. White Paper on Urbanization.

South Africa 1986. Report of the Department of Local Government, Housing and Works, 1984/5.

South Africa 1989. Report of the Department of Local Government, Housing and Works, 1988/9.

South African Institute of Race Relations, 1980–9. *A Survey of Race Relations in South Africa*. Johannesburg: South African Institute of Race Relations.

Watson, V. 1987. South African urbanization policy: past and future. *South African Labour Bulletin* 11: 77–90.

Wilkinson, P. 1983. Housing. In Southern African Research Service (eds.) *South African Review 1*, pp. 270–7. Johannesburg: Ravan.

5 *Class struggle over the built environment in Johannesburg's coloured areas*

MALCOLM LUPTON

J71 R11
South Africa

The suburban landscape of the Johannesburg Metropolitan Region is dominated by racially segregated townships, among which are the predominantly working-class coloured areas. According to Harvey (1985) the built environment of capitalist cities can conceptually be divided into fixed capital (composed of items used in production such as factories, highways and railways) and collective means of consumption (consisting of items used in the reproduction of labour power, such as houses, roads and health care facilities). South African cities, despite the specificities arising from race, are shaped by the processes of capitalist urbanization. The study of the urban process in South Africa should proceed from the accumulation of capital and its corresponding categories such as labour and capital, as well as the struggles between classes. The coloured townships in Johannesburg are largely the result of the provision of collective means of consumption by state initiative to a predominantly working-class population. That the built environment is also the site of class struggles (Harvey 1985) is clearly the case with the coloured areas of Johannesburg.

The number of coloureds in Johannesburg was estimated in 1987 to be 163,905, the vast majority of whom live in Westbury, Newclare, Coronationville, Riverlea and Bosmont in the west of the city and the Eldorado Park complex in the south-west (Fig 5.1). An estimated 9,500 or 6 per cent of the coloured population in Johannesburg live in the white, Indian and African areas of the city. The occupational structure of the coloureds in Johannesburg suggests that most of the economically active population are employed in wage-labour categories (Table 5.1). The social relations which structure the class position of coloureds are also organized across space. In Johannesburg the coloured working class is located mostly on the periphery of the city in racially defined dormitory townships. Although the spatial separation between the place of residence and the workplace is a feature of all capitalist cities, in South Africa the reproduction of labour power occurs primarily in racially segregated areas at great distances from the centres of production.

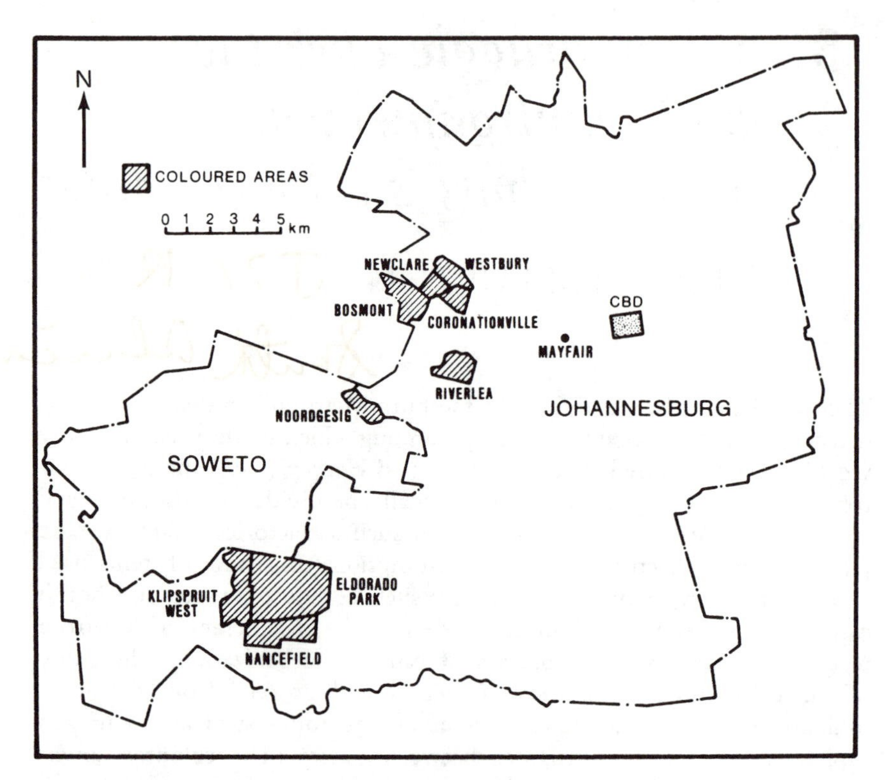

Figure 5.1 The coloured areas of Johannesburg.

Table 5.1 Occupations of coloured people in Johannesburg.

Occupations	Percentage of economically active coloured population 1960	1970	1980	1985
Industry	55.3	60.6	46.0	43.0
Clerical and sales	07.7	17.5	26.0	14.0
Services	13.4	07.5	05.6	07.0
Professionals and technical	03.7	04.4	07.0	07.3
Management and administration	00.0	00.2	00.7	01.9
Unclassified	19.6	09.7	14.5	12.8
TOTAL	100.0	100.0	100.0	100.0

Sources: Republic of South Africa, 1961, 1971, 1981, 1986 (figures subject to rounding errors).

The chapter is divided into three sections. The concern in the first part is with rent conflicts between working-class coloured tenants and the Johannesburg City Council. The second part is centred on struggle over urban renewal with specific reference to the township of Westbury. Part three is focused on an analysis of the interventions of capital, through state power, in class conflicts over urban space in the areas of study. In particular, the provision and management of collective means of consumption in Johannesburg's coloured suburbs is scrutinized as a determinant of urban politics. The social processes underlying the conflicts over the built environment in the coloured areas of Johannesburg offers a striking case of the interplay between labour, capital and the local state. Conflicting meanings and definitions that are attached to the built environment by different social classes and the local state form the central theme of the argument.

The coloured working class versus the appropriator of rent

To understand the content of struggles between coloured tenants, the Johannesburg City Council and construction capital, the underlying class nature of this conflict needs to be recognized. What appears merely as a fight over rent and urban renewal is essentially a struggle between capital and a predominantly working-class population.

Before 1950 coloureds were dispersed throughout inner city areas of the Witwatersrand, often living in dismal slum conditions. Coloured group areas were proclaimed for Johannesburg in August 1956 and as a result of the implementation of the Group Areas Act the coloured working class was relocated to more outlying areas. Also, in terms of the Housing Act of 1957 the provision of housing for low-income coloured workers was to be the co-responsibility of the Johannesburg City Council and the central state. The central state was to fund coloured housing in the newly proclaimed group areas through a Department of Community Development. From 1950 the state sought to create a few, large centres on the Witwatersrand where coloureds could be relocated. The policy resulted in the forced removal of many communities to racially segregated townships (Pirie 1983; Pirie and Hart 1985). In the coloured townships of Johannesburg, rent struggles became a conflict between the City Council and tenants since the role of landlord was to a large extent taken over by the Council.

The first major rent conflict between the coloured working class and the City Council occurred in August 1960 when the Council announced a rent increase for 1,000 of the 1,560 houses in Coronationville and Noordgesig. The increase in rent was also linked to family incomes at the urging of the National Housing Commission. The decision sparked off a sharp response from tenants. The general feeling among the Noordgesig and Coronationville residents was that the increase was 'unjustified' and it was argued that the increase would 'bring about hardship to scores of coloured families' (*The Star*, 16 August 1960). Typical of the perspective held among tenants was the contention that 'before there can be any increases in rents there must be increases in wages and the employment situation must be improved' (*The Star*, 16 August 1960). Residents interpreted the increase as an attack on their standard of living and began defaulting on payments. By 1963 Noordgesig and Coronationville tenants owed the City Council R21,000. The state commenced evicting defaulters and in 1964 required that employers made regular deductions from their wages to offset the outstanding rent. These measures were not very successful, because rent arrears stood at R65,000 by 1965.

By the late 1960s South African industry was concentrated in the major metropolitan regions despite the government policy of industrial decentralization. The passage of the Environment Planning Act in 1967 represented an attempt by the state to accelerate industrial decentralization as part of its broader apartheid policy. The Act was stringently applied on the Witwatersrand and in Johannesburg (Rogerson 1982). The aim was to coerce industry based in the Johannesburg Metropolitan Region to relocate to deconcentration points near or inside the bantustans as a means of stemming the flow of African labour into the cities. In terms of Section 3 of the Act ministerial permission was required before companies could increase the number of their African workers whereas coloured, Indian and white labour remained unaffected (Rogerson 1982). The application of the Act in the Johannesburg region meant that new employment opportunities were created for coloured workers which caused a substantial influx of work-seekers from across South Africa. Consequently pressure on the built environment was increased as the demand for housing, transportation and services grew, without a concomitant increase in supply. The housing stock actually decreased in 1972 with the demolition of houses in Westbury. Only in 1976 was there a significant increase of the housing stock when new houses were constructed in Eldorado Park.

Conflict over urban renewal in Westbury

The struggle over the definition and meaning of the built environment in coloured areas came strongly to the fore in Westbury during 1985 when working-class tenants confronted construction capital. Westbury was originally an African township called Western Native Township and renamed Western Coloured Township in 1962 when declared a coloured group area in terms of the Group Areas Act. The same year the City Council embarked on an urban-renewal scheme for the area. A complex three-way battle between tenants, the City Council and the LTA construction company (hereafter LTA) paralleled the renovation of Westbury. It took different forms and reached different levels of intensity from time to time. The renewal scheme was carried out in three phases. The construction of new houses involved the demolition of the original stock and the total alteration of the infrastructure and urban design of Westbury (Parnell and Payne 1989). LTA was requested to prepare a detailed feasibility study and awarded a contract of R20 million for the construction of the third stage of the renewal.

For LTA the renovation of Westbury was production for profit and the company was motivated by the need for capital accumulation. The requirements of profitability were clearly reflected in the urban design the company proposed for Westbury, a key concept of which was the trafficable playcourt where spaces for pedestrian and vehicular movement were integrated. The trafficable playcourts were extended right up to house frontages and represent an attempt by the developer to optimize the use of space thus reducing construction costs. This integration of public space (roads) with private places (play courts) violated the traditional concept of *die jaart* (township Afrikaans for backyard) in Westbury. The backyard accommodated day-to-day household activities such as laundry, secure playing for toddlers, storage and vegetable gardens. Westbury residents had a definite mental map of their community space ordered around a grid street pattern (PLANACT 1985). The trafficable playcourts and the irregular road pattern of the new urban design violated this mental map of community space.

The militant struggle of Westbury residents in 1985 over the redevelopment of their township was centred on the provision of collective means of consumption. Action against the construction project was swift. Foundations dug during the day by LTA workers were filled in at night by residents who protested against the restricted size of the houses. When it became apparent that construction would continue, building sites were entered during the day, walls knocked down, and building material damaged. The Westbury Residents Action Committee (hereafter WRAC) mobilized tenants through public meetings which culminated in a march to the Department of Housing in the city centre. Police suppressed the march with force and leading WRAC activists were detained. Students and unemployed youth of Westbury soon started to confront the police in the streets. Barricades were erected in the township and pitched battles were fought with security forces along the road separating Westbury from the white suburb of Newlands. After a bitter confrontation the City Council finally agreed to meet with representatives of residents and it was decided that the WRAC would be allowed to present alternative plans for the renewal of the area. In cooperation with a progressive architect's consultancy (PLANACT), the WRAC made new proposals for the renovation of Westbury which incorporated many of the demands of tenants.

Underlying the conflict over the renewal of Westbury was a deeper conflict over the very definition of urban space in coloured areas. For the coloured working class the built environment means items of consumption which are used in the reproduction of labour power. Workers in the affected areas have defined their urban space in terms of its use value to them whereas capital and the local state see the built environment in terms of its exchange value. For the construction interest represented by LTA the built environment provided a field for profitable accumulation. This is also the case where the local state such as the Johannesburg City Council assumes the role of landlord because not even the state can entirely bypass the logic and circulation of capital. Consequently the working class in coloured areas is engaged in a series of struggles over a variety of issues related to the creation, management and use of the city's built environment. Capital, as a class, can also not afford the outcome of conflict to be determined by the relative strength of labour, rent appropriators and the construction interest and therefore intervenes, primarily through state power, in urban struggles (Harvey 1985). The state itself is not a neutral force external to class relations but reflects class conflict and is also the site of class struggle.

The intervention of capital in urban struggles

Collective or socialized consumption refers to state provision of consumption items such as housing, transport and health care which are necessary for the reproduction of labour power and capitalist relations of production (Harvey 1978; Preteceille 1981, 1986). Capitalist states generally secure the social conditions required for capital accumulation by providing collective means of consumption which cannot be produced profitably by commodity production (Harvey 1978; Preteceille 1986). The construction of the Eldorado Park complex, which consists of Eldorado Park, its nine extensions, Nancefield and Klipspruit West (Fig.5.1) is an exemplary case of state provision of collective consumption to a predominantly working-class population.

Construction of Eldorado Park, located 15 km south-west of the Johannesburg central business district, started in the mid-1960s on cheap land without amenities. The isolation and distance from the central business district proved to be unattractive to private developers. Eldorado Park is today a high density area with a population in excess of 50,000. The agglomeration of housing units reflects the need to reduce the impact of ground rent on total housing cost. Eldorado Park was not planned as a unit, but expanded when necessary as a response to the growing housing crisis of the 1970s and 1980s, hence the long time-span of its construction.

The spatial form of Eldorado Park is characterized by standardized state built houses and large, uniform complexes of flats connected to the city by rail and road. There are insufficient amenities such as health care and cultural services, poor public transport and a shortage of shops, schools and day-care centres. Overwhelmed by the bleak skies of the Johannesburg region Eldorado Park was built in stages, and each phase named with immutable bureaucratic logic Eldorado Park Extension 1,2,3, . . . 9. Struggles developed in parallel with construction.

Two processes operated in conflicts over the provision of the collective means of consumption in Eldorado Park. On one level, the highly socialized nature of the production and management of Eldorado Park by the state created common interests around which residents could mobilize. Although the central state initiated the construction of Eldorado Park in the 1960s the Johannesburg City Council progressively assumed the role of developer, landlord and manager of the means of consumption. On another level, the harsh living conditions had an adverse effect on human interaction and social networks. The first inhabitants of Eldorado Park practically lived in the midst of a construction site with no urban amenities. Social anonymity and cultural alienation were especially acute for women who became isolated in their households. Many working-class coloured women were also forced into a dual role as wage labourer and domestic worker, with the geographic separation between work and home creating unique problems for them. These degraded living conditions were the outcome of broader social and economic processes. Eldorado Park is a state-provided concrete dormitory where labour power is reproduced.

Conflict between tenants and the City Council surfaced in 1977 over a rent increase announced for Eldorado Park. When some 200 people, mostly women, gathered at the rent office in the area to protest against the increase the demonstration was violently suppressed by riot police. That same year squatters from the Kliptown slums forcibly took possession of newly built houses in Extension 3 because of the acute housing shortage. A protracted struggle which followed between the local state and squatters was only resolved when the City Council provided alternative accommodation to the squatters. In 1980 the coloured education system went into an unprecedented crisis when 100,000 students began a nation-wide school boycott which was quickly transformed into a political conflict as the central state reacted with severe repression. The revolt of coloured youths against the state had a radicalizing effect on tenant struggles in Johannesburg.

The City Council announced in 1982 that it intended to increase rent and service charges in all coloured townships by up to 60 per cent. The decision was met with 'massive opposition' (*Speak*, November 1982). Tenants in Eldorado Park, Riverlea and Coronationville held mass meetings in protest against the increase. The WRAC already engaged since 1981 in a struggle with the Council over the urban renewal scheme for Westbury urged residents in all the affected areas to greater militancy. Action committees were formed in all the coloured townships and combined with the WRAC to create a Co-ordinating Residents Action Committee. The conflict peaked on 28 September 1982 in a confrontation between the Co-ordinating Committee and the Council when activists presented the demands of tenants to the city authorities.

Women held a placard demonstration in support of the committee in front of the Civic Centre in the CBD. Slogans on placards reflected the form of consciousness generated by the rent struggle. Demands for 'decent housing, decent rents' and 'rents we can afford' as well as accusations that 'the city council is robbing us' were linked to a strong protest against 'low wages, high rents' (*Speak*, November 1982). Clearly residents saw the built environment according to its use value whereas the Council viewed the increase in terms of the exchange value of the housing stock it owned and controlled.

Conclusion

The relative homogeneity of living conditions made it easier for residents to realize the commonality of the urban problems from which they were suffering. The socialization of consumption processes led to the socialization of protest. Urban politics in the coloured townships of Johannesburg was an effect of the high levels of direct state intervention in the urban terrain which is so characteristic of the South African city. What is important however is to understand why working-class tenant struggles over urban space in coloured townships did not expand their sphere of action beyond very specific and even economistic demands. Three factors appear to have shaped the form and content of struggles in these areas.

First, the absence of any developed mobilization can be explained as a result of weak or non-existent organization. This does not mean that the coloured working class have not developed organizations for promoting their interests. On the contrary, many tenants' and residents' organizations have been created from time to time in coloured areas. These organizations however never deepened their structures within the communities they were meant to serve and also failed to reproduce themselves because of weak or inefficient leadership, poor administration and insufficient resources. Second, state repression adversely affected the capacity of residents to mobilize on a massive scale. The detention and harassment of activists such as the leaders of the WRAC weakened the structures of community organizations. Third, the absorption of urban-related demands by the state as concessions neutralized some of the militancy. The demands of coloured working-class tenants can be described as reformist since they never articulated a vision for a fundamentally transformed social order and could thus easily be conceded to by the state. The presence of a reformist ideology among the coloured tenants should come as no surprise since there is a general tendency for the working class everywhere to spontaneously express reformist demands. The strong reformist ideology of coloured tenants reveals the absence of a highly developed working-class consciousness which can only be introduced by a political party guided by Marxist theory.

These three factors were clearly present, to a lesser or greater extent, in all the urban struggles in the coloured suburbs of Johannesburg. What these struggles reveal is a need to define research problems in spatial terms to gain a better grasp of urban questions and issues. Given that space is so obviously manipulated in the South African city by economic, political and ideological forces it is surprising that urban researchers in this country have not yet asserted the centrality of 'the spatial' in their analyses of the urban process under apartheid. The concepts of 'class struggle' and 'built environment' lend themselves particularly well for uncovering the social and political content of urban spatial forms. Space and society cannot be separated in urban research.

Note

This chapter is based on an honours research project supervised by Professor K.S.O. Beavon during 1989 in the Department of Geography and Environmental Studies of the University of the Witwatersrand, Johannesburg and he is in no way responsible for the views expressed here. The original research document contains detailed references to primary and secondary source material. Mr P. Stickler is thanked for the preparation of the diagram.

References

Harvey, D. 1978. The urban process under capitalism: a framework for analysis. *International Journal of Urban and Regional Research* 2: 101–31.

Harvey, D. 1985. *Consciousness and the Urban Experience*. Oxford: Basil Blackwell.

Parnell, S. and Payne, R., 1989. Contemporary housing struggles in South Africa: the politics of shelter in a Johannesburg coloured township. *African Urban Quarterly*, in press.

Pirie, G. 1983. Urban population removals in South Africa. *Geography* 68: 347–9.

Pirie, G. and Hart, D. 1985. The transformation of Johannesburg's black western areas. *Journal of Urban History* 11: 387–410.

PLANACT Architects Collective and the Westbury Residents Action Committee 1985. Western Township Redevelopment. (Unpublished report.) Johannesburg.

Preteceille, E. 1981. Collective consumption, the state and the crisis of capitalist society. In M. Harloe and E. Lebas (eds), *City, Class and Capital*, pp. 1–16. London: Edward Arnold.

Preteceille, E. 1986. Collective consumption, urban segregation, and social classes. *Society and Space* 4: 145–54.

Republic of South Africa 1961. *Population Census* Pretoria: Government Printer.

Republic of South Africa 1971. *Population Census* Pretoria: Government Printer.

Republic of South Africa 1981. *Population Census* Pretoria: Government Printer.

Republic of South Africa 1986. *Population Census* Pretoria: Government Printer.

Rogerson, C. 1982. Apartheid, decentralization and spatial industrial change. In D. Smith (ed), *Living Under Apartheid*, London: Allen & Unwin, pp. 47–63.

6 *The 'spatial impress' of the central and local states: the Group Areas Act in Durban*

BRIJ MAHARAJ

NA J71 R12 South Africa

Group Areas legislation is one of the key instruments used to reinforce the ideology of apartheid and has considerable spatial implications for the residential restructuring of South African cities. Conventionally, residential segregation is associated with the Group Areas Act of 1950, and the ascent to power of the National Party, and the entrenchment of apartheid. However, in this chapter it will be argued that the local state, the Durban City Council (DCC), played a critical role in the formulation, refinement and implementation of the Group Areas Act.

The prelude to the Act

The legal segregation of Indians in South Africa preceded that of urban Africans by more than 30 years. The whites of Durban perceived natives as a passive threat, but Indians were regarded as a 'sophisticated and active menace to their own position in colonial society, competing for space, place, trade, and political influence with the imperial authority' (Swanson 1983: 404).

In Durban agitation for Indian segregation commenced in the 1870s. In 1871 the Durban Town Council adopted a policy to create separate Indian locations. It represented the 'first concerted attempt at group area segregation in Durban and one of the first in a major South African town' (Swanson 1983:405). The Durban Town Council also attempted to control Indian residential expansion by invoking sanitary building regulations and vagrancy laws. In 1922 Durban Town Council requested the Natal Provincial Council to pass Ordinance No. 14, which introduced an 'alienation' clause into the title deeds so that ownership and occupation of property was confined to one race group. In practice this was an anti-Indian clause.

Anti-Indian agitation and allegations of penetration into white areas in Durban dominated the 1940s. Historical factors determined that 80 per cent of the Indian population, which comprised only 2.5 per cent of the total population of the Union, were concentrated in Natal. This was compounded by the fact that provincial boundary restrictions prevented them from moving into the other provinces. Changing occupational patterns, particularly the shift from primary to secondary activities, resulted in half of the Indians in Natal living in and around Durban. The Indian population was increasing fairly rapidly and there was a demand for more housing. The DCC developed thousands of acres in the newly acquired areas, outside the city's boundaries, for modern European townships. In the process thousands of Indians who had occupied these areas for decades were displaced. Hundreds of whites living in the Old Borough of Durban, in former elite areas in decay, sought to dispose of their deteriorated dwellings in order to purchase houses in the new townships. They found ready buyers in Indians and it was against these acquisitions that the Europeans agitated.

In the 1940s, largely as a result of pressure from the DCC, the Minister of the Interior appointed two Commissions of Inquiry, chaired by Mr Justice Broome, to investigate the problem of Indian penetration. Following the Second Broome Commission Report in 1943, the government was forced to bow to pressure from the Europeans in Durban, who demanded that legislative controls be imposed on the property acquisitions of Indians. In 1943 the 'Pegging Act' was introduced, which basically pegged the racial pattern of land ownership for the next three years in the Durban municipal area. It was replaced by the 'Ghetto Act' in 1946, which extended the control over the ownership and occupation of property of Indians throughout Natal and the Transvaal by creating controlled and uncontrolled areas. Controlled areas were reserved for European ownership and occupation, and all inter-racial property transactions were prohibited. In uncontrolled areas there were no restrictions on the ownership and occupation of property. With the ascent of the National Party to power in 1948, the stage was set for the Group Areas Act.

The Durban City Council and the Group Areas Act

The Group Areas Act, introduced in 1950, was a powerful tool for state intervention in controlling the use, occupation and ownership of land and buildings on a racial basis. The state also controlled all inter-racial property transactions. The Minister of the Interior, Dr T. E. Donges, emphasized that the aim of the Group Areas Act was to provide separate areas for the different race groups, and if necessary, this would be achieved by coercion.

The City Council of Durban, as the representative of the white minority of the city, historically played a significant role in the development and promulgation of Group Areas legislation, and worked in close collaboration with the National Party. In many respects the DCC was ahead of the National Party, agitating for the dispossession of the Indian community. In fact the genesis of the Group Areas Bill was presented in a secret memorandum submitted by the City Council to the Asiatic Land Laws Amendments Committee and the Land Tenure Act Amendments Committee (*Indian Views*, 7 June 1950). The Mayor of Durban, Percy Osborne, regarded the Group Areas Act as a lifeline by which Durban would remain a European city. He contended that if the Indian community cooperated with regard to race zoning they would find a very sympathetic city council and government. Percy Osborne was to later claim that apartheid 'was the traditional policy of the burgesses of Durban and their municipal representatives long before the Nationalists came to power' (*The Sunday Times*, 13 October 1957).

There were several indications that the DCC was the prime motivator for the Group Areas Act. The National Party claimed that the Group Areas Act was a response to the calls from Durban, Pietermaritzburg and practically the whole of Natal to act against Indian penetration and expansion. Replying to criticisms of the Bill from Opposition MPs in Natal, Dr Donges stated that they did not correctly represent popular opinion in their constituencies, and that he would be much more prepared to rely on the opinion of the Durban City Council. He also stated that the DCC had publicly supported the Group Areas Bill even before they had discussed the matter with him (*Hansard, Senate* 1950: col. 7823).

The DCC accepted the Group Areas Bill in advance of the parliamentary debate, and sent three representatives to Cape Town to assure Dr Donges of the Council's support, and to discuss suggestions to improve the measures embodied in the Bill. Further evidence of the close working relationship between the government and the DCC, and particularly the influence of the latter in the formulation of the Group Areas Act, is evident from a letter written by the Minister of the Interior to the DCC, thanking Durban's Town Clerk for his advice with regard to the Group Areas Bill, as well as representations made on behalf of the City Council. He welcomed the DCC's 'promise of co-operation in the administration of the Act in the interests of all the groups concerned' (Pather n.d.: 27–8).

Given its unanimous acceptance of the Group Areas Act, the DCC immediately set machinery into motion to implement the Act. A Technical Sub-Committee (TSC) was appointed by the DCC on 20 November 1950, which drew up detailed race zoning plans for Durban. The DCC submitted a number of amendments to the Group Areas Act, as a result of the recommendations of the TSC, for consideration by the government, which were ultimately incorporated in later amendments to the Act. The main aim of the amendements was to ensure that the Group Areas Act was implemented with the minimum displacement of settled population, and the provision of alternate accommodation for displaced groups.

The Technical Sub-Committee's race-zoning proposals

The TSC maintained that residential neighbourhoods should not only be clearly racially defined, but spillovers into another group area, or casual crossings of borders, must be reduced. Borders must be clearly delimited, and must serve as barriers. In addition, one race should not traverse the residential areas of another, as this was also likely to lead to conflict (Durban Corporation 1951: para. 20). In making these recommendations the TSC went beyond the legal requirements and was acting from conviction rather than from compulsion. The authority to determine border areas was only conferred on the Governor-General in 1955. The Act did not forbid members of one race group passing through the residential area of another (Kuper *et al*, 1958: 184).

It is important to note that, in presenting the principles that should govern residential segregation, the TSC was reflecting, with confidence, the views of both the local and central state:

> It had . . . the benefit of close and frequent consultation with His Worship the Mayor and of the attendance of some of its members at meetings between the Mayor and the Ministers of Interior and Native Affairs, whose advice upon matters of general principle made it possible to proceed with greater confidence. (Durban Corporation 1951: para. 5)

Thus the TSC had functioned in consultation with members of the National Party and the DCC, who were in complete agreement with each other with regard to the Group Areas Act. In the light of the fact that the TSC had received guidance from National Party cabinet ministers themselves, its plans justifying the principles of racial segregation influenced local authorities throughout the Union as they implemented the Group Areas Act.

The course of zoning favoured by the TSC was that of locating each race group as conveniently as possible in terms of its relation to other groups and access to places of employment. Planning in this context would lead to racial residential ribs or radii, extending from a spinal working area. This approach would utilize as far as possible existing natural topographical boundaries, as well as considering the type of development which had occurred in each area, the needs of each group in terms of employment, transport, and so on. The residential areas were then divided into different zones by natural or artificial boundaries: the major rivers, the South and North

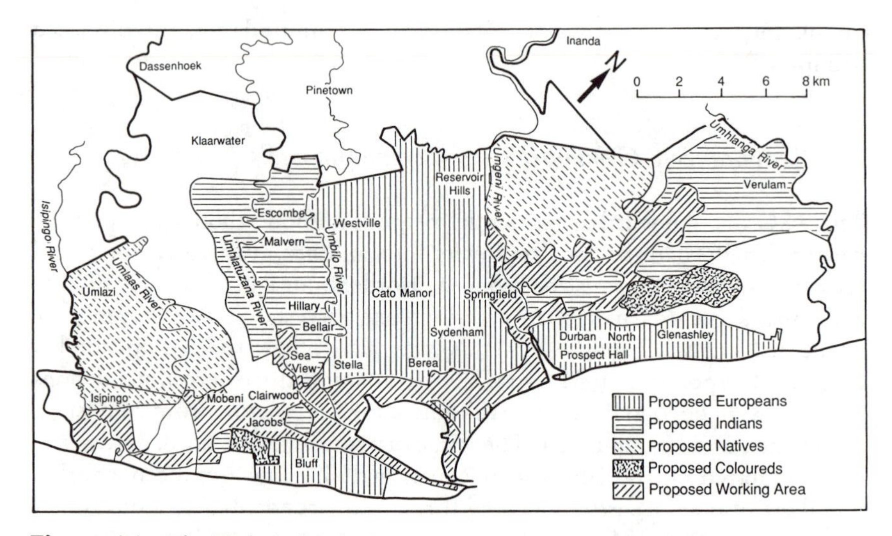

Figure 6.1 The Technical Sub-Committee's race zoning proposals.
Source: Durban Corporation 1951.

Coast railway lines and the existing and future working areas (Durban Corporation 1951: paras. 270–2). The TSC's race zoning proposals are presented in Figure 6.1.

The Durban City Council's response to the Technical Sub-Committee's proposals

Initially, the DCC supported the general group area proposals of the TSC. The principle behind the plan entailed neither evictions nor compulsory movement of population. The Mayor stated that racial zoning was intended to allay irrevocably the fears of insecurity which had existed among the different race groups in Durban for many years. He commended the TSC for adhering to the principles of foresightedness, justice and fair play in the course of its work, and expressed the hope that the Council would follow likewise.

According to the TSC plan about 119,249 blacks out of a total population of 145,744 would be dispossessed, however, and the future of the remainder was uncertain. Indians would be evicted from the mixed areas of the Old Borough of Durban, as well as from Sydenham, Springfield and Cato Manor, areas they established as pioneers almost a century ago. In comparison, the TSC plan envisaged the removal of between 7,000 and 12,000 Europeans.

Indian political and civic organizations argued that the TSC was not concerned with making modifications with regard to residential occupation, or sorting out mixed areas. The implications of its plans were far-reaching in that it envisaged the ultimate expulsion of all Indians from the city. Consequently, Durban would become a white group area, with the other race groups being located outside its boundaries.

In a letter to the Mayor protesting against the TSC's plans, the Natal Indian Congress contended that if the Group Areas Act were to be applied fairly, then Durban would have to be divided proportionately among the different races. However, the TSC's proposals condemned blacks to ghettos outside the city's boundaries, while the developed area was handed over to whites who comprised one third of the population.

The Natal Indian Organization argued that an evaluation of the TSC report revealed that its primary objective was to serve and entrench the white interests in Durban at the expense of displacing settled Indian communities. The Durban Combined Indian Ratepayers Association, representing 12 civic organizations, protested emphatically against the zoning plans of the TSC, maintaining that the proposals would uproot and displace settled Indian communities and businesses and was considered to be iniquitous and unwarranted.

Europeans generally were opposed to moving under the terms of the Group Areas Act, and believed that Indians should be forced to do so. At a Council-in-Committee meeting held on 23 January 1952, the DCC tabled objections to the TSC race zoning proposals from 15 individuals and organizations, the majority of whom represented European interests. The most vociferous opposition to the TSC's proposals came from whites in the Main Line suburbs (Sea View, Bellair, Hillary). The strident resistance of whites to the zoning proposals of the TSC, the Natal Indian Organization maintained, emphasized an important factor neglected by the Committee – the strong attachment which people have for their homes and established communities. The Natal Indian Congress maintained that whites were alarmed by the slightest possibility of being displaced, yet they approved of the Group Areas Act, and supported the uprooting of settled Indian communities.

The Durban City Council's amendments to the race-zoning proposals

The DCC was forced to amend the TSC's proposals in response to objections lodged by white residents to being displaced by the Group Areas Act. Predictably, the objections raised by Indians were totally disregarded. Under the circumstances attempts to implement the Act were likely to exacerbate racial friction. An amended race zoning plan was accepted by the DCC on 5 May 1952 (Fig. 6.2), and was viewed as the most acceptable ultimate grouping of races in the residential areas in terms of the Group Areas Act. As evident from Figure 6.2, it was

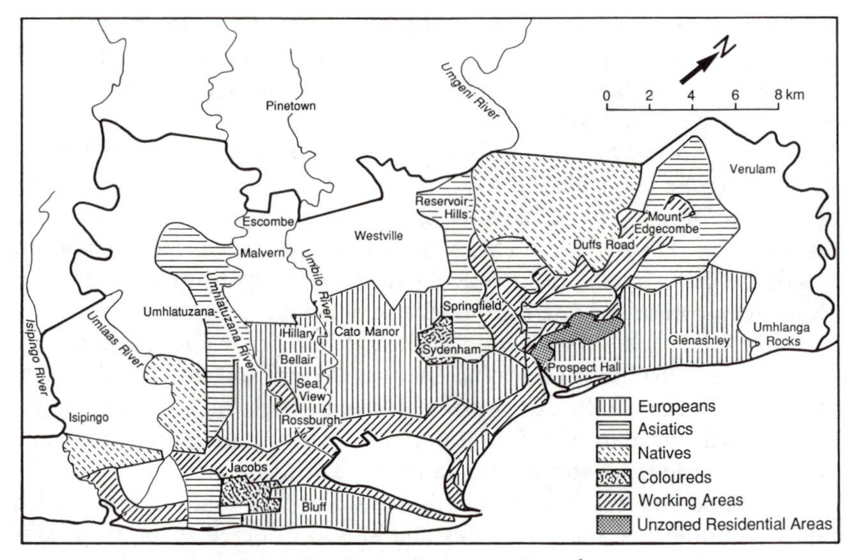

Figure 6.2 The Durban City Council's race zoning plans.
Source: Natal Regional Survey 1952, map n0. 11.

most desirable because it reserved the most suitable areas for, and envisaged no uprooting of, Europeans. The Main Line suburbs, for example, were zoned for whites. The zoning plan of the DCC, Councillor Bolton pointed out ominously, was immoral and financially impracticable because the Council could not bear the cost of relocating tens of thousands of people. In terms of this scheme, whites were able to expand their landholdings within the city, at the expense of the Indians. Residential areas of blacks were confined to undeveloped territories, outside the city.

As a result of the amended plans, 2,700 Europeans, 8,200 coloureds, 62,900 Indians, and 82,500 natives faced dislocation. European areas comprised 37,100 acres in total, and could sustain a population of 415,000 persons (Durban Corporation 1952). Within the Borough of Durban the Indian zones comprised 6,500 acres. Outside the City, the Indian section of rural Umhlatuzana, and to the north, the area encompassing the Fosa Settlement and Duffs Road, totalled 9,000 acres. Indians were thus allocated a total of 15,000 acres, which could accommodate 330,000 persons. It was evident that the largest proportion of the residentially developed area of Durban was proclaimed for whites.

The Group Areas Board hearings were held in Durban in 1953. In spite of objections from Indian political and civic organizations, the Board's recommendations to the Minister of the Interior, announced in July 1954, revealed that it had basically accepted the DCC's proposals. The first group area proclamations for Durban were announced in June 1958.

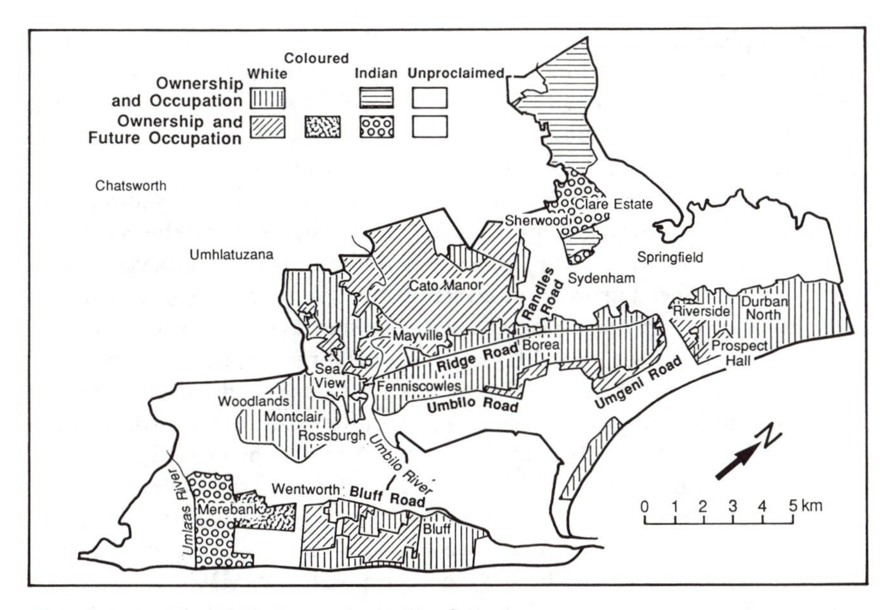

Figure 6.3 The 1958 Group Areas proclamations.
Source: Prentice Hall Legal Services, Durban.

The 1958 Group Area proclamations in Durban

The 1958 group area proclamations in Durban basically confirmed the recommendations of the Group Areas Board in 1954. The proclamations distinguished between group areas for immediate ownership and occupation, and for ownership and future (undated) occupation (Fig. 6.3).

The Board had estimated that about 25,000 properties would be affected by the proclamations. According to the Minister of the Interior, estimates based on the 1951 census revealed that 1,000 whites, 75,000 Indians, 8,500 coloureds and 81,000 Africans would have to move as a result of the proclamations (*Hansard, Assembly* 1958: col. 1472). In 1958 the figures were obviously higher.

The Indian community was devastated by the proclamations. They would lose property valued at millions of pounds. The proclamations galvanized the various political and community organizations into recognizing the need for a united front in opposing the Group Areas Act. The community was urged to participate in the mass protest action, which culminated in a rally held at the Curries Fountain ground on 26 June 1958, convened by the Natal Indian Congress, the Natal Indian Organization and Durban Combined Ratepayers' Association (which was referred to as the Sponsors' Committee), and attended by over 20,000 people. The meeting condemned the proclamations, and declared its total opposition to the Group Areas Act and its implementation (*The Graphic*, 4 July 1958). It was emphasized that the proclamations were not forcibly imposed upon Durban, as the government had merely rubber-stamped the proposals of the DCC. Most significantly, the DCC itself was beginning to have second thoughts about the group area proposals it had submitted to the Group Areas Board.

In 1957, the DCC's Planning and Development Control Committee, chaired by former mayor Councillor R. Carte, suggested that the possibility of rezoning Cato Manor for Indians should be examined, especially since it would not affect European interests in any way whatsoever. Indians had limited expansion possibilities, and with the shortage of land, Cato Manor was the logical choice. The Mayor, Percy Osborne, replied that the DCC had 'given the government an undertaking that we would cooperate in making Cato Manor an all-white area and nothing can stop its proclamation as such' (*Natal Mercury*, 18 May 1957). In the process he proved once again the close working relationship between the DCC and the government.

At a Council-in-Committee meeting held on 27 May 1958 the DCC agreed that the 1952 proposals were unfair, and assented in principle to revoke them, and to draft a new proposal. The DCC resolved at Council-in-Committee meeting on 5 June 1958 (ironically a day before the government's Durban proclamations) that a 'more realistic approach' be adopted with regard to race zoning, with emphasis on the 'minimum disturbance of the existing population', that Cato Manor be allocated for Indians, and that the government does not make any proclamations in Durban until the DCC has reviewed its proposals.

Three factors led to the review decision: the passage of time and the financial implications, as well as the fact that the composition of the Council had changed. Ultimately, the DCC was forced to review its race zoning plans because of the exorbitant costs of implementing apartheid. It recognized that there was insufficient land within the boundaries of the city to accommodate the displaced groups. There would be a need to develop land outside the city's periphery. However, the cost of land and services was astronomical, and without government financial assistance, would be beyond the capacity of the DCC. The government itself was unable to answer the question of who would foot the bill for relocating about 63,000 Indians.

The government argued that the proclamations of 6 June 1958 were based on the recommendations made by the DCC. The Minister of the Interior stated that in the five-year interim period between the Group Area Board's recommendations and the proclamations there had been no objections (obviously from whites) to Durban race zoning plans. He maintained that the greater community approved of the proclamations of June 1958. Under the circumstances, a very strong case had to be made for a review. The government gave the DCC a warning with the draft amendment to the Group Areas Act Development Bill, which made it quite clear that while the Group Areas Board would demarcate group areas, the DCC would be compelled to pay for its development, regardless of whether it approved or not (*Hansard, Assembly* 1958: cols. 3461–74).

The DCC asserted that rather than impeding the implementation of group areas legislation, the council had made every effort to make it feasible on the equitable and just basis which the government seemed to desire. It would therefore be far better for the government to cooperate with the council in regard to the problems which it had experienced, instead of threatening it with compulsion. Since 1951, the DCC had emphasized that it was the government's duty to provide the necessary funds for relocation costs incurred by people displaced under the Act. These facts did not justify the contention that the City Council was intentionally hindering the work of the Group Areas Development Board (*Daily News*, 12 September 1958).

The revised race zoning plan was accepted by the Planning and Development Control Committee on 14 August 1958, and the only concession was the zoning of Cato Manor for Indians. Immediately, Europeans in Durban began to clamour for Cato Manor to be retained as a white group area. They had the support of the government. The Minister of the Interior maintained that at the 1953 group areas hearings it was emphasized that Cato Manor 'was the main gateway to Durban from the inland and that it had to be kept in white ownership' (*Hansard, Assembly* 1958: col. 3473). Tensions between the local state and its constituency, as well as the central state had increased remarkably, and the municipal elections were due on October 1958.

The Durban City Council – an about-turn?

On 1 October 1958 the composition of the DCC was drastically altered by the white electorate. Almost all the councillors who were supporting the call for a revision of the June proclamations lost their seats, including Mr Carte. There was a total change in the composition of the Planning and Development Co-ordinating Committee, which was directly involved in making recommendations with regard to group areas, with Percy Osborne, a sympathizer of the National Party, appointed as Chairman. Clairvoyant powers were not required to forecast the future. Immediately upon his appointment as chairman, Mr Osborne declared that he supported the 6 June proclamations.

At a special meeting of the Planning and Development Committee held on 10 February 1959, the Chairman, Councillor Osborne, contended that the 1952 proposals would best serve Durban's interests. He strongly supported Cato Manor being zoned for whites. It was not suitable for Indian occupation as it was surrounded by European areas, and problems could arise if the Indian population expanded to the extent that the area would be unable to contain it. This could result in the area for Indian occupation being increased or else they would be forced to move out of this area.

The Committee unanimously accepted Councillor Osborne's resolution revoking the DCC's decisions of 5 June 1958, reaffirming support for the Group Areas Act, and allocating Cato Manor for whites.

Significantly, there was strong white opposition to the recommendation that Cato Manor should be zoned for Europeans. A mass meeting to protest against this, convened by the 'Citizens Committee', which consisted of concerned Europeans, was attended by more then 2,300 whites on 23 February 1959. The meeting resolved that the removal of Indians from Cato Manor was immoral, and was beyond the financial capacity of the City Council. An appeal was made to the DCC to make every effort to prevent this grave injustice. For the first time in the history of Durban, white conscience had been stirred into mass action and protest against the injustices perpetrated by the DCC against voteless and voiceless blacks. The concern expressed at the meeting was regarded as a reflecting of the views of most whites in Durban. It was envisaged that this opinion would eventually influence the DCC to reconsider its resolution of 5 June 1958.

The European protest affected the DCC, and at its meeting on 3 March 1959 it was resolved that the Planning and Development Control Committee give further consideration as to whether Cato Manor should be zoned for whites. This 'refer back' was regarded as procedural manoeuvre used by the DCC to temporarily shelve the issue.

Failure of compromise – the impress of the central state

On 2 June 1959 the DCC accepted a compromise proposal of the Planning and Development Control Committee that a portion of Cato Manor which was predominantly Indian-owned and occupied be zoned for this group. This would require an amendment to the June 1958 proclamations and the DCC agreed to send a delegation to discuss the matter with the Group Areas Board. The compromise proposal allocated 1,300 acres out of 5,000 to Indians. About 10,000 Indians would be displaced, and a large amount of property owned by them was excluded. Indians, therefore, rejected this plan.

There was a further change in the local politics of Durban in October 1959, with the election of liberal Councillor C. A. Milne as Mayor. Mayor Milne recalled Dr Donge's assertion that the Group Areas Act would be administered with justice, and contended that he expected the government to adhere to this policy. He also maintained that the DCC would not be party to the displacement of settled communities without the provision of suitable alternative accommodation.

The DCC delegation to discuss the rezoning of part of Cato Manor for Indians met the Group Areas Board in January 1960. The Board rejected the DCC's appeal. This meant the displacement of the entire

Indian community from Cato Manor, and the estimated cost of finding alternative accommodation (about £20 million) had to be borne by the DCC. Mayor Milne accused the government of grave injustice because while the city was trying to provide accommodation for thousands of Africans, it had to provide for Indians uprooted from Cato Manor who had some form of housing.

In June 1960, the DCC decided to appeal to the Minister of the Interior to reconsider the zoning of Cato Manor, as well as to obtain an assurance that the Group Areas Act would be applied with justice and fairness. In March 1962 the Group Areas Board announced that it would be making a decision on whether the proclamations of June 1958 should be implemented, and called for representations in this regard. The DCC argued that the proclamations were unjust and inhuman, and there was no alternative accommodation for those displaced. Indians viewed this as support for their stand that there should be no forced uprootings and displacements. In spite of receiving more than 10,000 objections to the zoning of Cato Manor for whites, the Group Areas Board merely affirmed the government's decision of 6 June 1958. The racial ecology of Durban today reflects the 1958 proclamations.

Conclusion

It is evident that the local state played a significant role in the formulation and implementation of the Group Areas Act. The DCC had been at the forefront of the calls for compulsory residential segregation, particularly against Indians, because this group presented a serious threat to whites in the competition for economic and social space. There was substantial evidence of close collaboration and complicity between the DCC and the National Party with regard to the formulation and implementation of group areas legislation. The TSC had worked in close conjunction with the government in drawing the 1952 race plan for Durban.

The different race zoning plans drawn by the DCC ensured that white interests would be entrenched, for they had the vote, and the councillors were accountable to their electorate. A good example was the rezoning of the Main Line suburbs for whites. In spite of being disenfranchised, the Indians presented a spirited opposition to the Group Areas Act and its proclamations via their political and civic organizations. Nevertheless, their objections were not considered by the central and local states.

However, eight years after the implementation of the Group Areas Act, and with the election of more liberal councillors to positions of influence, and with the financial costs of implementing apartheid becoming more evident, the DCC reviewed its support for the Act. The June 1958 proclamations fulfilled the worst fears of the council in terms of the costs of providing alternate accommodation for tens of thousands of Indians. The DCC appealed, in vain, to the government for a review of the proclamations on the basis of minimum disruption of settled communities, and the rezoning of Cato Manor for Indians. The government replied that it could force the DCC to bear the costs of providing alternate accommodation. The alliance between the National Party and the DCC with regard to the implementation of the Group Areas Act was revoked, and tensions between the central and local states increased markedly.

Conservative councillors objected to the rezoning of Cato Manor. However, influential whites supported the plan. The DCC accepted a compromise proposal, which zoned a portion of Cato Manor for Indians. This was rejected by the government, in spite of numerous appeals by the DCC. The spatial impress of the central state was finally imposed upon the local state.

Note

The main sources for this paper were Minutes and Reports of the Durban City Council, Hansard, and newspaper articles. Detailed references have been omitted because of space constraints.

References

Durban Corporation 1951. Report of the Technical Sub-Committee on Race Zoning, 22 June 1951.

Durban Corporation 1952. Factual Report on Alternate Race Zoning Proposals submitted by the Technical Sub-Committee, 24 March 1952.

Kuper, L., Watts, H. and Davies, R. 1958 *Durban: A Study in Racial Ecology*. London: Cape.

Natal Regional Survey 1952. *The Durban Housing Survey. Additional Report No. 2.* Pietermaritzburg: University of Natal Press.

Pather, P.R. n.d. *Seventy Years of Frustration and Unhappiness*. Durban: South African Indian Organization.

South African Institute of Race Relations 1958. *A Survey of Race Relations: 1957–8.* Johannesburg: South African Institute of Race Relations.

Swanson, M. W. 1983. 'The Asiatic menace': creating segregation in Durban, 1870–1900. *International Journal of African Historical Studies* 16: 401–21.

7 The destruction of Clairwood: a case study on the transformation of communal living space

J71 R11
South Africa

DIANNE SCOTT

Introduction

A large corpus of legislation has evolved in South Africa which is non-racial in letter but produces segregatory outcomes in urban space. The aim of this chapter is to examine how, in the Clairwood area in southern Durban, the local implementation of non-racial legislation facilitating capitalist urban development has had segregatory outcomes and has weakened place-based social relations upon which community and self-identity are based (Cox and Mair 1988).

As the formalized outcome of the political process, laws serve to recreate the structures of power in society (Thompson 1975). In capitalist societies laws have been instrumental in creating the appropriate form of urban society via a utopian set of rules and standards (Clark and Dear 1984: 110). Laws regulating the construction of urban space constituted an important means of establishing the hegemonic order in the colonial period in Africa (Cooper 1983), and were framed within the utopian vision of a racially segregated urban colonial society. Superimposed upon this mode of conceptualizing urban society was the powerful set of standards as to what constituted 'order' and 'disorder', derived from early urban administration ideology that had evolved to cope with the emerging industrial towns in Britain (considered further in Ch. 24).

A set of mental categories evolved in Natal, and other colonial contexts, for dealing with the experience of white settlers, living among the indigenous African people and later the immigrant Indian community, as they sought to maintain their dominance. The institutionalization of white dominance and the legitimacy of white norms and standards with regard to the structuring of sociospatial relations is evident in the labyrinth of inhibitive legislation that emerged in the colonial and post-colonial period. This legislation explicitly defined which categories of economic activity, urban form, family structure, and behaviour were deemed appropriate and 'legitimate' and which were inappropriate and hence illegitimate (Marris 1979).

In the 'struggle for the city', which began in earnest in the early twentieth century in South African towns, the mutually exclusive, and implicitly racial categories, of 'formal' and 'informal' formed a broad implicit framework within which municipal authorities devised policies and legislation for replacing informal settlements with formal planned developments which were appropriate to the emerging industrial order.

This case study presents an instance of the destruction of a predominantly Indian informal communal living space in the southern Durban corridor centred on the node of Clairwood (Fig. 7.1). The municipal authority's policy of industrialization was largely responsible for this destruction. This was executed via 'non-racial' housing, health and town-planning enactments, and implemented through the engineering and administrative functions of the municipality. The ideology of urban segregation implemented via non-racial legislation in this way was not therefore the product of the national state.

Indians as 'aliens' in colonial and post-colonial Durban

In Durban, the social relations between the dominant white group and the 'alien' Indian group evolved historically and were based on specific local experiences (see Ch. 6). The rapid growth of the Indian population in Durban gave rise to intense Anglo-Indian conflict at the turn of the century, resulting in the emergence of a corpus of legislation restricting the political, economic and social life of the Indian community (Meer 1969). Although explicit segregatory legislation was not sanctioned in the colonies by the Imperial government, municipal regulations relating to health and trading rights had succeeded in quartering the Indian population within the Borough of Durban in an enclave adjacent to the white business district (Pachai 1979). In the early twentieth century, the lack of cheap centrally located land resulted in the majority of low-income Indian families locating outside the Borough boundaries.

Both anti-Indian sentiments on the part of the white settler class and the encapsulation in legislation of the notion of the Indian as a temporary 'alien' in colonial society (Pachai 1979) were reflected in the spatial structure of the town with its predominantly white core and emerging peripheral 'black belt' of marginalized African and Indian informal settlements (Davies 1976).

The legislative constraints placed on Indians gave rise to the emergence of Indian political action in the form of 'passive resistance' as a strategy aimed at redressing the injustices against the Indian community in South Africa (Pachai 1979). Undeterred by the consequent elevation of the 'Indian Question' to an international forum, and the promise to uplift Indian living conditions made by the South African government in the Cape Town Agreement of 1926, the municipal authority in Durban continued to

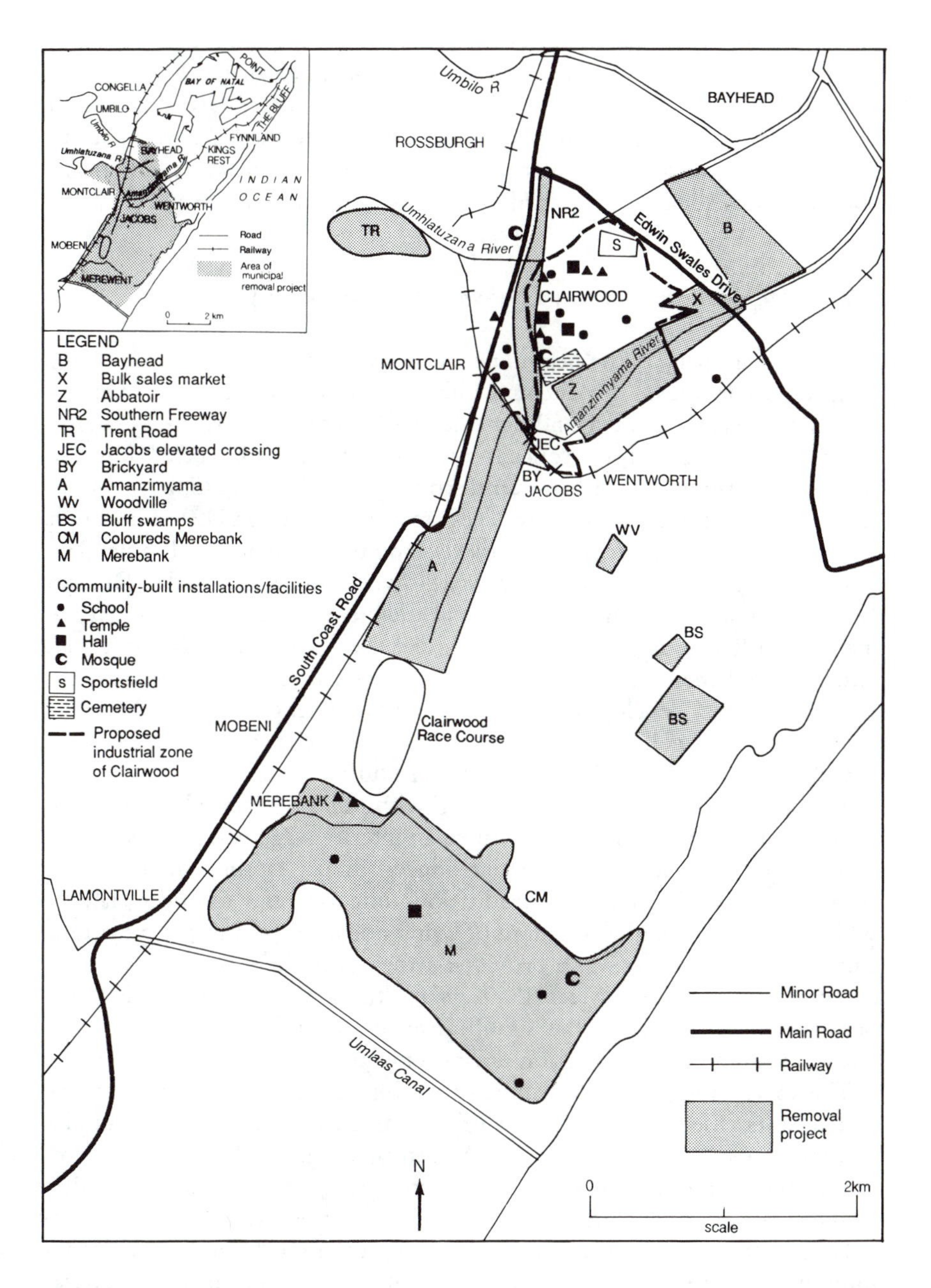

Figure 7.1 Municipal removal projects (1963–75) and community-built facilities in southern Durban prior to removals.

neglect the informal Indian settlements on the periphery of the town. Even when these areas were incorporated into the Borough of Durban in 1931, no attempts were made to provide infrastructure and facilities. The critical shortage of Indian housing in the period from 1920 to the 1950s was a result of earlier colonial conceptions that the Indian community was a temporary component of society and would eventually be repatriated to India.

The founding of the southern Durban Indian community

Durban's periphery at the beginning of the twentieth century consisted of a large population of low-income Indians and Africans living as tenants on both privately owned and state land in acutely impoverished living environments. The largest concentration of Indians in this belt lay in the southern corridor of Durban. Unrestricted in their migration to urban areas, Indian families found easy access to cheap land outside the borough boundaries where municipal regulations and rates were not applicable. The large areas of swampy, undeveloped state-owned land at the head of Durban Bay and Amanzimnyama constituted a suitable environment for intensive market gardening as well as for residential use. From the 1920s scattered informal settlement emerged on low-lying land that was undesirable for white residential development, such as the bay-facing slopes of the Bluff and around the small node of Merebank, as is evident from the 1931, 1949 and 1956 municipal aerial photography of Durban.

Market gardening in particular, along with other petty productive activities reliant on traditional skills, formed an interim stage between indenture and proletarianization, in that it introduced Indian families to the urban economy. The characteristic production unit was the extended family which permitted a higher standard of living due to the sharing of nutrition, clothing and possessions and provided a form of social insurance in times of economic hardship (Meer 1969). With the emergence of settled and permanent communities, the significance of the Indian family began to reassert itself after its disruption during indenture.

The early pioneer families initially established themselves as squatters and tenants, the more successful of whom purchased properties in the emerging node of Clairwood and Merebank (Fig. 7.1). Wood and iron houses predominated and the living environment was generally poor, due to the swampy terrain and the paucity of civic amenities.

Despite these conditions and the low incomes of most families, the communities that grew up around these agricultural activities in the peripheral areas were held together by a rich network of social ties based on the extended family. United by a common cultural and religious heritage (the predominant group being Hindu) and regulated by a set of ethnic and religious norms, the southern Durban Indian community presented a clearly identifiable culture and identity. Peripheralized by the mainstream society and in urban space, this culturally unified community evolved a high level of cooperation and reciprocity despite some differences in religious and language affiliation. (Most of the evidence on which this account is based was derived from oral sources via intensive interviews with past and present residents of southern Durban which were conducted between 1987 and 1990).

The first Hindu temple in southern Durban, the Shree Ambalavanar Alayam Subramanya Temple, was founded in the early 1880s. The Shree Shiva Subrahmanya, the Shree Luxmi Narayan and Jacobs Road temples followed, along with many other smaller sites of worship throughout the southern area (Fig. 7.1). These temples became the focus of social life defining a sphere of predictability and confidence within the community and acting as a source of self-identity for all residents. Through religious festivals and daily worship, the local community identity was affirmed and celebrated.

A wide array of communal facilities and institutions also became established through the investment of communal resources and initiative focused around the religious, social, cultural and educational dimensions of everyday life. In the Clairwood area the donation of privately-owned Indian land led to the concentration of schools, community halls, cemeteries, temples, mosques, sports grounds, and a clinic (Fig. 7.1). These institutions were administered by boards of trustees elected from the community.

With sacrifice and dedication the Clairwood community struggled to provide education for the growing school-going population through a series of community schools. The first of these was the Clairwood Boys School established in 1903. A strong emphasis was placed on vernacular education and the reproduction of the Indian cultural identity.

In addition, a multitude of clubs, cultural and social groups, sporting and welfare associations, and 'Old Boys' and 'Old Girls' associations proliferated in the first half of the twentieth century. There was an overlap in membership of these various groupings with prominent community members represented on a range of organizing committees and trusts.

Status in the community was not conferred according to the educational, income and occupational criteria of industrial society but rather in accordance with an individual's service to the community through the social, religious and welfare organisations. Even upon voluntary or forced relocation from the locality, many prominent ex-residents continue to serve their community from a distance.

The historical significance of the communally created space in southern Durban prior to the 1950s was that the intricate networks of kinship, religious, social and cultural relations within the community were directly place-related. Clairwood and district was both materially and imaginatively created by the Indian community of southern Durban. Clairwood itself, however, was the 'heart' of the area as it was here that the land was held in private tenure by Indian land owners. Although the surrounding informal shack settlements were established on undeveloped state land in the early part of the century, they formed an integral part of the larger Clairwood area.

The local economy of Clairwood had been established around the South Coast Road commercial node which was second only to Grey Street as an Indian trading area in Durban, providing speciality shops catering for Indian culinary, religious and clothing requirements. Linked into patron–client relationships with the community, the family businesses extended credit to low-income families in times of need.

With the growth of industry to the south and increasing population pressure on the land in the 1930s and 1940s, proletarianization of Indian labour was rapid. Market gardening and informal petty productive activities such as hawking provided a very low standard of living compared with the rewards of wage labour in industry. However, the communal support networks and the existence of the extended family which provided a refuge for the high percentage of unemployed in the community, and the continued patronage of informally skilled entrepreneurs, delayed the full proletarianization of the Indian workforce.

Easy access to the expanding employment opportunities in the surrounding industrial areas of Jacobs, Rossburgh, Maydon Wharf, Umbilo and Merebank (Fig. 7.1) made Clairwood in particular, and the southern Durban area in general, a desirable residential location for the Indian community. Long periods of residence by generations of Indian families in southern Durban, and the emotional investment in creating a place for themselves formed the material basis for a stable and self-sufficient community. Neglected and peripheralized by the municipal authority over many decades, and thrown back on their own resources, members of the southern Durban community created an informal space the parameters of which were antithetical to the visions of the municipal authority of what southern Durban 'could be'. However, unlike the African informal settlement at Cato Manor which exhibited a 'culture of resistance' – as did other urban African communities in South Africa – the Indian community of southern Durban presented to threat to the municipality in terms of a 'dangerous class' (Bonner *et al* 1989). It is contended that the culturally specific social relations of the Indian family and religion, which led to the emergence of a stable place-based community in Clairwood, gave rise to the development of a specific politically neutral response by the municipal authority in its attempt to transform the communal living space to the productive core of Durban.

The destruction of communal living space in southern Durban

The southern area of Durban, due to its location adjacent to the harbour and incipient nodes of industrial activity, was very early in the twentieth century pinpointed by local industrial capital and subsequently the municipal authority as the future productive zone of the emerging industrial metropolis. To accomplish the goal of reconstructing the space of southern Durban, the municipal authority had to create plans. The first step was to gain administrative control over the southern Durban area in order to be able to reconstruct it as an industrial space. This was accomplished by the incorporation of this area into the Borough boundaries in 1931.

In this early period, the municipal authority began to formulate a clearer vision of the future of Durban in order to justify municipal intervention on ideological grounds. The planned development and concomitant 'formalization' of southern Durban was envisaged as part of the broader planning goal of creating an efficient industrial city to both provide employment and to encourage investment in the Durban area, allegedly in the public interest. However, the primary goal of facilitating industrialization was couched within the municipal authority's broader agenda of maintaining white domination and creating a spatially segregated city.

The plan for the restructuring of the southern Durban corridor was essentially based on the principle of racial zoning, to be achieved through the application of Western standards of health and planning to control and formalize the use of space. In this manner the rationale of segregation was masked behind the benign machinery of non-racial legislation created to order various elements of physical space.

Even before the incorporation of the southern area into Durban to implement its plan, the municipal authority set about establishing the means of controlling land use in the area by acquiring 619 acres of land in southern Durban for the explicit purpose of creating peripheral planned African residential areas adjoining planned industrial estates.

In order to facilitate industrial growth to the south, a programme of engineering works commenced in 1930. The major works involved upgrading and extending the road system; providing water-borne sewerage and stormwater drainage; canalizing the Umbilo, Umlaas and Umhlatuzana rivers; levelling and reclaiming of large tracts of municipal land; and providing serviced sites for the housing scheme areas. Clairwood itself was the only residential zone not serviced with sewage facilities, nor upgraded in any other respect.

These early policies and actions on the part of the municipal authority began to structure sections of the city on a racial basis, in an ad hoc manner. In 1938 the policy of industrialization was officially sanctioned and formally expressed as a spatial plan in the municipal Post-war Reconstruction Plan in 1944. It defined southern Durban as a 'productive zone' surrounded by reservoirs of labour.

It was the Group Areas Act of 1950 and financial allocations for housing from the national state that enabled the radical and comprehensive transformation of the land-use patterns of urban areas in South Africa in the

1950s and 1960s. However, it is apparent that the creation of an ideological framework for racial zoning, and the piecemeal actions towards this goal – providing physical infrastructure and land resources – had been undertaken in Durban since the 1920s, laying the foundations for subsequent national imperatives in this direction (see Ch. 6).

Urban-based industrial capital accumulation underwent a period of rapid acceleration towards the end of the Second World War and the municipal authority perceived an urgent need to facilitate industrial expansion in southern Durban. To achieve this, two strategies were embarked upon in the post-1950s period. Firstly, the preparation of undeveloped municipally-owned and expropriated property for industrial and major infrastructural uses formed the chief strategy of municipal authority from 1950 until the mid-1970s. This strategy further necessitated the removal of large numbers of predominantly Indian families through the termination of their leases on municipal land or expropriation of private property. Related to the removal process in southern Durban in the early 1960s was the construction of a nationally-funded public housing scheme for Indians in Chatsworth to serve primarily as the destination for families disqualified in terms of the Group Areas Act.

The simultaneous implementation of development projects, clearance of shacks, and allocation of housing required the evolution of an efficient administrative procedure to be successfully achieved. The 'pegging' of shacks (i.e. the prevention of incursions of 'illegal' shacks), the execution of socio-economic surveys of the resident populations, and strict housing allocation procedures, provided formal housing for only those families that were bona fide residents in priority areas. (Data on removals projects was derived from the records of the City Estates Department of the Durban Corporation, and interviews with its officials.)

Underlying the predominant goal to industrialize southern Durban was the implicit objective of formalizing the character of the area to conform with Western standards of sanitation, health and physical order. The process of shack clearance, which commenced in 1963 and continued until 1975, therefore fulfilled the twin imperatives of providing land for industrial development and removing informal areas, within the ambit of the racial zoning plan. However, it was not via the Slums Act of 1934, nor the Group Areas Act of 1950, that removals were effected. The shack clearance schemes of the municipality were part of the administrative procedure involved in the physical planning of the area, enabled by apolitical planning and housing legislation, and implemented by the bureaucracy.

The second important strategy adopted by the municipality was to obtain legal sanction for the industrialization of Clairwood. The Mobeni industrial estate had been proclaimed in 1948, and preparation of the Amanzimnyama Estate was underway. Attention was then turned to Clairwood to make further industrial land available. This was pursued by lodging an application in terms of the Town Planning Ordnance of 1949, to rezone the specific area of Clairwood as an industrial area and to obtain powers of acquisition for the entire area.

Since this strategy involved an attack on the heart of the southern Durban Indian community and proposed the large-scale expropriation of private property, destruction of community-built facilities and sacred spaces, and the removal of about 40,000 people, it was strongly resisted through legal channels of objection by the representative civic body, the Clairwood and District Residents and Ratepayers Association. The protracted legal wrangle over industrialization continued from 1956 until 1986, when, with mounting resistance from the community and increasing pressure from the national government, it was proclaimed that Clairwood would remain residential and become subject to replanning by the Council.

Removals in southern Durban for municipal development projects commenced in 1963 in the Merebank, Amanzimnyama and Woodville settlements (Fig.7.1) resulting in the removal of 1,109 families. From 1964 to 1967 removals precipitated by the construction of the Southern Freeway and those from the smaller Trent Road and Brickyard settlements led to a further 655 families being relocated. After the removal of 226 families from the Bayhead area on behalf of the Railways Administration in 1968, attention was turned to the heart of the southern Durban Indian community – Clairwood. The largest removal project commenced here in 1969 and continued until 1975 with the relocation of 1,222 families for the construction of a bulk sales market and proposed abattoir. In addition to these development projects, the civic association of Clairwood vociferously opposed the municipal authority's attempts to rezone Clairwood as an industrial estate. During the early 1970s, 75 families were removed from Clairwood because of flooding, and 19 to make way for the construction of the Jacobs Elevated Crossing; 163 coloured families were removed from the Bluff Swamps and Merebank areas (Fig.7.1).

By the mid 1970s, Clairwood had become a slum due to the ban on residential development and the lack of civic amenities. Insecurity among residents had caused many to relocate voluntarily. Conditions of extreme overcrowding existed as people not qualifying for, or refusing the option of, formal housing had gravitated to Clairwood where cheap tenancies were available. The 1976 municipal census of Clairwood revealed a population of 11,680 people on 565 sites.

The municipal authority has relentlessly pursued the goal of facilitating industrialization in the southern corridor of Durban, ultimately overcoming local resistance, and restructuring the social relations of the resident community through relocation and intervention in the built environment. These actions resulted in the transformation, by 1970, of the southern Durban area into the productive core of the Durban metropolitan area, except for the small node of privately-owned Indian residential property in Clairwood.

After a mounting struggle by the residents of Clairwood in the 1980s, it is ironic that it was the intervention by the national apartheid state that resulted in the retention of a portion of Clairwood as a residential area. By 1988, the population had dropped to around 6,000 people, a shadow of the former community of an estimated 40,000 in Clairwood, and 80,000 in the larger southern area in the 1950s.

Action by the municipal authority has succeeded in weakening the complex network of social, religious and cultural ties that constituted the material basis of this community by the removal of the people who actively reproduced it. Relocated to far-flung corners of the city, ex-residents of the older generation nevertheless retain an identity with the place of Clairwood, perhaps returning to attend the annual Kavadi festival, or to bury a family member in the Clairwood Hindu or Madras Cemetery, but no longer to engage on a daily basis in the constitution of the place. But growth of the welfare state had already served to weaken local dependence on communal social relations. In 1946, the community schools of southern Durban became state-aided, and by the 1970s, the rich diversity of community organizations had largely become inactive, except for the ratepayers association and the temple communities.

The mapping out of the utopian vision of a continuous industrial zone south of Durban Bay represented a 'frozen space' (Vernon 1973) in which all aspects of experience that did not accord with the logic of this productive space were siphoned out. The plan for industrialization was therefore bereft of any considerations of the historical significance of the area, the emotional attachment to place, and communally created value systems.

As a result of community resistance, the physical reconstruction of Clairwood has begun after many decades of neglect. In consultation with the residents, the municipal authority is now engaged in replanning the area and providing the necessary civic amenities. By July 1990, installation of water-borne sewerage commenced. However, the social relationships which emerged in the southern Durban Indian community, and were strongly identified with the particular locality of Clairwood and its surrounding districts, cannot be materially reconstructed. The small node of Clairwood remains as a vestige of the former extensive area occupied by the southern Durban Indian community, serving as an 'anchor' to retain a sense of what Clairwood was and reproducing the ethos of the past community to which many displaced ex-residents still pay allegiance.

Forced removals in southern Durban executed to accomplish the goals of industrial development, formalization of space, and segregation, had served to accelerate the process of state intervention within a place-based community, diminishing the social relations centred around community, religion and the family in a community historically conceptualized as 'alien' within the mainstream colonial order and into the apartheid era.

Conclusion

The contestation of space in southern Durban represents one of many similar instances of state intervention in urban space in the construction of the apartheid city. District 6 in Cape Town, and Sophiatown in Johannesburg, for example, are also cases where place-based communal social relations have been destroyed in the implementation of the vision of the utopian segregated city.

The industrialization of land in the strategically located southern corridor of Durban required the removal of informal non-productive uses and their replacement with intensified uses that could contribute to capital accumulation in the industrial sector. This required the material and ideological transformation of the area. This was achieved through planning, rezoning and slum clearance effected through non-racial housing, slum clearance and planning legislation and justified by a vision of a planned, segregated and formalized city that was embodied in this corpus of legislation. Although the relevant planning and housing legislation purports to be non-racial, its definition of informal spaces as undesirable and illegal has resulted in racially discriminatory outcomes as it is historically the black people who squat and live in slum dwellings and shacks.

The relevance of understanding the creation of informal space is crucial. As South Africa moves into the twenty-first century, it is apparent that an increasing component of the urban fabric will be informal in character. The material basis of life in these informal settlements is constituted by the social relations of the communities that establish themselves and the ethnic, family, cultural and political structures and values within which they interact in localized spatio-temporal contexts.

Since it is the law through which society seeks to define what is right and just, it is instructive to understand how the law, in many different forms, can be instrumental in either creatively reproducing such contexts or defining them as 'illegal' and undermining them. Constructive efforts are required in both state and private sectors to create the legal machinery that allows for place-based informal communities to establish a sense of identity which could be harnessed as a constructive force in the creation of a democratic society in the post-apartheid era.

References

Bonner, P., Hofmeyr, I., James, D. and Lodge, T. 1989. *Holding their ground.* Johannesburg: Ravan.

Clark, G. L. and Dear, M. 1984. *State Apparatus: Structures of Language and Legitimacy*. Boston: Allen & Unwin.

Cooper, F. 1983. *Struggle for the City: Migrant Labour, Capital and the State in Africa.* Beverly Hills: Sage.

Cox, K. R. 1981. Capitalism and conflict around communal living space. In M. Dear and A. J. Scott (eds), *Urbanization and Urban Planning in Capitalist Society*, pp. 431–55. London: Methuen.

Cox, K. R. and Mair, A. 1988. Locality and community in the politics of local economic development. *Annals of the Association of American Geographers* 78: 307–25.

Davies, R. J. 1976. *Of Cities and Societies: a Geographer's Viewpoint*. Cape Town: University of Cape Town.

Marris, P. 1979. The meaning of slums. *International Journal of Urban and Regional Research* 3: 419–42.

Meer, F. 1969. *Portrait of Indian South Africans*. Durban: Avon House.

Pachai, B. 1979. Aliens in the political hierarchy. In B. Pachai (ed.), *South African Indians: the Evolution of a Minority*, pp. 1–68. Washington: University Press of America.

Thompson, E. P. 1975. *Whigs and Hunters: the Origin of the Black Act*. New York: Pantheon.

Vernon, J. 1973. *The Garden and the Map: Schizophrenia in Twentieth Century Literature and Culture*. Urbana: University of Illinois Press.

8 *Urban (mis)management? A case study of the effects of orderly urbanization on Duncan Village*

CLAUDIA REINTGES

The White Paper on Urbanization (South Africa 1986) was presented as a wide-ranging reform document. Examination of the document and the processes of its implementation reveal a series of contradictions that have far-reaching implications for the ultimate creation of an 'orderly urbanized' environment. This chapter brings to light some of the contradictions that are emerging on the ground in Duncan Village, a Black township in 'white' East London (Fig.8.1). Since the rapidly changing political climate has created new forms of mismanagement, their effects both on a specific community, Elexolweni, and their implications for future urbanization will also be considered.

Orderly urbanization – the policy

After the publication of the White Paper there was widespread speculation that the policy that it espoused, orderly urbanization, was simply a new and more effective form of influx control. This observation arose because, while influx control legislation had been done away with, the state argued that some form of indirect control over influx had to be exerted 'in order to prevent undesirable social conditions' (South Africa 1986: para. 5.3.1). The form of control accepted by the government was that new arrivals to the city should have access to both a site and housing that are approved for living purposes by a government authority.

The shortage of land on which Blacks in urban areas can settle, together with the estimated national housing shortage of approximately 2 million units, excluding the 'independent' homelands (South African Institute of Race Relations 1989), suggests that insisting that people have access to a site before allowing them to settle in an urban area could provide a very effective means of influx control. That such an application might have been intended is reinforced if one recalls that the White Paper stresses deconcentration of people away from metropolitan areas.

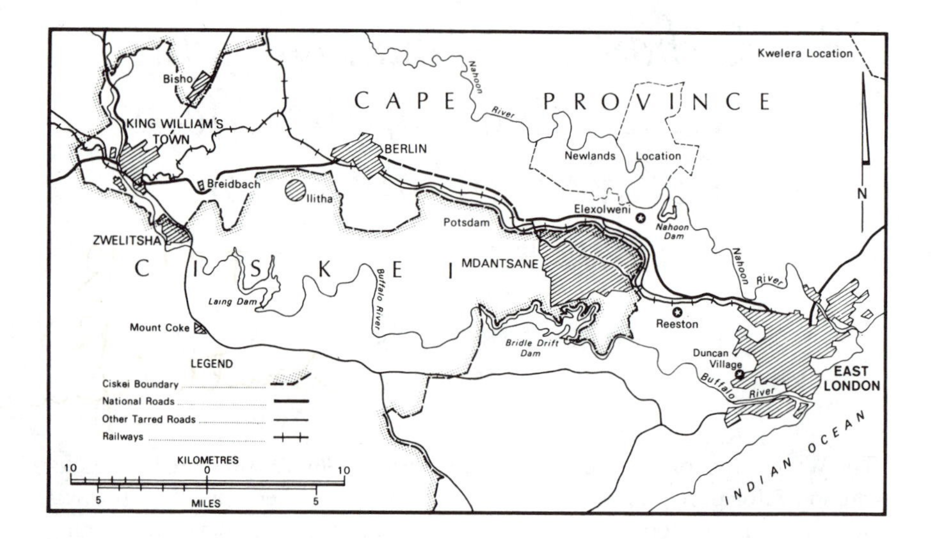

Figure 8.1 Case study locations, in relation to East London and the Ciskei.

Another issue which emerged from the White Paper, forced removal, had resulted in the displacement of approximately 3 million people (Platzky and Walker 1985) over the last three decades. The White Paper stated that no more forced removals would take place and the relevant legislation was withdrawn from the statute books. However, extensive amendments to the Prevention of Illegal Squatting Act (1986 and 1989), forced removals of some 42,000 people from Langa in 1986 and evictions at Crossroads did little to win the confidence of people in this regard. The withdrawal of removals legislation was also announced hand in glove with policy which set out to upgrade the existing townships which are close to 'white' urban areas thus creating 'acceptable' middle-class areas. Since there will be insufficient space to accommodate everyone in the upgraded townships, the urban poor will be removed to the urban periphery, where they will be provided with serviced sites. A factor exacerbating the minimal space available in the upgraded areas is the cost recovery basis on which upgrade is to be effected.

The cost recovery nature of the upgrade programmes are linked to the establishment of 'acceptable' Black Local Authorities. Previously, Black Local Authorities (BLAs) had only certain administrative and consultative powers, and were essentially controlled by the indebted and unpopular Administration Boards (Bekker and Humphries 1985). BLAs had little power and money, were imposed by the state without any consultation, and were rejected by those whose interests they were supposed to represent (Grest and Hughes 1984). The new BLAs are touted as autonomous bodies with wide-ranging powers. They are answerable to the administrator of the province, should be financially self-sufficient and have voting rights on

Regional Services Councils. These BLAs are part of the state's programme of the devolution of power to local levels. It is these bodies, whose political legitimacy and financial viability is questionable, that are to be credited for the upgrade that is undertaken.

Making BLAs responsible for housing provision and administration of their 'own' areas, together with the privatization of housing, must be seen in the context of state withdrawal from its previously direct and control-oriented involvement with Black housing (see Ch. 4). This withdrawal, it can be argued, came about as a result of both political and economic crises prevalent in Black urban areas in the 1980s. Withdrawal of direct control can be understood as an attempt to de-politicize the housing crisis, making the local rather than central state responsible for the implementation of unpopular policy. Furthermore, privatization absolves the state from its financial commitments regarding housing, while simultaneously, so the state argues, boosting the economy by enabling an expansion of the construction sector. While privatization will cater for relatively wealthier people's housing needs, economics prevail against investment by developers in the low-income sector given its lack of profitability. The state has, to a very limited degree, acknowledged this by admitting its responsibility to those Blacks earning less than R150 per month in the form of constructing peripheral site and service schemes. The state's focus at present is, however, on the upper market housing sector. The developers can obtain low-interest loans, and most of the people who can afford to purchase these houses are state-subsidized employees. It is interesting to note that the Urban Foundation's recent loan guarantee initiative (Swilling 1990) extends and makes more viable the state's privatization programme. In essence, however, these programmes have potentially devastating effects on the poor as they are not dissimilar to the process of gentrification (Swilling 1990). The situation that could arise is that the poor, by virtue of their inability to pay for privatized urban goods, will be forced to move. The problem with these removals is not only that they could be widespread, given the vast numbers of impoverished Blacks in urban areas, but that they will be difficult if not impossible to resist since they will occur on a highly individualized basis (Swilling 1990).

The above résumé by no means covers the full extent of the White Paper. The concern here has been to create an awareness of the nature of the issues that have arisen. What needs to be borne in mind is that the White Paper was central to the state's reform package in the mid to late 1980s in that it recognized the permanence of Blacks in urban areas, supposedly devolved power to local levels of government, privatized housing and set out to improve the living conditions of some urban Blacks. As such it represents, at face value, the relinquishing of absolute control over certain elements of urbanization, while creating less direct means of management over the urban environment (Sutcliffe *et al.* 1989).

The history of Duncan Village

Duncan Village today represents an amalgamation of problems arising from previous state policies and the manner in which they were implemented. The entire history of the area is one where the problem of urban Black housing was dealt with either by removing the problem elsewhere, thus 'sanitizing' the problematic areas, or by creating legislation and policies, which enabled both local and central authorities to dismiss the 'problem'.

Duncan Village was established in 1879 as a result of the forced removals of Blacks from three small townships within East London itself. These townships were overcrowded and their removal to Duncan Village was seen by the municipality as a means to 'clean the areas up' (Reader 1961). Having removed the problem little effort was exerted to construct the number of dwellings required to accommodate everyone, which resulted in overcrowding and a lack of facilities and services. This neglect was entrenched through the 1923 Native (Urban Areas) Act, its amendments, and the 1952 Native (Abolition of Passes and Coordination of Documents) Act which was central to effecting the temporary nature of Blacks in urban areas given that its strict definitions of legal urban dwellers formed the basis of the infamous influx control. So-called illegal residents in urban areas never formed part of census-counts nor influenced forward planning since they were theoretically not there. The neglect engendered by this legislation gave rise to the development in the 1950s and 1960s of areas such as Cato Manor (Durban), Sophiatown (Johannesburg) and Duncan Village. These areas became uncontrollable and unacceptable to the apartheid state who solved the 'problem' by removing the areas and locating their legal populations in state-controlled satellite townships. The homeland ideology created by Verwoerd also made it possible for townships to be located within homelands, or close to their borders for later incorporation, thus almost totally eliminating urban Blacks from 'white' South Africa.

The removal of approximately 112,000 from Duncan Village to the peripheral township of Mdanstane, now part of Ciskei (Fig. 8.1), began in 1964 when all development and improvements in Duncan Village were frozen. Between 1964 and 1983 approximately 80,000 people were removed to Mdanstane. The removal was never completed and in 1985 Duncan Village was granted a reprieve: it would remain a Black residential area. Over a period of 20 years no development or improvement had taken place in Duncan Village. It was appropriate then that shortly after the reprieve the state announced that Duncan Village was to be upgraded.

Upgrade proposals for Duncan Village

The planned end result of the upgrade is to create a middle-class environment for some 23,000 people. Since Duncan Village accommodates between

90,000 and 140,000 people in approximately 15,000 shacks and 4,500 houses, between 67,000 and 117,000 people will be forced to move to peripheral land (see Table 8.1).

Table 8.1 Potential densities of people per site at Reeston.

Estimated overflow	Densities per 4,000 sites	Densities per 5,000 sites
12 739[a]	3.2	2.5
37 000[b]	9.2	7.4
67 000[c]	16.7	13.4
117 000[c]	29.5	23.4

[a] The Minister of Constitutional Development quoted the population figure of 35,739 (−23,000 = 12,739).
[b] Setplan (1987) calculated a population figure of 60,000 (−23,000 = 37,000).
[c] The lowest estimate provided by the Duncan Village Mayor and the Chief Engineer is 90,000 (−23,000 = 67,000) and the highest is 140,000 (−23,000 = 117,000).

Duncan Village (Fig. 8.2) is divided into several areas. Upgrade proposals for each area vary. In Ziphunzana south of the Douglas Smit Highway the existing houses will be retained as they are in good condition. There are no shacks in the area and upgrade will focus on the improvement of services and the construction of houses in vacant areas via a job creation scheme that the council has initiated. These houses are being built by labourers earning between R5 and R10 per day and they are sold at their cost price of R7,000 which must be paid in cash. The vacant land in Ziphunzana north of the Douglas Smit Highway is being used for elite housing development. The two phases are complete and house costs range between R50,000 and R120,000.

In the remaining areas development is more complex since a variety of formal housing types and more than 15,000 shacks are to be found. Plans call for the redevelopment of the whole of C section which means that all the structures will be demolished (mostly shacks), and new housing will be constructed. Formal dwellings that are not too deteriorated will be upgraded with their sites enlarged and services upgraded. Since sites are to be increased in size, and the single accommodation (which currently houses families) is to be turned into family dwelling units, many people currently living in formal accommodation will be forced to move.

In order for the upgrade to be undertaken, land must be available on which to accommodate the overflow of people. It was stressed by the Chief Engineer responsible for Duncan Village that people could not be moved until alternative land had been made available. While people initially feared that they would be forced to move to Mdanstane in the Ciskei, land at Reeston (Fig. 8.1) was identified in 1988. As Kruger (1987) argues, the upgrade plans epitomize the aims of the orderly urbanization policy. If the

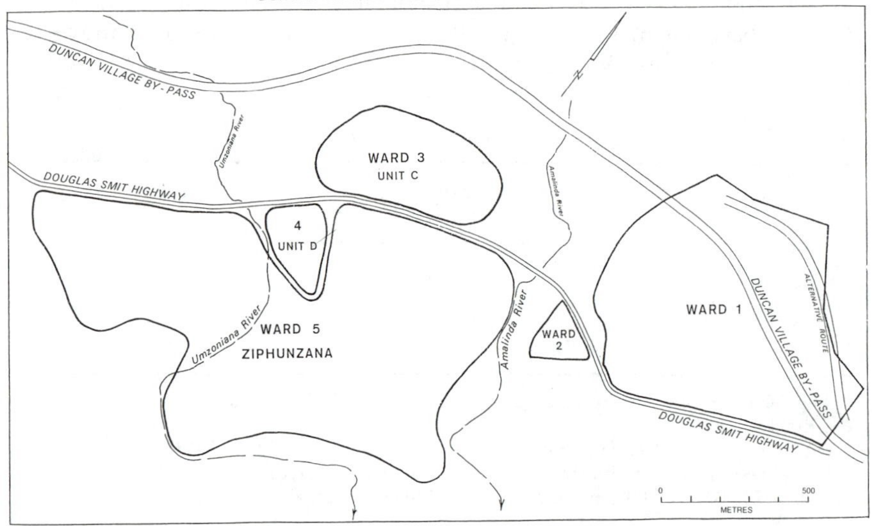

Figure 8.2 Divisions of Duncan Village.

upgrade plans in Duncan Village are ever completed the area will become, as a result of cost recovery and privatization, an upper income suburb and the poor will be displaced to peripheral land.

Underlying this discussion is a disjuncture between the idealism of apartheid policy and the reality of effecting the plans. Understanding these disjunctures in a broader context will shed light on the contradictions that are emerging.

An assessment of the upgrade proposals

While officials envisage that 5,000 sites will be developed at Reeston, an East London planner argues that the area will only accommodate 4,000 sites. Table 8.1 indicates the potential densities of the population at Reeston. Only the obviously unrealistic official figures indicate an improvement in living conditions as a result of low densities. Unreliable population data are, however, not the only problem. The slowness with which the bureaucratic structures function exacerbates the already serious housing crisis. Only a general plan for Reeston exists and the land is still being surveyed. While the policy of orderly urbanization calls for the rapid identification and development of land for settlement, this is taking place at a pace that is unable to have any real effect on the housing crisis given the exceptionally rapid rate of shack development.

Land availability is a problem since full upgrade cannot take place until people are moved, people cannot be moved until land is available, and it seems that land will not be available for some time. The upgrade that is currently being undertaken thus focuses on developing vacant areas and maintaining, in the other areas, a minimally liveable environment.

A further problem with the upgrade is that the land at Reeston is insufficient to accommodate the Duncan Village overflow, as shown in Table 8.1. No land has, as yet, been identified for future Black residential development. While the guide plan for East London should identify such land, it has not yet been completed.

The final point that relates specifically to the upgrade proposals is the fact that the Mayor of the Gompo (the official name for Duncan Village) Town Council has stated that no site and service or squatting will be tolerated in either Duncan Village or Reeston. This is a very short-sighted stance given the extent and nature of the housing problem. In reality the policy emphases indicate that the lower end of the market is, to a large degree, being ignored. A recent survey indicates that only 19.5 per cent of the people living in Duncan Village (that is those earning over R800 per month) can afford privatized housing (Levin and Du Plessis 1989). This will have drastic long-term consequences: the people living in shacks will not simply disappear, rather, their numbers will continue to grow and one wonders how the people earning between R150 and R800 per month will accommodate themselves without some form of state assistance.

Ridding a place like Duncan Village of its shack dwellers and creating an exclusive middle to upper income environment, while being a pleasant dream for some is, in reality, unattainable. The lack of employment that is prevalent in this impoverished region, the nature of the policy, the short-sightedness of planning, the cumbersome and time-consuming procedures involved in making land available and the false scarcities of land due to the existence of Group Areas, give rise to a situation where even those involved in the implementation of upgrade do not believe that the area will ever be fully upgraded. Shacks will remain prevalent and once the land at Reeston is developed shacks will appear there too.

Assessment of urbanization policy

The state has provided a neat package for how urbanization and the management of the urban environment should proceed, which has been handed to local authorities to implement. The state, having divested itself of the responsibility of effecting the policy, has failed to provide the tools with which the policy can realistically be carried out. The resultant lack of development on the ground, in Duncan Village, demonstrates this.

Privatization is also problematic given the lack of available land. The Mayor of the Gompo Town Council's short-sighted stance on site and service is to a large degree symptomatic of the fact that the council has no power to overcome this impasse. It cannot proclaim land for Black residential purposes. If it could it would have no funds to purchase and develop the land. This leads to a situation where, while low-income site and service schemes should, according to the White Paper, be on the planning agenda for Reeston, the land will probably be made available to private developers since state funds allocated for this development are only sufficient to purchase the land. Yet again low-income people will not be provided for.

The BLA also cannot effect any real control over urban influx, given the extensive shacks. Thus using access to accommodation as an indirect means of controlling access to the urban area is inappropriate. The BLA fails with regard to controlling urban influx and cannot, given the limited power and resources at its disposal, effect extensive upgrade. It is further frustrated by the lengthy bureaucratic procedures that are being utilized to proclaim land which are contradictory to the provisions for land identification and rapid development made in the amendment to the squatting legislation.

All of the above points to a realization that central state's (mis)management of the urban environment is creating a situation where it cannot achieve the aims of its own policy. Understood in a history whereby legislation and policies have been created which allow central and local state to ignore, or remove, the housing problem, this conclusion does not really surprise one. The housing crisis is getting worse daily and the White Paper as it is currently being implemented is exacerbating rather than relieving the crisis. What is required for the disjuncture of policy and reality to be resolved is not only a new policy but a major restructuring of the state such that democratically elected local actors can develop practically implementable and appropriate solutions. A brief discussion of an East London community, recently settled at Elexolweni (Fig. 8.1) will develop this line of thought. Discussion of this example allows one to move beyond a mapping of the mismanagement that has arisen as a result of the orderly urbanization policy since the freeing up of the political climate at a national level is not yet being followed by any developments with regard to policy. Furthermore, while the politicians' attitudes are substantially altered the bureaucracy is still firmly intact, as is the major legislation that upholds apartheid reality. While the implications are not yet clear, and will remain fairly dynamic as the politics change, suffice it to say that the kinds of situations developing on the ground are very interesting.

A sample of the future?

The people currently living at Elexolweni originated from Blue Rock, a 'black spot' between East London and Mdanstane. The people were forcibly removed to Potsdam, Ciskei in 1983, and have had many problems with harassment and tax payments since their arrival in Ciskei. In two separate attempts to obtain rights to live in South Africa the people of Potsdam literally walked out of Ciskei and camped on the roadside in South Africa. On both occasions they were moved back to Ciskei. In November 1987 they applied to the court to have their 1983 removal to Ciskei declared illegal. After two years the court ruled in their favour and they gained permanent residence rights in South Africa. The court, however, failed to identify land on which they could settle. After several unsuccessful petitions and deputations they again left Ciskei in April 1989 and settled on land owned by the South African Development Trust and administered by the Department of Development Aid (DDA). The regional branch of the DDA has, after an initial period of negotiation, taken a fairly harsh stance against the people of Elexolweni. They have threatened the community with eviction under the Prevention of Illegal Squatting Act, they have tried to get people to move to small 150-square-metre sites and have generally been uncooperative and unwilling to negotiate. The community have an elected residents' association which administers the area. This body has consulted a town planner who has drawn up plans which accommodate the people on 2000-square-metre plots in several villages. The community argue that this is the minimal size on which they can practise basic subsistence farming. The plan caters for all the people previously resident at Potsdam. There is sufficient land at Elexolweni to implement this plan and there is land that can be used for communal grazing, shops, clinics and schools.

The community has on several occasions tried unsuccessfully to get the local DDA to consider these plans. Given the more encouraging tone of national politics in recent months the people of Elexolweni decided in February 1990 to set up a meeting directly with the Minister of the DDA. They perceived this as essential since people were becoming more permanent and this was not happening as set out in the plans. They were thus anxious to finalize matters so that they would not find themselves in a position where they had to remove people in order to develop the planned settlement. The people also refused to move to the 150-square-metre plots that were being developed for their settlement purposes.

The Elexolweni Residents' Association met with the deputy Minister of the DDA on 26 February 1990. At this meeting it transpired that while the DDA discouraged ad hoc planning they would consider the community's plans. Furthermore, the deputy Minister overruled the eviction notices of the local representative of the DDA and assured people that the Prevention of Illegal Squatting Act would not be used against them. He further stated that the people would not be moved to the 150-square-metre plots, that the development of these would be suspended and that people could remain on the land unhindered but that no more people should settle there while the state considered the plans. They further clarified that their consideration of the plans would take place within the context of the East London guide plan.

One must bear in mind that the people are continuing their lives while the state thinks. Since the people at Elexolweni cannot be expected to live in temporary plastic shelters until the state makes its decisions, they have marked out their plots and more people are settling on the land. The residents have also begun construction of a school and clinic. To date the state has not responded to their proposed plans. The contradictions which emerge are:

(a) Within one state department regional bureaucrats are implementing old-style apartheid while their seniors, in the spirit of current politics, are prepared to negotiate.
(b) National-level ministers of departments, having overridden the traditional apartheid response to a problem like Elexolweni, are left in a dilemma. They must operate within old-style apartheid structures and legislation since these are the only ones available. As such they can only inform the people to wait since there is no alternative policy to apply.
(c) While the central state officials seem, in principle, to agree on the proposed plans they also insist that the plans must be considered in the context of the as yet incomplete guide plan.

What becomes obvious is that the current 'glasnost', if not rapidly translated into policy and if not communicated to all levels of bureaucracy, is creating a degree of mismanagement which, in the near future, could have severe consequences.

Conclusions

This chapter has considered the implications of a specific state policy on a particular residential area in East London. It has also been pointed out that discussion of orderly urbanization is no longer as pertinent as it might have been in the late 1980s given that the policy has apparently been superseded by the exceptionally dynamic nature of politicking at a national level. This was highlighted by the situation of the people at Elexolweni. The disjunctures that are arising both from policy and politics need urgent attention. In both cases the disjunctures affect people and whether there is policy or not, whether decisions are taken or not, people will try to get as close to sources of employment as they can. They will, given the devastated nature of the Ciskei economy, consider seriously their chances in and around urban areas such as East London. People will not wait for the state's permission to settle on land. This is clear both from the extensive shack development at Duncan Village and the growing permanence of Elexolweni. The state is not achieving anything positive by waiting for guide plans to be published, nor by leaving its apartheid bureaucracy intact while it enchants international audiences with its recently enlightened politics. It is, for example, very likely that the people at Elexolweni, if they remain in their current state of limbo, unserviced and unfacilitated, will take to the most lucrative farming available to them, shack farming. The existence of thousands of shacks in Duncan Village demonstrates this possibility, one that the state decision-makers and planners are aware of. Perhaps, given the degree of poverty of the people in the Eastern Cape and Border region the Elexolweni option, even though it is sometimes referred to as scratch patch poverty, is one that needs serious consideration. People are hungry for land and they need employment. The relative closeness of Elexolweni's 2000-metre-squared plots to East London marry these needs. If scarcity of land for residential development in South Africa's cities remains a problem when Group Areas are no more, and if unemployment remains as high as it is, the points raised in this chapter concerning the future nature of urbanization demand serious thought.

Acknowledgements

To Oakley West and John Keulder for the maps, people interviewed for their time and information, and the Grahamstown Rural Committee for information and stimulating discussion.

References

Bekker, S. and Humphries, R. 1985. *From Control to Confusion*. Pietermaritzburg: Shuter & Shooter.

Grest, J. and Hughes, H. 1984. State strategy and popular response at the local level. In South African Research Service (ed.), *South African Review Two*, pp. 45–62. Johannesburg: Ravan.

Kruger, F. 1987. *Upgrading Duncan Village*. Unpublished information pamphlet, East London.

Levin, M. and Du Plessis, A.P. 1989. *Socio-economic study of Gompo*. Vista University: Employment Research Unit.

Platzky, L. and Walker, C. 1985. *The Surplus People: Forced Removals in South Africa*. Johannesburg: Ravan.

Reader, D.H. 1961. *The Black Man's Portion*. Oxford: Oxford University Press.

Setplan 1987. Greater East London Metropolitan Development Study, report prepared for the Urban Foundation. East London.

South Africa 1986. *White paper on Urbanization*. Pretoria: Government Printer.

South African Institute of Race Relations 1989. *Race Relations Survey*. Johannesburg: South African Institute of Race Relations.

Sutcliffe, M., Todes, A. and Walker, N. 1989. Managing the cities: an examination of state policies post 1986. Paper given at the Conference on Forced Removals and the Law in South Africa, Cape Town.

Swilling, M. 1990. The money or the matchbox. *Work in Progress* 66: 20–5.

PART THREE

Informal settlement

The significance of informal (spontaneous, squatter or shack) settlement has emerged from some of the earlier chapters. Those brought together in this part of the book have been chosen to reveal something more of the nature of these settlements, and of their variety, based on research in different parts of South Africa.

First (Chapter 9), Horn, Hattingh and Vermaak examine Winterveld, within the 'independent' state of Bophuthatswana but functionally part of the Pretoria metropolitan region. They describe it as an 'interface' settlement where people experience transformation from rural to urban life, and explain some of the problems arising from the tangled racial and tribal geopolitics imposed by apartheid. Next (Chapter 10), Cook provides an account of the nature and significance of Khayelitsha in the Cape Peninsula, where the state first sought to relocate Africans from Cape Town squatter settlements such as Crossroads, subsequently using the project to legitimize informal settlement as a solution to the problem of Black housing – albeit in a location peripheral to the metropolis. A vivid picture of life in the shack cities is provided. Then (Chapter 11), Crankshaw, Herron and Hart report the results of research on a small peri-urban squatter settlement on the Witwatersrand. Using quantitative and qualitative methods, they show how people came to the settlement, illuminated by individual case histories of ex-farmworkers in the process of urbanization. Finally (Chapter 12), Boaden and Taylor draw on experience with upgrading projects in KwaZulu/Natal to consider why there has been so little improvement in the informal settlements. They attribute this to a series of misconceptions, or myths, using local case studies to reveal the realities of the problem in different environmental and social contexts.

9 *Winterveld: an urban interface settlement on the Pretoria metropolitan fringe*

ANDRÉ HORN, PHILLIP HATTINGH & JAN VERMAAK

J71 J61 R11
South Africa

Introduction

Government policy has in numerous ways played a significant role in structuring and regulating urbanization in South Africa (Davies 1986). In adaptive response to restrictions many Blacks settled in peripheral places allowing access to towns and cities to fulfil a dream of better living. Winterveld, an informal settlement on the fringe of the Pretoria metropolitan area, exemplifies such a response.

Winterveld also reflects the constitutional and functional complexity that complicates urban form and living in contemporary South Africa. The Pretoria city system (Fig 9.1) comprises several Group Areas components: Pretoria, Verwoerdburg and Akasia, three white-controlled urban areas, are at the core with several non-white dormitories such as Atteridgeville and Mamelodi (Black towns), Eersterus (coloured town) and Laudium (Indian town) adjacent to it. In conjunction with these local manifestations of racial apartheid, grand apartheid (the division of Blacks into ethnic entities constitutionally linked to homelands) was also superimposed on the Pretoria metropolitan system, 'internationally' dividing it into a sector under the control of the Republic of South Africa (RSA) and a sector in the Republic of Bophuthatswana. The latter includes the Black towns of Ga-Rankuwa and Mabopane as well as the Winterveld settlement, as functionally part of the metropolis.

Although the existence of the Winterveld squatter settlement as it is today can mostly be ascribed to apartheid practices, its diverse ethnic composition has made it a geographical anomaly in the system of grand apartheid which placed it in the 'homeland' of the Tswana people. As such, it complicated the acceptance of 'independence' by Bophuthatswana, leaving the RSA with a legacy of responsibility. In fact, paradoxically, it became the testing ground for bilateral cooperation between the two governments and development efforts were consequently mostly experimental and superficial. It is therefore no surprise that the position of Winterveld was over the years much

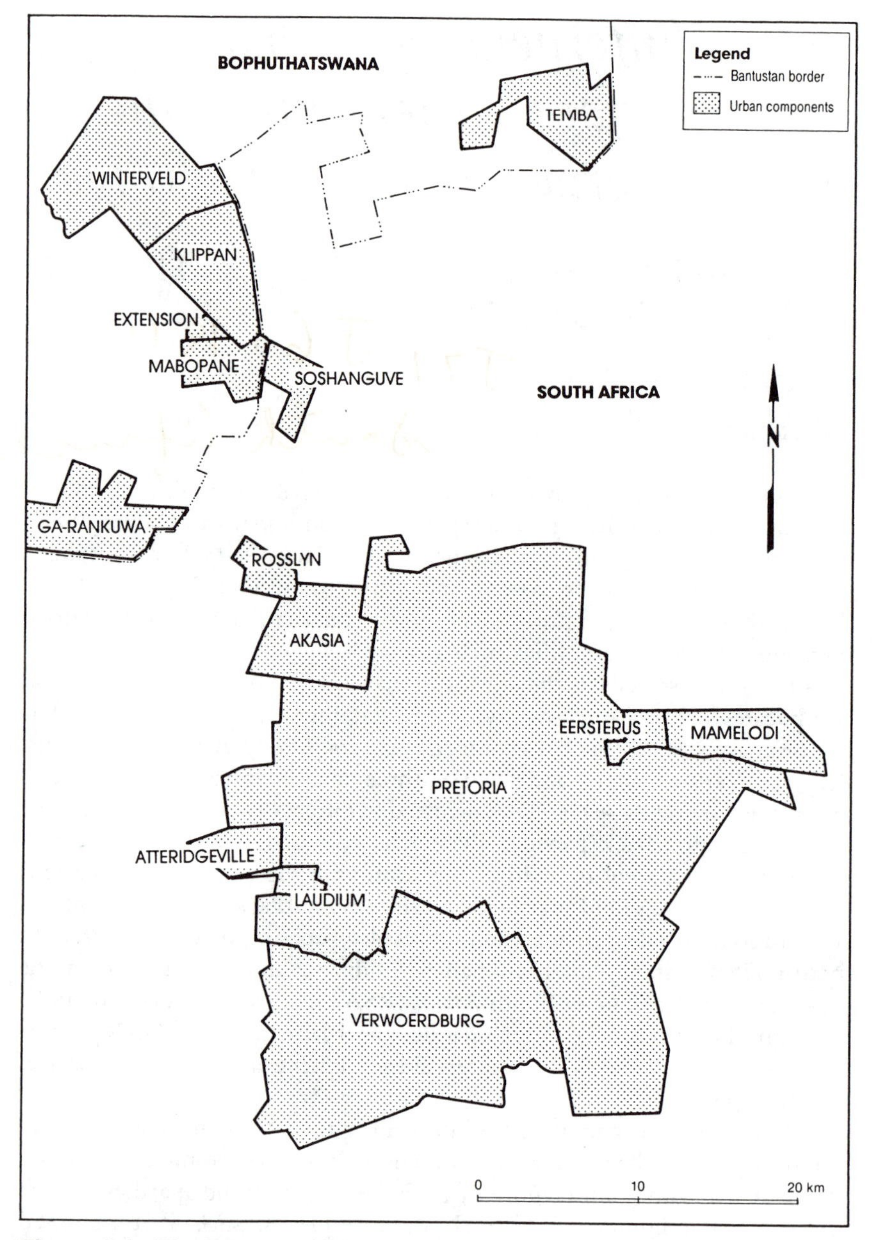

Figure 9.1 The Pretoria urban system.

politicized and even sensationalized. It was only after the bringing about of an institutional framework which includes resident representation and the obtaining of urban status that worthwhile development got underway.

Any presentation of the current situation in the settlement should take cognizance of the sociopolitical absurdities that gave rise to it, the poor quality of life and the accompanying institutional manoeuvres, as well as the sincere development efforts of late.

Hence the purpose of this chapter is: (a) to describe the spatial and functional evolution of the Winterveld settlement in the apartheid era; (b) to report qualitatively on living conditions within it; (c) to determine its status in the contemporary period of political transition; and (d) to contemplate its role in a future metropolitan system.

Spatial and functional evolution

Spatial history The Winterveld informal settlement, a sanctuary for people drawn into a process of economic and cultural transformation but caught amidst political crossfire, is situated in the Odi and Moretele Districts of the Bophuthatswana homeland some 30 km north of South Africa's capital city, Pretoria. Located on a plain in the Transvaal heartland, the area was formerly inhabited by African tribes, later used by white pioneers as winter grazing (still bearing the name) and was eventually divided into white-owned farms. The settlement originated between 1938 and 1945 when firstly the farm Winterveld and later also Klippan, a total of 10,386 hectares, were subdivided into 1,658 agricultural holdings varying from 4.3 to 8.6 hectares and sold over a period of time to individual Blacks.

The area commonly known as Winterveld was sparsely populated by 1950 with only 464 dwellings identifiable (South Africa, 1950) but a dramatic increase in population was to come. According to Hattingh (1975: 45–52) this was the result of: resettlement of Black squatters in and around Pretoria (Group Area legislation); removal of 'Black spots' in proclaimed white areas (grand apartheid); and voluntary influx enhanced by the lack of any influx control in spite of strict policies.

Despite the area being earmarked for agricultural landuse, owners of holdings, especially in the southern Klippan area, soon embarked on 'shack farming' (tenants being granted permission to erect their own makeshift dwellings) as a means of generating income. This practice relates to the 'lotsha' system – a traditional Tswana custom whereby a chief could grant a family the right to live and farm on a piece of tribal land. The exclusiveness to tribal members was relaxed under changing circumstances. The demand for land induced subletting and this became financially profitable (Hattingh 1975: 51).

The familiarity with the traditional lotsha system as encountered in the tribal areas, in combination with the push and pull factors pertaining to population movement and urbanization, facilitated the creation of an informal nucleated settlement neighbouring formally-structured Black townships on the metropolitan fringe.

Built-up area: expansion and density The growth of Winterveld as an urban-oriented settlement ties in with the general development of apartheid planning in South Africa (Smith 1983: 71), more specifically the Pretoria region. At the start of the century there were approximately 350,000 Blacks in urban areas in South Africa (Smith 1983: 55). Early days saw Pretoria displaying the pre-apartheid pattern of ethnic diversity with some segregation in the city and a large and scattered Black population in the countryside. Blacks within the city were accommodated in a variety of urban forms including the ghetto-like Marabastad, the freehold township of Lady Selborne, and Bantule, a partly developed location, as well as a number of squatter settlements.

With the coming of apartheid under the National Party government there were about 120,000 Blacks living in the city (Junod 1955: 75) and a clear need for some other scheme to cope with their constantly growing needs. While the city council developed three townships, namely Atteridgeville, Saulsville and Mamelodi, central government in conjunction with the council embarked on their policy of slum clearance and resettlement. According to Hattingh (1975: 49) some 155,000 people were uprooted and resettled in Mamelodi and Atteridgeville–Saulsville while others found their way to the proclaimed Black trust area (land reserved for Black occupation) north of Pretoria. Although the border of the trust area was still fairly insignificant, and appropriate accommodation was available in Black border towns founded after 1960 such as Ga-Rankuwa and Mabopane, the area was already earmarked for a Tswana nation state, implying the reservation of housing for Tswana people. Outcast by both white and Tswana establishments, many of the unaccommodated surplus labelled as 'throwaway people' (*Pretoria News*, 1990) and new arrivals found an opportunity for settling in Winterveld.

The general control of urbanization of Blacks (influx control) was exacerbated by the cessation of housing provision in Atteridgeville and Mamelodi from 1969 onwards. As a result of natural growth in population in the two towns the superfluous were directed to the homeland north of the metropolis. The 'Black spot' removal programme added population from Walmansthal and places as far as Heidelberg (105 km) and Nylstroom (97 km). Voluntary in-migration resulted from primarily economic pull factors in the white urban areas, for example the industrial decentralization programme dating from 1960, and push factors such as high population pressure, unemployment and poor farming conditions in the rural areas.

In accordance with the general pattern of homeland urbanization, with concentrations in the areas bordering White South Africa, Winterveld had higher densities in the south, becoming more prominent over time as the population increased. According to Vermaak (1987: 6), more than 60 per cent of the population lives in the southern Klippan area. There is some confusion over how many people are really residing in Winterveld. A

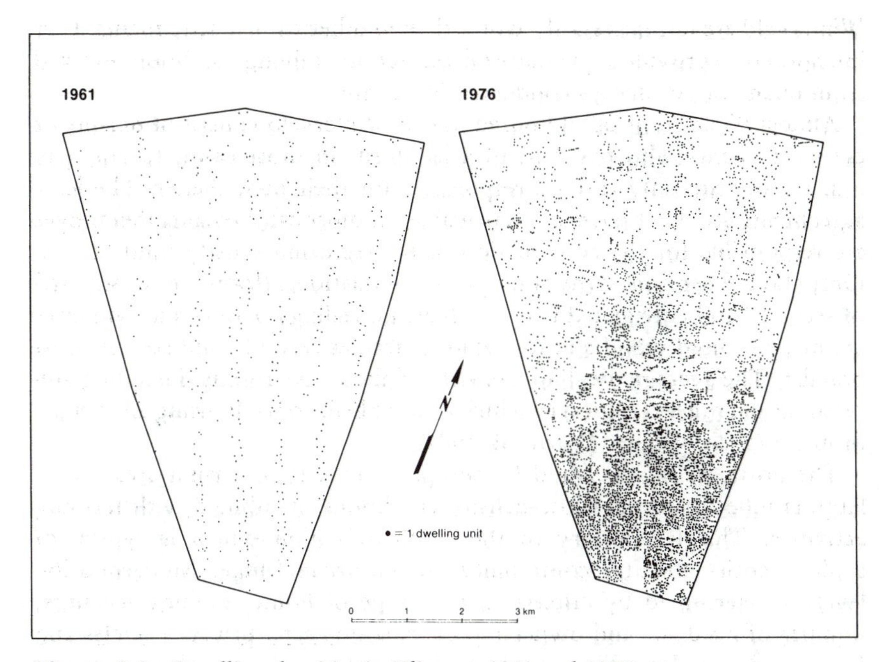

Figure 9.2 Dwelling densities in Klippan, 1961 and 1976.

decade ago Smit and Booysen (1981: 42) estimated that there were 300,000 to 400,000 people in the area. Other estimates are even higher, for instance 400,000 (*Pretoria News* 1984), 500,000 (*Beeld*, 1982), and 750,000 (President Mangope of Bophuthatswana, cited by Vermaak 1987: 2), as against the official figure of 102,000 (Bophuthatswana 1985). We estimate, based on dwelling density and occupancy, that there are currently 220,000 to 280,000 people living on the original farms of Klippan and Winterveld. There is general agreement that the bulk of inflow took place between 1960 and 1980 (Vermaak 1987: 13).

The expansion and density in the southern Klippan area are illustrated (Fig. 9.2) by data obtained from the official aerial photo series of 1961 and 1976 (South Africa, 1961, 1976). From this it becomes evident that dwellings were originally constructed alongside the major roads providing a linear appearance to the settlement but distorted afterwards. In 1961 the density of dwellings was on average one per 10 hectares in the Klippan area. By 1976 this had changed dramatically to nearly two dwellings per hectare. In total, there were 419 houses in the Klippan area in 1961, increasing nearly twenty-fold to 8,178 in 1976.

Living in Winterveld Slum conditions and squatter settlements as a result of the high rate of urbanization are a common feature in African urban areas (Obudho and Mhlanga 1988a: 3). Although appalling, conditions in

Winterveld are not necessarily worse than in other similar settlements. Our intention is to provide a qualitative presentation of living conditions derived from quantitative surveys conducted in the area.

Almost 90 per cent of all households in Winterveld consist of tenants or even subtenants who pay rent to a landlord. In most cases, tenants pay rent for the site only and are responsible for their own shelter. The lease agreements are short term and renewable. In most cases tenants themselves are responsible for the construction of houses using mostly mud blocks. Only about 8 per cent of the houses have foundations. Roofs are constructed of second-hand corrugated iron and have no ceilings. Doors and windows are made of used wood and for window frames second-hand steel is most popular. The average dwelling consists of about five rooms (including any washing and toilet facilities) of which two are bedrooms, housing an average of just over six people. (Vermaak 1987).

The northern region could be compared to a typical rural area where large families live under low-density conditions, combined with farming activities. This is contrary to the southern region which is typical of a high-density squatter community on the urban fringe. Modernization levels – determined by criteria such as type of home, savings accounts, number of residents and ownership of television sets, private vehicles and fridges – decline laterally from the major south–north arterial (Olivier and Booysen 1985: 567).

The age structure of the residents reveals that close on half are under 20 years (Bophuthatswana 1985) which is similar to the general Black South African situation (Sadie, 1988). The predominant economic activity in Winterveld seems to be that of selling labour, mainly in the Pretoria metropolis. Of this offering some 55 per cent are unskilled, 25 per cent semi-skilled (e.g. carpet fitters, barmen) while the remaining 20 per cent can be regarded as skilled (e.g. teachers, plumbers, electricians). Hence it is no surprise that unemployment, excluding the informal sector, is high and that at least a third of the households are probably living under the household subsistence level which is currently being estimated at R570 per month in the Pretoria area.

The economic base of the area is fundamentally the generally low income that households earn from outside Winterveld. Apart from that there seems to be a leakage or outflow of money. Accommodation in the area is relatively cheap (with most people spending less than 5 per cent of their income on it) so that there is a limited multiplier effect in the local economy.

The lack of water and water-borne sewage is one of the most vexing problems in Winterveld. Water-borne sewage is non-existent and without exception households are dependent on pit latrines. The water table is close to the surface and easily polluted in this way. Until recently the only sources of water were boreholes and wells, a situation which has led to numerous outbreaks of cholera. A programme of chlorinating boreholes resulted in the appearance of water vendors who started selling this chlorinated water. On the one hand it seems to be a profitable venture but on the other hand results in the vandalization and boycotting of schemes to bring piped water from Pretoria to within the settlement. Efforts to introduce electricity have also fallen prey to vandalism, a sociopathological phenomenon rather than politically related act. Stoned and burnt-out power boxes are a common feature in the southern area.

In certain areas the accumulation of refuse is so bad that access to properties can hardly be found. It is, however, obvious that residents usually try to clean up in the proximity of their own houses but they have nowhere else to dump refuse than in the streets. Some landlords provide a refuse removal service at a cost, while others try to burn it.

Apart from cholera emanating from drinking polluted water, the incidence of deficiency diseases such as kwashiorkor and scabies is high. Infant mortality is around 10 per cent but on occasion (e.g. in 1976) it has risen to as high as 50 per cent. Other common afflictions are tuberculosis and gastroenteritis. The only hospital is at Ga-Rankuwa some 10 km away and residents have to rely on three clinics in Winterveld and one in the adjacent Mabopane.

Although both public and private schools are packed to capacity, less than 60 per cent of children of school-going age attend school. This results from the obvious deficit of educational facilities and also from economic realities forcing pupils to look for employment. Another problem is that the Tswana language is the compulsory medium in government schools whereas the majority of residents are non-Tswana.

Hence, for the outsider, Winterveld may seem like a hell-hole imprisoning society's throwaways. To a certain extent this is true and subsequently we will address the sociopolitical tangle which lies at its foundation. It is quite appalling to learn of the existence of a despairing community where people are born and buried without documentation.

Yet, on the other hand, it is a settlement of its own making, and any form of superimposition is treated with suspicion as threatening the freedom of undocumented existence and bringing the burden of responsibility. It would, however, be quite wrong to imply that this inferior lifestyle is adequate, but for those who are in the process of getting an urban foothold this is what is preferred, affordable and what they are prepared to take responsibility for.

Settlement features: interface, imposition and intervention

By the concept of an 'interface settlement' we refer to the informal settlement as locating the encounter between different lifestyles as well as different political systems. This interfacial function of the informal settlement, of course, takes place within a normative framework which also channels intervention in the form of development efforts.

With reference to the first type of interface that takes place in an informal settlement (the social encounter between feudalism and industrial capitalism), it is our impression that Winterveld is not significantly different from other informal settlements in South Africa or elsewhere in the world. We share the opinion that informal settlement reflects a way of urban living and should be accepted, promoted and fully incorporated into metropoles. It offers a foothold for those who are in the process of urbanising and furthermore functions as an intermediary school where an urban lifestyle can be acquired. In the final section of this chapter we shall focus on this issue when addressing the future of the Winterveld settlement.

The informal settlement also accommodates the political encounter between the tribal system and contemporary Third World government and it seems that as an introductory response in order to survive, traditional community relations are mobilized into an effective lever to lobby for adequate living conditions (McCarthy 1988: 295). It will be argued that the real trauma of Winterveld is not its existence as an informal settlement but the constitutional tangle that was superimposed on it.

As stated earlier, the area in which the Winterveld settlement is located was concurrently occupied by various African tribes and later proclaimed as farms owned by White settlers. Although certain heartlands for exclusive Black occupation, based on existing settlement patterns, were scheduled by the Natives Land Act of 1913, it did not include the Winterveld area. Land earmarked later to expand these areas was proclaimed as 'released areas' (Bantu Trust and Land Act of 1936). This resulted in the transactions referred to previously whereby the farms in Klippan and Winterveld were divided into smallholdings and sold to Blacks. However, ethnicity (a cornerstone of the apartheid system that was to come) was of no importance. Blacks from all over Southern Africa settled in the area.

It was only in 1959, with the promotion of the Bantu Self-government Act, which resulted in the demarcation of homelands, that the Tswana as an ethnic group *inter alia* received recognition. The apex of the new system of self-government for the Tswana people was the Tswana Territorial Authority of 1961. The Bantu Homelands Citizens Act of 1970 granted citizenship for nationals in homelands with certain exclusive privileges such as the right to own land. That put many of the inhabitants of Winterveld in an unfortunate position since only about 10 per cent of them and about 45 per cent of the inhabitants of the Odi District are Tswana (Bophuthatswana 1985). The Tswana acquired internal self-government in June 1972 and gained sovereign independence from South Africa on 6 December 1977.

The position of Winterveld was problematic during the independence process and remained so afterwards. At first a part of the Mabopane township (subsequently renamed Soshanguve (Fig 9.1) and divided into ethnic sectors according to strict apartheid ideology) was incorporated back into South Africa in order to make provision for the non-Tswana. Despite the costly construction of an attractive town which offered a variety of residential options and which is currently occupied to capacity, the anticipated outflow of people from Winterveld did not take place. As a second option, the Bophuthatswana government accepted the incorporation of Winterveld into the homeland whilst the South African government has taken responsibility for the general upgrading and development of the area including infrastructure. The sum of R188.7 million was budgeted for this purpose.

It was because of this legacy that official attitudes towards the fate of the Winterveld inhabitants began gradually to change. In the period prior to the independence of Bophuthatswana, the South African government simply ignored the urbanization process in Winterveld. In the terms of Obudho and Mhlanga (1988b: 327) this was a typical colonial response. During and shortly after independence (1977) the *laissez faire* policy was replaced by one of eviction: while South Africa was choreographing its ethnic experiment in Soshanguve several raids to discourage the non-Tswana were executed in Winterveld by Bophuthatswana police on the basis that agricultural land was being used for urban practices (*Financial Mail* 1982: 1011).

Constant media coverage kept Winterveld in the political limelight and official attitudes had to change. The passive and negative policies of the past were replaced by the belief that the establishment of a local management framework was necessary to enhance positive development in Winterveld. Such a framework comprising a Development Board, Executive Management, an Advisory Committee and a Residents' Committee was gradually phased in. The Development Bank of Southern Africa was appointed as a development agent.

Despite the positive changes in attitudes, development efforts for some time seemed to be ad hoc, experimental and superficial and thus not addressing the core of Winterveld's problems. The harsh practices of eviction were replaced by the creation of new towns and alternative housing outside Winterveld. Apart from Soshanguve, the most significant of these projects was the creation of an 'idyllic' housing scheme on a piece of no man's land wedging in between Winterveld and the adjacent Mabopane (Extension, in Fig. 9.1). This was a joint venture between the two governments as part of the Winterveld Development Programme, an overall development strategy. The aim with Mabopane Extension, as it was introduced, was not only to provide additional housing in the area but also to aesthetically link the squatter settlement and the formal town. In the first two phases, 1,900 dwellings were constructed. Despite the incorporation of cute planning concepts Mabopane Extension (nicknamed Little Beirut) does not seem to be an overall success. Apart from possible blunders concerning layout and architectural design the question arises whether the general lack of appreciation regarding low-cost housing schemes in South Africa is the result of inadequate social and community development back-up.

A further step, however, making development an undeniable reality, was the proclamation in 1986 of the southern Klippan area as 'urban', and the introduction of an urban council in 1990. A second draft urban structure plan has also been tabled and in the last few years considerable progress has been made in the provision of public amenities and basic urban infrastructure. Uncertainty still prevails on the form of residential development and amongst residents there is still a conspicuous attitude of mistrust in and resistance to intervention. The full acceptance and incorporation of Winterveld in the urban system, however, seems to have ended the bickering about its status.

Future metropolitan role of Winterveld

The future role of Winterveld in the metropolitan system will be closely linked to the degree of development that is to take place in the area. The general state of the man–environment encounter in the settlement left no option but to develop. There exists, however, the danger of enforced upgrading, overregulation and overdocumentation.

In order to qualify this warning we must point out that the majority of people currently in the settlement are there by their own free will (Vermaak 1987: 15), that land and home ownership is not such a priority with them as employment is, and that they are generally happy with their makeshift dwellings in terms of affordability (Vermaak 1987: 47, 65). Principally, we support the general improvement of living conditions in the settlement but wish to emphasize that in metropolitan context its prime function is interfacial, that is, to operate as a cultural and political transformer. As such it also offers for the rural superfluous people an urban alternative to the socially undesirable system of migratory labour. The characteristics of Winterveld today, including irregular layout, secondary building materials and insecurity of tenure have evolutionary significance. The danger of conflicting priorities between the urban newcomer and the urban regulator thus exists. Overdevelopment and the associated responsibility and liability will inhibit the initiative of the newcomer in obtaining an urban foothold and will, furthermore, have the same outcome as eradication, namely alternative sites of squatting.

Furthermore, the future significance of Winterveld is that of an urban partner in the Pretoria metropolitan system. Close to 80 per cent of

Winterveld's economically active inhabitants are currently employed in the core of the metropole and its function as a labour reservoir will ensure its participation in the urban alliance. With its urban status recognized, its political presence as voiced from within will also be felt on different levels.

In retrospect we believe that Winterveld as a political catalyst is also of historical significance. While many squatter settlements in South Africa exemplify the manifestations of apartheid and the struggle against it, the position of Winterveld and its role within the political system in particular has grotesquely enlarged the latent cracks in that system.

References

Beeld 1982. Vordering in plakker-stryd. 28 November: 9–10.

Bophuthatswana 1985. *Population Census Report No. 2*. Mafikeng: Department of Economic Affairs.

Davies, R.J. 1986. When! – reform and change in the South African city. *South African Geographical Journal* 68: 3–17.

Financial Mail 1982. Winterveld residents squeezed. November: 1011–15.

Junod, H.P. 1955. The bantu population of Pretoria. In S.P. Engelbrecht *et al.* (eds), *Pretoria 1855–1955*, pp. 61–85. Pretoria: City Council of Pretoria.

Hattingh, P.S. 1975. Nedersetting en verstedeliking. In J.H. Moolman and G.M.E. Leistner (eds), *Bophuthatswana: hulpbronne en ontwikkeling*, pp. 25–58. Pretoria: Africa Institute.

McCarthy, J.J. 1988. Poor urban housing conditions, class, community and the state: lessons from the Republic of South Africa. In R.A. Obudho and C.C. Mhlanga (eds), *Slum and Squatter Settlements in Sub-Saharan Africa*, pp. 295–308. New York: Praeger.

Obudho, R.A. and Mhlanga, C.C. 1988a. The development of slum and squatter settlements as a manifestation of rapid urbanization in Sub-Saharan Africa. In R.A. Obudho and C.C. Mhlanga (eds), *Slum and Squatter Settlements in Sub-Saharan Africa*, pp. 3–30. New York: Praeger.

Obudho, R.A. and Mhlanga, C.C. 1988b. Planning strategies for slum and squatter settlements in Sub-Saharan Africa. In R.A. Obudho and C.C. Mhlanga (eds), *Slum and Squatter Settlements in Sub-Saharan Africa*, pp. 327–48. New York: Praeger.

Olivier, J.J. and Booysen, J.J. 1985. Moderniseringsvlakke binne die Klippan-Winterveldse informele nedersettingsgebied: 'n voorlopige verkenning. In L.A. van Wyk (ed.), *Development perspectives in Southern Africa*, ABEN: Research Papers 85–1, pp. 551–71. Potchefstroom: Potchefstroom University for Christian Higher Education.

Pretoria News 1984. Winterveld . . . a slum housing 400,000 folk in squalor. 6 June: 12–13.

Pretoria News 1990. Winterveld: the festering disgrace on Pretoria's doorstep. 1 February: 11.

Sadie, J.L. 1988. *A Reconstruction and Projection of Demographic Movements in the RSA and TBVC countries*. Research Report No. 148, Bureau of Market Research. Pretoria: University of South Africa.

Smit, P. and Booysen, J.J. 1981. *Swart verstedeliking proses, patroon en strategie*. Cape Town: Tafelberg.

Smith, D.M. 1983. *Update: Apartheid in South Africa*. London: University of London.
South Africa 1950. Aerial photographs Job 218/48–50.
South Africa 1961. Aerial photographs Job 453/61.
South Africa 1976. Aerial photographs Job 769/76.
Vermaak, J.L. 1987. *Urbanization Trends based on Winterveld Surveys and Reports*. Pretoria: Institute for Development Research.

10 *Khayelitsha: new settlement forms in the Cape Peninsula*

J61 R11
South Africa

GILLIAN P. COOK

The 1983 official announcement that a new town, Khayelitsha, was to be created for African residents of Cape Town came as a surprise. Since 1955 only a limited number of Africans had legal right to remain in the Western Cape and from 1966 a deliberate policy of exclusion and harassment of Cape Town residents was implemented. Whether Khayelitsha heralded a real change in policy or was a pragmatic response to gross overcrowding, unsuccessful removals and escalating violence is not clear (Cook 1986). At first it was planned that all African residents in Cape Town, starting with those in the squatter settlement of Crossroads (Fig. 10.1), would live there, but without any rights to land. With this in mind a 3,220-hectare site was cleared and building started on the first of four towns each to comprise four villages housing 30,000 people in 25–32-square-metre core houses on 160-square-metre plots. Included was space for higher-income private housing, land for educational and social services (but none for industry), a central area, a spinal green heartland and ultimately a rail link to Cape Town.

Almost immediately (June 1983) people from squatter settlements who were prepared to live in Khayelitsha were temporarily housed in 14.4-square-metre flexicraft huts adjacent to Town 1 (Fig. 10.1). In the face of refusals to be moved, representation through official channels, demonstrations and rioting, first 99-year leasehold was introduced in Khayelitsha (September 1984), then site-and-service plots, to which even 'illegal' squatters would be allowed temporary access, were incorporated in the plans (November). Finally in February 1985 the idea of moving all Africans in Cape Town to Khayelitsha was dropped. A site-and-service scheme commenced at site C, and land was allocated for the technicon, hospital and stadium (March). By October 5,000 core houses were rented out to 13,000 Africans legally in Cape Town while 8,300 squatter families from Crossroads occupied 4,150 site-and-service plots. In 1986 the private sector became slightly involved when government encouraged firms to build houses or hostel accommodation for employees. More significantly official acceptance of the orderly urbanization strategy meant that the site-and-service

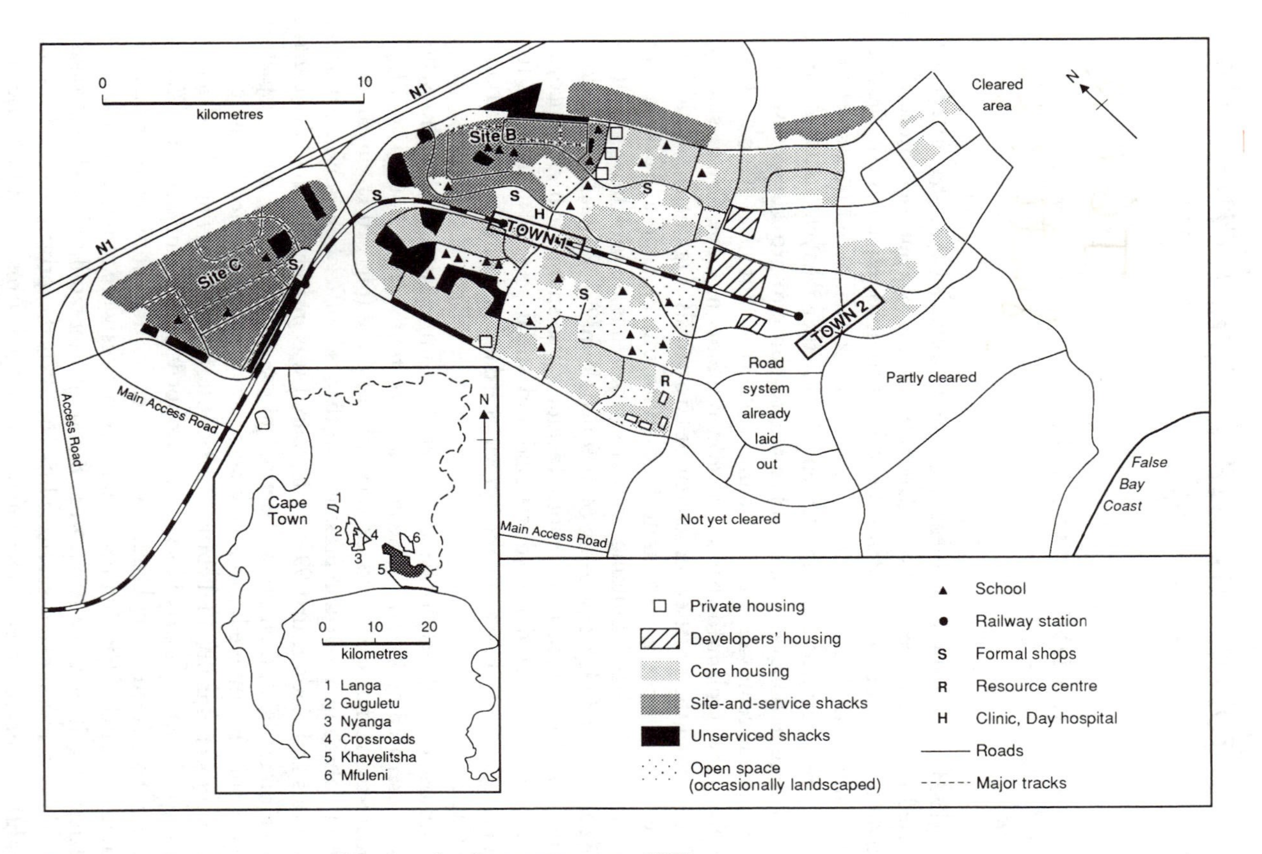

Figure 10.1 Khayelitsha, showing developments up to 1990.

component in Khayelitsha ceased to be regarded as a temporary expedient and 4,600 plots of 78–90 square metres were developed to a higher standard for informal housing at site B under 99-year leasehold (Dellatola 1987).

Residential character

In line with the original conception of a well laid out town in which householders would upgrade core houses, the largest proportion of the site is laid out to residential streets in a curvilinear pattern. Occasional small paved and landscaped corners fail to relieve the monotonous sprawl of uniformly low quality 27.8-square-metre houses set behind vibracrete fences against which sand is driven by fierce southeasters. Large prefabricated schools (mostly junior) have some grassed areas but they are behind high fences and strong gates. Tall floodlights provide street lighting and although schools, crèches, clinics and community centres have electricity few houses are connected to the service. Near the new railway line hoardings attract attention to show houses of property developers. Here building activity is concentrated in pockets overlooking and in some cases adjacent to dense clusters of informal settlement.

It is this latter form of settlement that makes the greatest impact by virtue of the degree (unusual in the Western Cape) of density of both housing and people. Two types of informal residential areas can be recognized. There are serviced areas of 'controlled squatting' (sites B, C and Greenpoint) which together form an arc on the north-east perimeter of Khayelitsha (Fig. 10.1). Here conventional planning constructs have been applied to include access roads, subdivisions and service hierarchies. The remaining housing is not preplanned and increasingly regularity is broken down by unserviced shacks erected wherever there is open space. Residents of informal areas, some in traditional dress, thread their way between shacks carrying goods, water and piles of firewood to their homes. Four out of every five people in Khayelitsha live in two-or three-roomed shacks made of corrugated iron, plastic, cardboard and soft wood in some combination.

Housing

Formal public housing The change in official policy regarding the form of settlement in Khayelitsha has meant that only a limited number of formal houses are now being built. Demand for rental houses is reflected in a long official waiting list. This may be bypassed when contact is made between a household wanting to move in and one intending to vacate a house. A financial arrangement, which varies around R1,000, is made between the two parties before they approach the council (to whom a 'gift' may be made) who then authorize the official transfer. Some racketeering may occur because there are cases in which several houses are rented by the same person under different names. Rentals are between R20 and R23 per month and include full municipal services but there is a high level of arrears and defaults which probably reflects a genuine inability to pay rather than a boycott. Only 14 per cent of houses on offer were sold by 1990, primarily because the R560 required to qualify for a mortgage is beyond the means of most households. Fully serviced sites can be purchased for R7,000 to enable residents to construct their own homes.

Anticipated upgrading of two-roomed core houses has not taken place. Given that houses have neither plastered walls nor ceilings, no bath/shower or electricity and an average of between six and ten occupants in less than 30 square metres, it is significant that even in Village 1 only 44 per cent of houses provided visual evidence of alterations after four years (as observed in a survey by Nigel Measures in 1988–9, on which the substance of this paragraph is based). The primary reason is lack of money coupled with an unstable work situation which together make it impossible either to get loans or to contract the work out. In the case of single heads of household the difficulty is compounded by lack of time and the physical inability to carry out the necessary tasks. A survey recorded 4 per cent of houses upgraded according to official specifications, as opposed to a figure of 10 per cent predicted at the planning stage. Some 9 per cent of houses had minor alterations (door changed, window put in). Internally priority is given to plastering walls to reduce the cold and damp. Significantly one-third of all houses had more living space created by erecting a makeshift room either attached to the house or in the grounds. In general the more alterations made the more likely that the head of household is not single, is permanently employed and has either bought the house and/or received financial assistance to upgrade. More than half the householders interviewed felt that the houses were below standard and therefore should be upgraded by the state, some arguing that by doing nothing they might pressurize officials into extending homes or moving the tenants into larger places. Gas is used for cooking and heating water and appliances run off batteries as very few core homes are connected to the electricity supply. In an effort to encourage electrification a limited supply is offered to residents unable to afford the full cost of wiring their houses. A low-cost solar water heater appropriate to Khayelitsha conditions has also been developed but at R1,200 is beyond the means of all but a few residents.

Formal private housing Private developers are introducing another form of housing to African residents in the Cape Peninsula and their impact is greatest in Khayelitsha where most land for development is available. Three large construction syndicates built an average of 80 houses a month in Khayelitsha during 1990. They advertise with slogans 'All costs included – no deposit' for people to invest in developments with evocative names like Gracelands. The cheapest homes on offer are three-roomed prefabricated houses of 50 square metres costing between R10,000 and R16,000 plus land. Most houses are conventional brick-built two-bedroomed homes available at 'affordable' prices (R30,000 with repayments of R262 per month) covering land, house, bond and transfer costs. Schemes are aimed to attract the small number of upwardly mobile residents earning above R900 per month: less than 5 per cent of all government employees, 7 per cent of squatters, 4 per cent of sites B and C residents and at least 10 per cent of core house residents (Dewar and Watson 1990a). The largest development is Futura Park with 1,000 units on plots ranging between 115 and 500 square metres and requiring a deposit of R250. Prices range from R30,000 to R79,000 including 'carpets, gutters and toilets'. A slightly more up-market development at Khanya Park (the 64 homes from R45,000 upwards sold out immediately) and its 600-unit extension Elitha Park is located conveniently near a railway station although there are no first class carriages on trains serving Khayelitsha.

Informal housing Site B has reticulated water and sewage systems with an outside toilet for each 78–90-square-metre site. A tap is shared by every two sites. On average shacks are built on the 25-square-metre cement floors provided, by dividing them into two or three 'bedrooms' and a 'kitchen' comprising a table and paraffin stove (sometimes with two plates) or a wood fire. The official rental is R15 per site to which access is strictly controlled by the local headman.

Greenpoint started as a tent town where refugees from the 1986 Crossroads crisis, people on the official waiting list and squatters removed from other areas in the Peninsula have been housed at different times. Roads are compacted sand, two taps are located at the end of each row and bucket toilets are shared by two to four families. At first only interiors could be improved and residents were largely transient. Today there is little chance of moving, tents have been replaced by shacks and some residents have been relocated to a serviced extension with slightly larger sites and some tarred roads in Town 2.

Site C has water reticulation with one tap for four sites and a bucket toilet for every two sites. Water-borne sewerage is intended to be installed but the estimated cost of R30 million makes it unlikely. Gross densities of 73 units per hectare rise to between 85 and 95 units net because the road structure has been altered with minor roads in the north-west obliterated and larger routes narrowed by unofficial shack encroachment. Official sites can be bought for R3 per square metre plus a service charge or rented for R10 per month. Residents have not paid for services since 1985 but rates increased by 10 per cent in 1990 and new charges are due to be promulgated.

Truly informal squatter areas have no services of any kind and water is obtained from taps at serviced sites. Many sites are unsuitable for housing and the location of shacks changes in response to flooding or drifting sand. Mid 1989 estimates were of 12–13,000 families averaging 6.5 persons all squatting informally.

As shells, cores and standard housing are put up by large contractors there is neither stimulation, training nor opportunity for small builders to participate and develop skills. Therefore help in shack construction is difficult to obtain and must be paid for at commercial rates. Second-hand shack-building materials are expensive and prices of cardboard and plastic reflect cost of transport from Cape Town. Used corrugated iron is almost as expensive as new and is obtained primarily from builders suppliers who deliver to so-called informal outlets located on the periphery of Khayelitsha. A building material depot at Philippi is the only alternative source but is far from Khayelitsha. Floors of most shacks are hardened earth which may be covered with plastic, cardboard or pieces of carpeting. The amount and type of wall covering and existence of ceilings depends on available finance. Inevitably shelters are generally inadequate in that they are badly ventilated, too hot in summer and too cold in winter, and suffer from interior condensation and rising damp due to the high water table and the materials used. Construction materials are all highly inflammable. Candles and to some extent oil lamps are used for lighting and the lack of space necessitates that cooking is done indoors on paraffin stoves or open wood-burning fires. Therefore shacks are frequently destroyed by fire which, on occasions, results in serious injury or loss of life.

The people

The population of Khayelitsha has increased dramatically. An official survey carried out late in 1988 estimated that between 72,000 and 84,000 people lived in the 5,329 core houses of Villages 1, 2, 3 and 4 which go to make up Town 1. There were also between 38,000 and 79,000 in the site-and-service areas of sites C and B (including Greenpoint) as well as somewhere between 8,000 and 26,000 in informal shelters. This made a total of 110,000 to 189,000 people. In 1989 only 250 new formal houses were built despite 5,000 families on the official waiting list, but 305,323 people were now officially resident. Rapid growth continued and by late 1990 an estimate of 450,000 people, living on less than a third of the land originally planned to house 600,000 at most, is not unrealistic. Of these, 14 per cent are in formal housing and 54 per cent and 32 per cent in serviced and unserviced shacks respectively.

Unemployment is estimated to be as high as 80 per cent and poverty is endemic. Market research shows that 76 per cent of households in sites B and C had earnings below the household effective level at that time (Dewar and Watson 1990a:45). People in formal housing are among the better off though over 52 per cent of households earn less than the household effective level. This is not surprising because low wages are paid to Africans in Cape Town; in addition a survey showed that heads of household in Town 1 were predominantly female and over half were single.

Most Khayelitsha residents are Cape Town born or have lived in one or other of the townships, often in backyard shacks or hostels, for a long time. There are links with the Transkei though in many instances they are more tenuous than is generally supposed. Women in site C are probably typical of shack residents and interviews confirmed that they were generally young (average age 32), married (86 per cent), usually by customary unions, and had come to join husbands or at their husbands' advice to make use of medical services for the children. The level of education is low as 47 per cent functionally illiterate (less than Standard Two). That horizons are limited is clear because 36 per cent had never been to Cape Town in three years. Daily cooking, washing and cleaning appears to leave ample time only to rest and visit friends nearby.

Social and administrative organization

With the introduction of Regional Services Councils in 1987 Khayelitsha became a single representative body separate from other African areas in Cape Town, when the Lingalethu Town Council was created. As an African residential area there is no direct representation and the Town Clerk is white as is the representative on the Regional Services Council. White attitudes towards Khayelitsha and its residents are reflected in the anonymity of names commonly used by officials. It is Town 2 not Bongweni; Village 3 not Ekupumleni; neighbourhoods are simply listed alphabetically as sectors and street names kept to a minimum. The mayor on the other hand is African, a local headman and strong leader who played an important role in Crossroads and the early moves to Khayelitsha. He and his town council are the officially accepted representatives of Khayelitsha and have their offices in site C. Their position was confirmed in October 1988 when 43.28 per cent of residents voted in the Black local elections as opposed to the national average of only 5 per cent. In the eyes of some residents their credibility is suspect however. They have been accused of a degree of patronage, of a lack of forward planning, of failing to respond to local grievances and they are indirectly associated with violence and crime. A Khayelitsha Residents' Organization has come into existence but support is limited and prominent members have had their homes damaged in petrol-bomb attacks. In the wider context political support for the Pan African Congress is said to dominate and be sufficiently strong to prevent the ANC from taking over the structural organization of Khayelitsha.

Social control in Khayelitsha is achieved through a strict, unofficial network of informal courts operating in a three-tier system (Burman 1989). At the lowest level there is the community council (equivalent to the street, section or headman's committees in other townships) of local members who are democratically proposed and serve a limited number of streets or a maximum of about 100 houses. They are generally well known, trusted and operate at grass-roots level to settle disputes and attend to daily affairs of the neighbourhood. Meetings are held monthly in private homes but when necessary contending parties are brought together in a hall or on open ground. Khayelitsha is unlike other townships in that the community council is sufficiently powerful to use physical punishment on adults and parental consent is not required for such punishment of children (Burman 1989: 159). Access to sites is in the hands of the community council which controls allocation and can evict any resident who disobeys their court decisions and then does not pay the double fine that follows. There are also community police or home guards who operate as vigilantes and work closely with the community council. Their membership is largely drawn from informal housing areas and they are said to be independent of the community council to whom they may report problems when necessary. When this is done the council normally notify them of any action required or taken. The next higher level of control is the executive committee (civics in some townships) to whom cases can be referred and who are required to ratify certain decisions. In Khayelitsha the executive were sufficiently strong to be unaffected by the Comrades' Courts and operated as usual in contrast to the situation elsewhere in Cape Town. Finally there is apparently an umbrella body providing overall control and direction, about whom there is little available information.

Services

Road transport plays an important role in Khayelitsha. Both the formal bus service and particularly the informal taxi services are heavily used. At first they were the only means of mass transport, but completion of a 10.5-km rail link with three stations in December 1987 provides an alternative for commuters. Distance from employment centres means that commuters average two hours and forty minutes daily and spend 11 per cent of income on travel. In Khayelitsha lack of any focus and of links between villages separated from one another by highways does not encourage local interaction.

There are no firefighting services, fewer than a dozen telephones, only two resident doctors and the two general clinics and one nutrition centre operate during the day. This is serious undersupply given that TB reached epidemic proportions in 1987 and clinic surveys show that half the children are nutritionally stunted and suffer chronic malnutrition. A day hospital followed a 16-bed maternity obstetric unit opened at site B in 1988 but it can take up to two hours to walk there and as much as half an hour by taxi. With 17 midwives, 15 student nurses and 6 assistants instead of a staffing norm of 95 members, it deals with 550 maternity admissions alone each month. This allows only 21 hours, day and night, per bed for labour, delivery and recuperation before returning home. This is less than half the average for other clinic deliveries in Cape Town (Rip and

Hunter 1990) and in reality a figure of 8 hours is commonly necessitated by pressure on facilities. Alternative maternity facilities are 10 and 20 km away by road while the nearest of three hospitals serving the area is one hour's journey away. Although new mothers in Khayelitsha and their infants have satisfactory weights, infant mortality increases dramatically when they return home reaching 31.7 per 1,000 in formal and 62.9 per 1,000 in informal housing areas where they are the highest in metropolitan Cape Town (Rip *et al.* 1988). There has been great pressure to establish community health centres but money only became available for this purpose in 1990.

Economic activity

Local employment opportunities are very limited and are largely confined to low-paying council jobs. Little support is provided to develop entrepreneurial skills or potential and to integrate Khayelitsha with the economy of Cape Town. For the majority of residents shopping must be done locally and most people complain that there are too few shops, but as prices are high they are underused. All outlets are geared to the low economic status of the clientele and sell single cigarettes, sugar by the cup, bread by the slice. The range of goods is also limited and the operations themselves are tenuous.

There are very few formal shops. Despite having been in operation since 1987, the two local shopping centres each comprise only a single supermarket together with one or two fruit and vegetable outlets. Some traders are assisted by the Small Business Development Corporation but capital remains limited and in all cases the scale of operation is such that profit margins are low and most entrepreneurs cannot afford an economic rental (Thomas 1987). Nevertheless similar outlets in the third shopping centre have been sold on a tender basis and there have been allegations of racketeering. Mobile residents shop in Mitchell's Plain 10 km away and intend to patronize retail outlets (10,000 square metres) in the large new shopping and industrial complex being erected on the route interchange providing access to Khayelitsha. This development by a white consortium anticipates that Africans will participate in management and fill all jobs at lower levels. Multiracial tenancy is also being considered. The first Cape off-course tote to locate in a Black housing area did so in Khayelitsha and the single petrol station is a hive of activity at all times. The Council put this site up for sale and despite an exorbitant asking price there was fierce competition among African entrepreneurs to purchase it.

As there are no trading restrictions local residents become hawkers or run shops (known as 'spazas') from accessible front rooms in their homes. These provide credit, open at all hours and are fairly evenly distributed throughout the formal and informal housing areas, but the frequency is greater in the latter. Some traders set up informal outlets at sites where the potential demand is greatest, for example near the clinic, pub or station and along main pedestrian access routes, particularly in informal housing areas. Retailing is confined to items with slightly higher profit levels, primarily fruit and vegetables in high demand (one-third of outlets) or the provision of perishable foodstuffs; a quarter sell raw or cooked meat, entrails and other prepared foods (Dewar and Watson 1990b). Almost 28 per cent of stalls are only operational on Friday evenings and weekends which are also the times when second-hand furniture, household goods and clothes as well as wood and other building materials are traded along arterial routes used by taxis and buses. Discotheques, shebeens and survival entrepreneurs (concerned with prostitution, protection and other crime) flourish and service activities proliferate. Dressmakers (who made up 60 per cent of applicants to the Small Business Development Corporation), childminders and traditional healers are all home-based. Backyard mechanics, panel beaters and repair workers operate on a small scale (one or two employing up to 12 people) in spaces adjacent to shacks.

Conclusion

The landscape of a portion of the Cape Peninsula has been transformed, but it is unlikely that the new town reflects either planning or policy objectives. Khayelitsha has failed to become an attractive local alternative to residents of overcrowded townships in Cape Town. That its growth has been extraordinarily rapid is largely a reflection of the number of Africans unofficially resident in Cape Town, coupled with people returning to the city in which they had spent a large proportion of their lives. Neither formal housing nor site-and-service development have kept pace with household formation and an increasing proportion of residents live at high densities in collections of shacks without access roads, running water, sewage or refuse removal. Despite relatively attractive offers only 700 homes have been purchased and few have been upgraded. There is little real security of tenure in the face of warlording, rent racketeering and corruption. Yet Khayelitsha is very much alive and residents show great initiative in the use of space, construction of homes and creation of communities which together make up a new settlement that will remain an integral part of the urban environment in the Peninsula.

Note

Except where specific citations are made, the information included in this chapter is based on a continuation of unpublished surveys (some confidential), anecdotal evidence, field investigation and interviews.

References

Burman, S. 1989. The role of street committees. In H. Corder (ed.), *Democracy and the Judiciary*, pp. 151–66. Cape Town: David Philip.

Cook, G.P. 1986. Khayelitsha – policy change or crisis response, *Transactions, Institute of British Geographers* N.S. 11: 57–66.

Dellatola, L. 1987. Khayelitsha, homes for all, *South African Digest* 45: 3–4.

Dewar, D. and Watson, V. 1990a. *An overview of development problems in the Cape Town Metropolitan Area*. Cape Town: Urban Problems Research Unit/Urban Foundation (Western Cape).

Dewar, D. and Watson, V. 1990b. *The structure and form of metropolitan Cape Town: its origins, influences and performance*. Cape Town: Urban Problems Research Unit/Urban Foundation (Western Cape).

Rip, M.R. and Hunter, J.M. 1990. The community perinatal health care system of urban Cape Town, South Africa – II: geographical patterns. *Social Science Medicine* 30: 119–30.

Rip, M.R. *et al* 1988. Perinatal health in the peri-urban township of Khayelitsha, Cape Town, Parts I and II. *South African Medical Journal* 75: 629–34.

Thomas, W. 1987. Business development and job creation in Cape Town. Address given to the South African Institute of Citizenship, Cape Town.

11 *The road to 'Egoli': urbanization histories from a Johannesburg squatter settlement*

OWEN CRANKSHAW, GAVIN HERON & TIMOTHY HART

Introduction

The popular interpretation of the growth of squatter camps on the Witwatersrand is that it is the direct result of recent immigration from the rural areas of South Africa's 'homelands'. Evidence from surveys conducted in 1990 (Crankshaw 1990) does suggest that this road to 'Egoli' (the place of gold) is becoming an increasingly significant cause of squatting. However, up until very recently, squatting on the Witwatersrand has generally been the result of natural urban population growth in the face of a chronic housing shortage, low wages and high unemployment. This is true for most of the free-standing squatter camps that have become established on vacant land adjacent to formal townships. For example, surveys conducted in 1988 and 1989 amongst squatter communities adjacent to Alexandra and Soweto and in white urban areas reveal that the vast majority of squatters were either born on the Witwatersrand, or had lived there for at least a decade (Crankshaw 1990).

This study examines a rather different sort of squatter settlement. In contrast to many that are located near existing African townships, Vlakfontein is in peri-urban farmland some 35 km to the south of Johannesburg. As such, it is a catchment area for peri-urban farmworkers as well as for township dwellers. If we examine the birthplaces of our sample of 95 household heads, we find that these people originate from the townships and farms of the Witwatersrand itself, non-metropolitan towns, white farming areas, and rural areas in the homelands and neighbouring states. Despite this diversity of birthplaces, however, most (54 per cent) of the sample of squatters were born on white-owned farms. About half (51 per cent) of these ex-farmworkers in the sample were in fact born on farms in the southern Witwatersrand, between Johannesburg and Vereeniging. The other half were born on farms in the northern Orange Free State and south-eastern Transvaal (30 per cent), and in the eastern and western Transvaal (12 per cent). In comparison with other squatter settlements that are found adjacent to formal African and white townships, Vlakfontein has ten times as many ex-farmworkers amongst its residents (Crankshaw 1990). So, the high proportion of ex-farmworkers in Vlakfontein makes this squatter settlement unique.

As our research amongst the community progressed, it also became clear that the forces behind squatting at Vlakfontein were unique to the peri-urban character of the area, and to the social relationships on the smallholdings. The story of Vlakfontein is, for the most part, the story of ex-farmworkers and their urbanization. For this reason, we have chosen to focus on the ex-farmworkers in the sample; to tell their story of urbanization and homelessness.

Research method

This study is based on both qualitative and quantitative data. After consulting with the community leaders, we undertook a pilot survey on the basis of which a standardized questionnaire was designed. A market survey company then administered the questionnaire to household heads during a weekend in October 1988. The sample size was 95 household heads, providing a 19 per cent sample. The information collected by the survey was presented to the community and also made available to their lawyers in the hope that the evidence would assist the community's struggle for secure accommodation. Later, we returned to conduct in-depth interviews in order to provide qualitative data on urbanization patterns. Other in-depth interviews were also conducted with Josi Adler, a member of the Black Sash, who has worked with the Vlakfontein community and is familiar with the area.

The pattern of urbanization from white-owned farms

The urbanization routes taken by ex-farmworkers in our sample were quite distinct from those of squatters who were born in the rural areas of the homelands or neighbouring states or in small towns outside the Pretoria–Witwatersrand–Vereeniging complex (PWV). Since this study is based on a survey of squatters, it is impossible for us to draw any conclusions about the urbanization of farmworkers in general. We cannot know, from this evidence, the extent to which farmworkers have become integrated into the urban economy. Nor can we establish how successful they have been relative to groups urbanizing from other rural areas. However, we can begin to understand how and why the farmworkers in the peri-urban areas south of Johannesburg have been marginalized from urban accommodation and employment.

The most obvious feature of the urbanization routes taken by ex-farmworkers in our sample is how few actually moved from farms outside the PWV directly into the formal African townships. Many ex-farmworkers (11 out of a total of 31 cases) moved onto smallholdings and farms in the peri-urban areas of the southern Witwatersrand (Table 11.1). A further 8 ex-farmworkers found accommodation and employment as domestic servants and gardeners in private households in white, coloured and Indian residential areas in the south of Johannesburg. Another 2 found accommodation in hostels, and 2 more moved directly from farms into the squatter settlement. So, only 8 out of a total of 31 ex-farmworkers moved from a farm outside the PWV directly into formal urban accommodation.

Table 11.1 Summary of migration routes of Vlakfontein squatters.

	Migration route from urban accommodation to Vlakfontein					
Migration route To the PWV	From African town-ship	From domestic servant's rooms	From peri-urban farm	From farm	From hostel	Total
1. Born on farms outside the PWV						
From a farm:						
to African township	2	1	4	0	1	8
to domestic rooms	0	7	1	0	0	8
to peri-urban farm	1	0	9	0	1	11
to hostel	1	1	0	0	0	2
to Vlakfontein	0	0	0	2	0	2
Total	4	9	14	2	2	31
2. Born on farms within the PWV						
From a farm:						
to African Township	3	2	1	NA	0	6
to Domestic Rooms	0	0	2	NA	0	2
From a peri-urban farm to Vlakfontein	0	0	16	NA	0	16
Total	3	2	19		0	24
3. Not born on farms	13	9	13	NA	0	40
Grand Total	20	20	46	2	2	95

A similar pattern is evident amongst those ex-farmworkers who were born on farms within the PWV. Out of a total of 24 peri-urban farmworkers, 16 never moved from the smallholdings into formal urban accommodation of any sort; they moved directly from smallholdings into the Vlakfontein squatter settlement. Of the remainder, 2 moved into domestic servant's rooms in non-African townships and 6 moved into the formal African townships (Table 11.1).

What emerges from this analysis of urbanization patterns amongst squatters born on farms on the Witwatersrand and further afield, is that ex-farmworkers tend to remain in employment either on peri-urban farms or in domestic service. Clearly, both these forms of employment are characterized by a marginal urban status. In the first place, the work is of an unskilled nature and is poorly paid by urban standards. These jobs are also subject to high unemployment and high turnover rates. Secondly, farmworkers and domestic servants experience very insecure tenure precisely because their accommodation is provided by their employer. When they become unemployed they automatically forfeit their accommodation.

Before we go on to examine the social relations and changes in the peri-urban areas that underlie squatting, there is still another group of peri-urban farm-dwellers that needs to be examined. Although almost half of the squatters at Vlakfontein were not originally farmworkers, a significant proportion (33 per cent) left the townships to live on the peri-urban smallholdings before they moved to Vlakfontein. This is also true for many squatters who were born on the peri-urban farms but who managed to secure accommodation and employment in the urban areas. Long before they resorted to squatting at Vlakfontein, they had been forced to return to living on the smallholdings in the south of Johannesburg.

So what are the general patterns of movement amongst farmworkers in our sample? There are farmworkers who have migrated to PWV and those who were born on smallholdings within the PWV. Most of the immigrant farmworkers from outside the PWV found employment-linked shelter on peri-urban farms to the south of Johannesburg and in domestic service rather than accommodation in the African townships. Squatters who were born on the peri-urban farms tended to remain on the farms rather than find urban accommodation. Finally, there is a third category who were born in the PWV townships or urbanized from small towns or rural areas in the homelands to live in the PWV townships. Many of these people moved out of the townships to live on the smallholdings in the south of Johannesburg where they became subject to the same forces that pushed other farmworkers into Vlakfontein.

Causes of squatting amongst peri-urban farmworkers

Farmworkers share a number of characteristics that are quite distinct from other squatters in our sample. Squatters who were born on white-owned farms did in fact have a marginally lower level of education than that of their homeland-born and urban-born counterparts. On average, ex-farmworkers had received two years less education than other squatters in the sample. The reason for this lies in the fact that there is no system whereby the state can initiate the provision of education on the farms in the Republic of South Africa. There is only provision for a state subsidy, should farmers choose to provide schooling facilities for their workers (Ardington 1986: 71). Farmers themselves have little interest in educating their farmworkers beyond primary school. A secondary school education is not only unnecessary for farm work, but also provides farmworkers with an escape route from the farms and into urban employment.

Prior to the abolition of influx control legislation in 1986, farmworkers were also denied legal channels to find urban work. Although labour bureaux in rural areas were empowered to place farmworkers in urban employment, it was seldom that farmworkers were legally qualified for urban employment (Roberts 1959: 126). After all, the influx control system was designed, at least in part, to ensure an adequate supply of African labour to the farms.

These factors could explain why ex-farmworkers relied upon securing employment and accommodation on peri-urban smallholdings where their presence was perfectly legitimate under the old influx control laws. Further, although domestic service offered no legal protection from the pass laws, it was nonetheless still possible to find such employment without detection by the authorities. In addition, since farmworkers have limited education and skills, their work experience as unskilled labourers and domestic servants on the farms confined them to these categories of work.

On the basis of in-depth interviews, we have discerned the following social dynamics amongst peri-urban farmworkers. Most often the head of the household worked for the farmer in return for accommodation for the rest of the family. The other members of the family were then free to find work in the nearby urban areas. In spite of the fact that farmers are not permitted to have tenants who are either not employed by them or are not dependants of the farmworkers (Hathorn and Hutchison 1990), this practice is nonetheless rife.

Some farmworkers were forced to leave their accommodation because the farmer died or because the farm was bought up for development. Others were dismissed or resigned over disputes regarding wages and working conditions. However, other conflicts between the farmworkers and farmers revolved around the control of the labour that the farmworkers' offspring could provide. For example, when the head of the household became ill, grew old or was otherwise unable to continue working, the farmer demanded the labour of the younger generation. Since they were often employed in urban work, the farmworker's children preferred to move into urban accommodation or to squat, taking their elderly parents with them. Farmers sometimes insisted that more than one member of the family should work on the farm in return for their accommodation. In some cases the farmer demanded that the farmworker's children should leave school in order to take up work on the farm. These are the most common reasons why ex-farmworkers and their families chose to leave the smallholdings. However, not all ex-farmworkers had household members with urban employment. Almost as often, ex-farmworkers and their wives lived alone on the smallholding. When the husband or wife lost their employment on the farm they were forced to find employment and accommodation elsewhere. Since many farmworkers found themselves in rural employment precisely because they were unable to find urban alternatives, they had little choice but to squat.

The following residential case histories illustrate the impact of these peri-urban processes and relationships on the lives of farmworkers.

Case 1. Josephine Morure. A farmworker who was born on a farm in the Orange Free State, and subsequently migrated to live on a smallholding near Johannesburg.

Josephine Morure was born on a white-owned farm near Amersfort in the Orange Free State in about 1950. She worked there as a domestic servant in the farmhouse. Sometime in the mid 1970s the farmer died, and she had to leave the farm in order to find work and a place to stay. She moved to live with her aunt who was employed as a domestic servant on a white-owned smallholding in Eikenhof. Here she met her husband who worked (and still does) as a sales assistant in a small car-spares retailing business in nearby Grasmere. Josephine found work, again as a domestic servant, in a nearby white township of Johannesburg. When the squatter settlement at Vlakfontein began to grow, they decided to move there because it meant that they could have a place of their own which was conveniently situated next to schools, transport and their places of work.

Cases 2 and 3. Magdalena Matlong and Makhulu Skosana. Farmworkers who were born on farms within the PWV.

Magdalena Matlong was born on the farm of Hasenbekfontein near Evaton in 1940. Her parents both worked for the farmer. When she was still very young her parents were able to move into Evaton where they rented a home. Magdalena attended school in Evaton, and after completing her schooling, married in 1958. Since her husband was not legally qualified to work in an urban area, the couple moved onto a farm in the region. They lived on various farms, moving regularly, in some cases because of disputes with the owners. The couple eventually settled at Elandsfontein farm where Magdalena worked as a domestic servant. Over the years they raised a family on the farm. Three of their sons (one with his own family) remained on the farm with them. Although they lived on the farm, their sons did not work for the farmer. The oldest son worked for a panel-beating company in the Nancefield industrial township and the younger sons worked as casual labourers on nearby building sites. After her husband died in the early 1980s, Magdalena started to experience difficulties in holding down her job. She fell sick on numerous occasions and could not continue to work on the farm. With the help of friends Magdalena made contact with John Mkhize, the community leader at Vlakfontein, and moved with her sons to the squatter settlement in 1987.

Makhulu Skosana was born on the Tamboersfontein farm near Heidelburg in 1958. He attended a farm school up to Standard Four, after which he worked on the Tamboersfontein farm and other farms in the nearby Koppieskraal district as a tractor driver. In 1986 he was fired from his job. Since he had difficulty finding another job, he came to live with his two brothers who had previously settled at Vlakfontein. Since then he has been unable to find employment except casual work as a truck driver for a peanut retail company in Lenasia. He and his family therefore rely upon meals provided by 'Operation Hunger', a charitable organization.

The most striking characteristic of ex-farmworkers who did find work in the urban areas is that most of them were employed as domestic servants in white, coloured and Indian townships in the south of Johannesburg. Out of 26 cases, 11 had lived in domestic servant's rooms prior to squatting at Vlakfontein (Table 11.1). These domestic servants were either forced to leave their accommodation because their employment was terminated or because of the restrictions associated with their accommodation. Most servant's rooms in private homes are designed to accommodate only one person, so that a domestic servant cannot comfortably live with his or her family. Even if the couple have no children, or are prepared to live in cramped conditions, their employer and landlord usually restricts the number of people who may live in the room as well as the frequency and duration of visits by friends and relatives. Domestic servants are also subject to low wages and very long working hours (Gordon 1985). Given these conditions, it is easy to understand why domestic servants in our sample chose to leave their urban accommodation in order to live in a squatter settlement. For some, it meant not having to share cramped quarters or tolerate restrictions on their social life. For others it meant the opportunity to live with their families.

Similar restrictions on social and family life also obtain in single-sex hostels and rooms on industrial sites. The couple of ex-farmworkers in the sample who moved into Vlakfontein from such accommodation did so for reasons similar to those given by domestic servants.

Case 4. Ashley Mkhize. A farmworker who moved to Johannesburg where she found employment and accommodation as a domestic servant.

Ashley Mkhize was born on a farm near Heidelburg in 1926 where her parents worked as unskilled labourers. When she was 19 years old she decided to leave the farm to work in Johannesburg. She found work as a domestic servant and lived in the servant's room of a private home in Jeppe. She later married and raised a family of four children, some of whom shared her cramped quarters. The family were evicted from this accommodation by local authorities on the grounds that only employed domestic servants were permitted to occupy servant's rooms in white townships. Ashley managed to find work and accommodation with another household in Bezuidenhout Valley, but the family was evicted once again by the authorities. In desperation, the family moved to live on a smallholding in Eikenhof, where Ashley was able to work as a domestic servant. In 1986, at the age of 60, she fell ill and could not continue working. Consequently, the family was evicted, and this time they moved to live in Vlakfontein. She and her husband now live with their grandchildren and a son. Her three daughters, who are employed as office cleaners in Johannesburg, moved to live in a backyard shack in Soweto.

The small number of ex-farmworkers who moved directly from formal and informal accommodation in the African townships to squat in Vlakfontein were largely influenced by conditions associated with overcrowding. Since for these people, employment was not directly linked to accommodation, only a couple were obliged to leave their homes because of unemployment. Most township dwellers exercised a certain degree of choice about leaving their homes. All the ex-farmworkers who secured township accommodation, in fact either shared a house with other lodgers, or lived in backyard shacks. Invariably they decided to move to Vlakfontein because they were tired of sharing. Squatters who came from backyard shacks complained about landlords who placed restrictions on their social life by preventing guests from socializing and drinking on the property. Finally, backyard shack dwellers as well as those who shared formal houses also cited quarrels with fellow tenants (often relatives) as a reason for leaving for Vlakfontein. These accounts reveal that the overcrowded conditions in the townships make the option of squatting relatively attractive. Although squatters forfeit services such as running water and water-borne sewerage, they gain access to relatively uncrowded living quarters and freedom from landlords and even rent.

Case 5. Frans Majoro. A farmworker who was born on a farm in the Orange Free State, and who urbanized to live in formal African townships on the Witwatersrand.

Frans Majoro was born on a farm near Bloemfontein in 1934. He moved to Johannesburg in 1959 and found work on a peri-urban farm as a gardener.

He later left this job when he found work as a semi-skilled manual worker in a boiler-making firm in the south of Johannesburg. While working in Robertsham he and his wife stayed with his brother's family in Soweto. In 1976, as a result of a quarrel between his wife and sister-in-law, Frans and his family moved from his brother's house to live in a converted garage, also in Soweto. In 1983 the boiler-making company went out of business, and Frans found a similar job in Evaton. Soon thereafter he lost his job and has been unemployed ever since. Because he was unable to pay rent, Frans searched the informal settlements for a place to live. He chose Vlakfontein above other settlements in the area because he found it to be uncrowded, sanitary and free of crime.

Locally specific causes of peri-urban squatting

These case studies raise the important question of why peri-urban squatting has become feasible since the mid-1980s, and what the particular attractions of Vlakfontein are. The high proportion of ex-farmworkers who recently lost their peri-urban employment does suggest that township development in the Grasmere area is having some impact on farming activity. Since the mid 1980s there has been considerable expansion of coloured and Indian townships in the area, especially the region surrounding Vlakfontein itself. This development has been preceded by state purchase of farmland and the cessation of farming activities on large tracts of land in the area. The development of this farmland has had two effects. On the one hand, where farming activities have been stopped but the original landowners or tenants still occupy the land, it has contributed to unemployment amongst farmworkers who are then driven off the farms. On the other hand, where the land is vacated by its owners or tenants, the unsupervised farms (such as Vlakfontein and Weiler's Farm) have been settled by squatters.

Given these developments, why has the state tolerated squatting? The growth of the Vlakfontein settlement took place at a time when state urban policy and its administrative structures were being reorganized. Although West Rand Administration Board officials and the police acted twice against Vlakfontein squatters, on both occasions the squatters made representation to higher authority and succeeded in putting a stop to arrests and demolitions. With hindsight, it is clear that the state was in the process of developing legislation to allow for legalized squatting on designated land in the area (Crankshaw and Hart 1990). Given the sensitivity of demolitions and forced removal in the international media, state officials no doubt decided to turn a blind eye to the squatting until a more acceptable solution was possible. The provision of formal site-and-service plots in the area only became available in 1989, by which time the Vlakfontein settlement had grown to over 1,200 shacks.

In the course of their resistance to state intervention, the Vlakfontein squatters became fairly well organized. This resulted in the presence of a legitimate leadership who was able to control crime and regulate overcrowding and sanitation. Through its contact with other organizations the community also established a crèche and a clinic, and instituted a feeding scheme for the unemployed. These are desirable services, even in comparison with formal townships. Although Vlakfontein is further from Johannesburg than Soweto, it is located alongside a major route to the city which is well served by a bus service from Evaton. So, although many squatters stressed that they had little choice about living at Vlakfontein because there was simply nowhere else that they could go to, a number also claimed that they chose to live in Vlakfontein because the settlement was reputed to be free of crime, is near to transport and employment opportunities, and because accommodation is rent-free.

Conclusion

Vlakfontein shares many characteristics with other squatter settlements in the peri-urban areas of the southern Witwatersrand. What sets it apart from the settlements that have sprung up alongside formal African townships is that it is home to a high proportion of ex-farmworkers. An analysis of the residential histories of these ex-farmworkers reveals that the farms and smallholdings in the Grasmere area have served as a staging post for urbanizing farmworkers moving to the Witwatersrand from farms in the Orange Free State and the eastern and western Transvaal. The peri-urban farms provided a refuge for urbanizing farmworkers because they offered legal and cheap accommodation while at the same time providing access to better paid urban employment. However, urbanizing farmworkers were not the only people who used the peri-urban farms as a refuge from unemployment and a shortage of affordable urban accommodation. They were joined by significant numbers of ex-township dwellers.

It is clear from our analysis of residential histories that many farmworkers have struggled to establish themselves in stable urban employment and accommodation. This is demonstrated by the high proportion of ex-farmworkers who have only been able to find urban employment as domestic servants. In the face of township development in the peri-urban areas of Grasmere which is both displacing farmworkers as well as providing the opportunity for squatting, many ex-farmworkers have chosen to squat in the peri-urban area rather than to move to the townships. The choice to squat at Vlakfontein appears to have been informed by two factors. On the one hand, these ex-farmworkers are mostly unskilled workers who are subject to low wages and high rates of unemployment which act to exclude them from rental accommodation. On the other hand, the formal townships are not only expensive to live in, but are also overcrowded and crime-ridden.

The phenomenon of squatting in South Africa is as yet imperfectly understood. This study has focused on one settlement and a specific set of routes to it. As a sub-theme in the general area of migration studies, squatter settlement poses particular methodological challenges because it is complex, and because individual strategies are so idiosyncratic. A complete understanding of squatting in South Africa has to merge the broader structural changes with patterns of household strategies. This study has attempted to mix such a brew.

References

Ardington, L. 1986. Farm villages: A relief measure for rural poverty. *Indicator South Africa* 4: 71.

Crankshaw, O. 1990. Unpublished summaries of squatter surveys conducted in association with the Black Sash.

Crankshaw, O. and Hart, T. 1990. The roots of homelessness: causes of squatting in the Vlakfontein settlement south of Johannesburg. *South African Geographical Journal* 72: 65–70.

Gordon, S. 1985. *A Talent for Tomorrow: Life Stories of South African Servants*. Johannesburg: Ravan.

Hathorn, M. and Hutchison, D. 1990. Labour tenants and the law. In C. Murray and C. O'Regan (eds), *No Place to Rest: Forced Removals and the Law in South Africa*, pp. 194–213. Cape Town: Oxford University Press.

Roberts, M. 1959. *Labour in the Farm Economy*. Johannesburg: South African Institute of Race Relations.

12 *Informal settlement: theory versus practice in KwaZulu/Natal*

BRUCE BOADEN & ROB TAYLOR

J61 R11
South Africa

Introduction

The nature and extent of informal settlement in the KwaZulu/Natal region are well documented. It is the predominant housing delivery system in the metropolitan region and represents the actions of the very poor and not so poor to house themselves without subsidy from government or employers. It is the response of people to the reality that alternative, affordable housing is not available. Whether it is illegal squatting on land or the building and occupation of illegal structures, the informal settlements of the Durban and Pietermaritzburg metropolitan areas are an inevitable consequence of the housing shortage.

Although people have been housing themselves in this way for a long time, it is only in recent years that the seriousness of the situation has been recognized and efforts made to address the problem. Private-sector organizations, government agencies, academics and church groups have all been drawn in. Despite the concern and efforts to improve housing conditions and despite the goodwill that has been generated amongst the general public towards the plight of the urban poor, informal settlements continue to grow rapidly and continue to be the most common means of poor people acquiring shelter.

There is also a growing acceptance on the part of the authorities that standards and methods of construction need to be reviewed in order to find realistic solutions. At a recent housing conference (Institute for Housing of South Africa 1989), a paper by the Minister of Housing in the House of Representatives advocates the use of self-help methods that ten years ago would have been unacceptable in government circles. In the same vein, a paper delivered by a member of the executive committee of the Transvaal Provincial Administration suggested a totally new way of viewing the problem – one sympathetic to the use of appropriate building methods and standards.

This raises the question: why is there so little improvement occurring in the informal settlements when such a positive attitude exists within the public and private sectors? Why, with all the knowledge and experience, both here and overseas, is so little progress made? It is reported, for example, that none of the R1.2 billion recommended for upgrading by the KwaZulu/Natal Planning Council in 1986 has been spent to date (*Natal Witness* 1989). It is argued that this is partly a result of a mythology (Boaden 1990) that has developed which prevents results being achieved on the ground. The intention is to highlight some of the misconceptions standing in the way of real progress. These views stem from involvement in three upgrading projects while working for The Urban Foundation, a private-sector developmental agent. It is hoped that this will lead to more realistic approaches being adopted by implementing agencies in the future.

The meaning of shack 'upgrading'

Shack 'upgrading' is a concept that is interpreted in many ways. Usually it is understood to mean the physical improvement of an existing, often temporary, structure. For our purposes, this view is too restrictive. Shack upgrading, in the context of this chapter, refers to the process of improving both the living conditions and the economic well-being of the present and future residents of informal settlements. Upgrading would include creating or reinforcing community social structures to ensure local participation in the project; legitimizing the land, that is, ensuring that residents will be able to own the site on which they live or wish to settle; ensuring that settlement is in terms of some acceptable layout plan; providing training and advice on building matters, including services; giving people access to building materials at reasonable prices; attending to problems of finance; and creating employment and entrepreneurial opportunities through the housing process.

It should be clear that these activities require the on-site presence of an implementing agency. Too often well-intentioned agencies are unable to make an impact on the problem because they do not have personnel in the field or are hampered by the actions of community groupings or other agencies engaged in the project area.

In this chapter three case studies with some interpretative material are presented. Problem areas or 'myths' are then identified. The 'reality' presented in relation to each of the 'myths' reflects the substance of the case study material. The observations made are the result of the personal experiences of the authors and their colleagues over three years in shack upgrading projects in the Greater Durban area. Conclusions reached, therefore, do not have universal applicability with respect to place and time. In fact, shack areas vary considerably in their social, physical and historical contexts which makes the validity of a model solution questionable.

Case studies

St Wendolins St Wendolins lies within the Durban metropolitan area, about 20 km from its central business district. The Catholic church is a major landowner and has been dominant in the affairs of the community for more than 100 years. There was a threat of removal in terms of racial legislation until 1984 when the permanent settlement of Blacks in the area was acknowledged. This came after negotiation spanning a period of over 30 years. During that time there was a deterioration in the condition of the limited formal housing stock. There was also an increase in the number of informal dwellings in the area. Densities are generally low with only a few pockets of dense informal settlement which tends to coincide with a system of tenure which is largely leasehold in nature with an abundance of tenants and lodgers.

The fact that residents were given the right to remain in St Wendolins was interpreted by many of them as an indication that the area was ready for redevelopment and formalization. This assumption has proven itself to be a negative factor in the drive towards upgrading. The community could be seen to have what can best be defined as a rural mentality while occupying an area close to an urban centre. This mentality has had an important bearing on attempts to upgrade the area. Many residents lay claim to large areas of land on the pretext of wanting to use it to grow crops. There is, however, little evidence of farming activity. These people, understandably, can be very obstructive in attempts to formalize settlement on that land.

There is also the question of self-help as an assumed delivery process. Self-help is defined here as a process of people building permanent structures with the support of the implementing agent on demarcated sites. There has been no evidence of self-help as a delivery process of any kind. Experience over two years did not result in the delivery of a single formal house using self-help. It also did not result in the production of any building components on a self-help basis. This phenomenon is not confined to St Wendolins but its interpretation could have several dimensions. The first would be a sense of helplessness born of years of dependency fostered by the political system. A second interpretation in the context of St Wendolins could be a dependence on the church as the purveyor of goods and services over time. This relates to the general philanthropic orientation of the Catholic church in its dealings with its parish since the time of the establishment of the mission at St Wendolins in 1882. Such an orientation has created an atmosphere of expectation in relation to the further development of the area. A third interpretation which also bears upon, and probably originates from the experiences of mission life, could be the expectation of subsidy which would deliver large numbers of sites and formal houses at virtually no cost to the end-user. In this regard subsidy mechanisms tend to be misplaced and subject to abuse.

They are also geared in favour of formal conventional housing in formal townships as a norm where this norm is known to be non-sustainable in the longer term.

The Community Committee in St Wendolins tends to comprise an elite group which derives its status from its relative affluence and only pays lip-service to the needs of the poor in order to attract ever greater subsidies to the area whilst striving for the construction of larger, conventional houses as the general norm. In this way access to housing for the very poor was denied by vehement opposition to the construction of any form of appropriate, or technologically innovative, housing notwithstanding affordability levels and need. The manipulation of the community by its leadership became problematic, as evidenced by the inability of those most in need of housing to gain access to even the most rudimentary of formal houses. Acquisition of sites on a site-and-service basis for the erection of informal housing was discredited by the leadership. The presence of a development agency made it possible to gain credibility for the leadership in terms of attracting resources. The fact that those resources were not used appropriately was not considered to be the fault of the leadership. The peculiar phenomenon of development resources being underutilized on the one hand and the clamour for appropriate development on the other represented a paradox which only clarified over time.

Another complicating factor was the involvement of so-called interventionist agencies in St Wendolins. An interventionist agency is an ostensibly independent agency whose role is usually reduced to one of positioning itself between the community and the implementing agent. While purporting to represent community interests in their actions, agencies of this nature tend, over time, to begin to pursue their own agendas. These agendas are supported by surveys and reports which delay progress, but which the implementing agent cannot ignore, except under threat of community alienation. Agencies of this nature, which carry no financial responsibility and little responsibility for the success or otherwise of the development project, must always retain community credibility at whatever cost. The cost is usually time delays and inflation, placing housing which would otherwise have been affordable beyond the reach of many or making what is affordable ever smaller in accommodation terms.

In this regard, St Wendolins has been over-serviced and over-surveyed. Because of the ability of an interventionist agency to hold the confidence of the community it was possible for it to become the only credible vehicle for information transmission to the community and vice versa. This enabled distortion of information flow and provided barriers to direct consultation. In this manner the community ceased to drive or manage the process of development which became effectively driven by an outside agency whose agenda was not entirely developmental in orientation. Development was significantly delayed.

Richmond Farm Richmond Farm, also within the Durban metropolitan area, was settled through a process of allocation on the part of individuals holding de facto (rather than de jure) power in the area during the period 1985 up to the present. Many of these informal leaders are regarded as warlords in the sense that they protect their territories with armies (or 'impis') drawn from the local residents. Although located within the formal proclaimed KwaZulu township of Ntuzuma, the area is entirely informally settled with substantial barriers to formal settlements occurring resulting mainly from political unrest in the area and extremely low levels of affordability.

The community obtained access to the land as a result of a political and economic subservience to warlords who exact allegiance from the inhabitants. The interests of the community at large are therefore forcefully and often violently subsumed beneath those of the local warlord.

The warlords derived power from control over the land. In order to retain power following settlement, the warlord must retain control over the people who settle. The warlord would pay lip-service to the principles of development and improving the lot of the people but also recognize that any improvement by way of independent land ownership would erode his power base. It would therefore be acceptable for services to be provided to the community at large, for roads to be improved and for infrastructure to be put in place. It would not however be acceptable for land to be alienated or for home ownership to occur on scale. Any move towards formalization would also result in dedensification meaning the erosion of political power base as people move away into adjacent areas.

The consequences were high densities where informal settlers were entirely dependent on the largesse of the warlord. The tendency was for these densities to increase in order to improve revenues to the warlord. Such densities overload the service infrastructure and over time create a breakdown of communal services.

In Richmond Farm, there was a strong desire to protect territory. This again related to the protection of the economic power base but also gave rise to certain anomalies which could be seen in the coexistence of extremely dense areas of settlement with less dense areas nearby. This could be explained in terms of differential territorial control making it impossible for individuals to settle in a rational way. Territory was vigorously and often violently protected. This militated against the possibility of formalization and improvement occurring across territorial boundaries. It also militated against integrated development where dedensification and resettlement would normally occur in a manner which subscribed to planning logic rather than territorial or political logic.

Qadi Tribal Area Qadi Tribal Area is situated at the inland end of Inanda near Durban. The land falls under tribal control and cannot be alienated.

The land forms the logical growth area for Inanda in terms of its expansion inland. The land is not densely settled and could provide an abundance of space for new urban development. Access to the land is bedevilled by its traditional form of ownership.

The community in this instance could be defined to be the chief and the tribal executive. The chief is custodian of the land. The land therefore forms his power base. Alienation of the land represents an erosion of that power base. It also represents an erosion of the economic base and therefore alienation was not desirable. In development terms it was impossible for persons living under a system of communal land ownership to gain access to finance in the formal sense of the word. The upshot of this was land ownership in the hands of the tribe but limited development or access to urban amenity.

Myth versus reality

The three cases presented are illustrative of some of the causes of development inertia in different environmental and social contexts. The remainder of this chapter serves to present a broad range of perceptions, reinforced by the specific, but not exhaustive, case studies. These perceptions are presented as a series of misunderstandings or myths.

Myth No. 1: It is purely a technical problem A widely held view is that the problem of informal settlements is a technical one that can be overcome by introducing innovative and cheap methods of construction together with affordable levels of services. The difficulty experienced in St Wendolins in getting self-help housing started was indicative of the fact that we were dealing with more than a technical problem.

Academic and commercial researchers have put a great deal of time and effort into finding technical solutions to the provision of 'appropriate' housing and services (appropriate being affordable and meeting acceptable performance standards). A plethora of 'solutions' exist ranging from labour-intensive systems using easily obtainable materials to the high-tech, prefabricated, quick-to-build systems that are advertised daily. These systems rarely find their way into shack communities. It has been shown (Boaden 1986) that a relatively narrow range of building materials and house types are to be found in Durban's informal settlements.

While there is clearly a place in the upgrading process for these technical developments, we would suggest that efforts in this direction are unlikely to make a significant impact on the problem for a number of reasons.

Firstly, the residents of informal settlements generally desire formal houses – usually of concrete-block construction – and find 'appropriate' or non-conventional forms of construction unacceptable. Most families would therefore prefer to continue living in their shacks while waiting, or pretending to wait, for a conventional house.

Secondly, the concrete block house is extremely difficult to improve upon from both a cost and an efficiency point of view, particularly if the blocks are made by the owner on site.

Thirdly, there is no point in a resident putting money, time and effort into a house – regardless of whether it is of 'appropriate' or conventional construction – if there is no security of tenure. The question of the appropriate house type is only of relevance once the land situation has been formalized, which is very rarely the case.

Fourthly, very little attention is given to appropriate marketing methods on the part of implementing agencies. In the literature one finds very little material on how to communicate alternative methods of construction to shack dwellers in order to obtain general acceptance of these methods.

Finally, conventional financing generally favours conventional forms of construction. Although elsewhere in the country there are examples of building societies and government agencies making long-term loans available for non-conventional housing, this is rare in Natal.

Myth No. 2: The formal housing system will eventually succeed This view, which does not come out clearly in the case studies, holds that the formal housing delivery system (including township establishment procedures) will, given the money and time, overcome the problem of squatting.

This technological fantasizing has been with us for many years and has created an attitudinal barrier to creative solutions being found. The shack housing delivery system as presently operating in the Greater Durban area is the predominant way in which people obtain housing and will continue to be so for years to come. It has an inertia of its own which will be extremely difficult to change. Although no figures are available, it is probably true that out of every ten houses built in the area last year for all race groups, eight were self-provided shacks. Instead of referring to formal and informal housing, one should perhaps be talking about conventional (informal) and non-conventional (formal) housing. It is unrealistic to think that this situation will change overnight – which suggests that new ways of doing things need to be found. The 1986 Black Community Development Amendment Act and the 1988 Prevention of Illegal Squatting Amendment Act appear to be an attempt on the part of government to face this reality.

People involved in the provision of housing for low-income people, such as planners, lawyers, builders, developers, architects, financiers and administrators, despite showing great concern for the problem, are nevertheless reluctant to come to terms with the informal settlement process. We are still trying to apply the conventional, formal township development model rather than learning to work with and to improve the de facto one. Pleas, such as that made by Merrington (1988), for more appropriate planning and engineering skills, go unheeded. An underlying assumption in much of the literature dealing with planning issues with regard to low-income housing is that the land is vacant. (See, for example, Urban Planning, Housing and Design, 1986). In the KwaZulu/Natal context this is simply not true – informal shacks are scattered over most of the vacant land at different densities.

Myth No. 3: Community consultation is uncomplicated Consultation with the shack community is often regarded as a straightforward process leading to a clear-cut course of action. This is rarely the case. In both St Wendolins and Richmond Farm major problems were experienced in terms of having to deal with shifting community views.

Most organizations involved in upgrading projects quite rightly adopt a participative approach involving consultation with the community. Long experience in South America and Africa indicates the importance of participative upgrading (Urban Foundation 1987). In practice, however, this process often breaks down, leading to a less than optimal solution being found. There are a number of reasons for this.

In many cases people in shack areas have lived this way most of their lives and have very little understanding of the township development process. This often leads to serious misunderstandings arising even when great care is taken to explain the process. The concepts of bond financing and home ownership are, for example, particularly difficult to communicate. Community leaders, also unfamiliar with these matters, are reluctant to advise their people for fear of being wrong. This can result in long delays in community decision-making.

The implementing agency is often unable to place clear-cut options on the table due to the many unknowns, particularly with respect to the availability of finance. Discussing hypothetical scenarios serves to further confuse the people and, more seriously, to create false expectations which are extremely difficult to dispel and which lead to resentment.

Community opinion is often divided, particularly between the haves and have-nots. To be able to identify the 'real' feeling of the people is not always easy. In addition, the leadership through which one works may, for various reasons, be relating a different message either from the people to the agency or vice versa.

Myth No. 4: Success of project is of primary concern It is usually assumed that all actors in a shack upgrading project have the interest of the shack dwellers as their primary concern, and that all their actions are aimed at improving the living conditions of these people. Problems experienced with other interventionist agencies in St Wendolins as explained earlier contradict this assumption. This statement needs to be revealed as a myth since it leads to many of the difficulties encountered in upgrading projects. Everyone, whether they are conscious of it or not, has 'hidden agendas' which may or may not be in the best interests of the project.

It is not the intention to raise specific examples of this type of behaviour; suffice it to say that when one examines the broad range of people typically involved in projects – private consultants, government officials, donors, community leaders, land owners (de jure and de facto), academics, field workers, commercial developers, financiers – it should be clear that many different interests are involved, some of which are potentially harmful to the overall objective of the project.

Myth No. 5: Informal settlements are all similar Those involved in development policy issues often regard informal settlements as a homogeneous, generic type of urban form for which a model solution can be found (e.g. see Institute for Housing of South Africa 1989). The three case studies indicate how different informal settlements can be.

Myth No. 6: Squatters are law-breakers That informal settlements, with their illegal squatters, are dens of iniquity resulting in increased crime in surrounding areas is a common assertion. Together with this comes the supposition that all people living in shack areas are squatters. General experience as well as that in Richmond Farm and Qadi in particular has shown the falseness of this assumption.

Myth No. 7: Upgrading is in everyone's interests It would seem reasonable to assume that it is in everyone's interests to formalize or upgrade existing informal settlements. Surprisingly, this is not always true, which accounts for difficulties that sometimes occur in undertaking upgrading work. Richmond Farm and Qadi are two very good examples of development not being in the interests of certain residents.

The shortage of affordable formal housing leads to the creation of informal settlements. This in turn results in land suitable for settlement becoming an instrument of power. An example of this is the protector or warlord syndrome referred to earlier, where the power to allocate land, or the right to settle on that land, is vested (quite illegally) in an individual who has taken on this role. Attempts to interfere with this role are bound to meet with resistance.

Similarly, but distinct from the protector or warlord syndrome, is the case where land in shack areas is being sold illegally by individuals. Here again, attempts to formalize the area, and therefore put an end to this practice, will give rise to problems.

The tribal land ownership system is also not conducive to the township establishment process. It is not in the interest of the chief or the tribe itself to alienate the land in this way. Shack areas on the urban periphery often fall on tribal land which makes it difficult to upgrade in the normal way, as in the case of Qadi referred to earlier.

Upgrading almost invariably results in housing becoming more expensive for existing residents. Although the quality of housing improves, this often means that those who cannot afford the improvements have to leave the area or reduce their budgets on other items. Very often shacks exist, particularly on privately owned land, because there is a demand for this type of housing at this price. Upgrading is clearly not in the interests of these people.

Myth No. 8: Informal settlements are a result of migration A commonly held view is that informal settlements are a direct result of the removal of influx control legislation – that this has led to a flood of poor people migrating from the rural areas. This may be true elsewhere in the country but is not true of the three shack areas discussed here or of metropolitan Durban as a whole. Research indicates that only about one third of heads of households in informal settlements were born in rural areas (Tongaat-Hulett Company 1989). Furthermore, we have not come across any studies indicating a radical reduction in rural population growth rates or the depopulation of rural KwaZulu.

Conclusion

In this chapter a number of commonly held misconceptions about the informal settlements problem have been identified. It is contended that those misconceptions have contributed to the lack of action, or ineffective action, with regard to the problem. It is hoped that attitudinal changes will result in more practical approaches being adopted by all those concerned with improving the living conditions of people presently housed in our informal settlements. Experience in the field leads to the following recommendations:

(a) A great deal more research needs to be done with respect to the communication process between the suppliers of housing and the users. This is particularly true where shacks dwellers on the urban fringe are being drawn into the urbanization process.
(b) Those directly involved in the planning and upgrading of shack areas need to develop new techniques for the establishment of townships that take cognizance of the situation on the ground – particularly the fact that the land is usually occupied and a social structure exists that has to be incorporated into the planning.
(c) Greater emphasis needs to be placed on the legitimization of the land as part of the upgrading process.
(d) Ways of bringing about attitudinal changes on the part of all those

involved in the housing process and the general public need to be developed. This would include discouraging the use of demotivating negative terms such as squatter, septic fringe and so forth.

(e) The question of 'hidden agendas' needs to be addressed. Any upgrading project should seek to identify these and to lay them on the table as a first step in the process.

(f) It is clear that the process of upgrading informal settlements is insufficiently determined by the economic, social and political forces that operate within these settlements. A balance needs to be struck between a top-down and a bottom-up approach.

(g) People involved in the technical aspects of housing need to become more aware of the non-technical context within which the upgrading process takes place. The constant pursuit of technical solutions, with little or no consideration of the economic, political and social environment, can only result in failure.

References

Boaden, B. G. 1986. The Imijondolo: a study of the shack housing delivery system in metropolitan Durban. Special report for the Urban Foundation, Natal.

Boaden, B. G. 1990. The myths and realities of shack upgrading. *Urban Forum* 75–84.

Institute for Housing of Southern Africa 1989. *Proceedings of the 9th Biennial National Housing Congress*. Sandton: Institute for Housing of Southern Africa.

McIntosh, A. 1986. Urban informal settlements in Natal/KwaZulu. Special report for the Urban Foundation, Natal.

Merrington, G. J. 1988. Finance – the pathway to housing. Paper presented at a conference at the School of Business Leadership, University of South Africa, Pretoria.

Natal Witness 1989. R1.2 billion for Durban's squatters lies unused. 5 April.

Tongaat-Hulett Company 1989. *The Durban Functional Region – Planning for the 21st Century*. Durban: Tongaat-Hulett Company.

Town and Regional Planning Commission 1988. *The Durban Functional Region: Statement of the Problem*. Pietermaritzburg: Town and Regional Planning Commission.

Urban Foundation 1987. Unpublished reports on informal settlements in South America and Africa. The Urban Foundation, Western Cape.

Urban Planning, Housing and Design 1986. Proceedings of International Convention. Singapore: Institute of Planners.

PART FOUR

Servicing the cities

Attention now turns to aspects of the broader functioning of the cities, in serving people's needs beyond shelter. In the first essay (Chapter 13), Rogerson considers the informal sector of the economy, which provides income, goods and services for large numbers of the black (especially African) population. He points out that some activities which for a long time were officially discouraged and suppressed are now being viewed as positive sources of employment and enterprise for black people. Intra-urban segregation and the bantustan policy have created peculiar patterns of transportation, as explained by Pirie (Chapter 14), imposing great cost in terms of time and money on those least able to afford it. Yet despite the dangers as well as discomfort of travel under apartheid, Blacks have at times turned the situation to their advantage, for example using trains as venues for social mobilization. Khosa (Chapter 15) examines the Black taxi industry, the recent expansion of which is often taken to exemplify the opportunities arising from initially informal activity. The state's stance has recently shifted from repression to tolerance and even promotion of the Black taxi business, from which major corporations serving the taxi operators now gain substantial business.

The major innovation of the 1980s reforms, with respect to the provision of urban infrastructure, was the Regional Services Council, planned to cover entire metropolitan regions whatever the (racial) designation of their component parts. Pillay (Chapter 16) examines the obstacles encountered in applying this strategy to the Durban metropolis, including the position of the KwaZulu homeland and conflict between national and local government. Next (Chapter 17), Butler-Adam and Grant turn to tourism, as a major element in the Durban regional economy. They argue that 'sustainable and equitable' development requires changes in the present pattern of tourist activity which mainly serves the white well-to-do; there needs to be more sensitivity to local culture and to the basic needs of the poor in the role of both consumers and producers. Part Four is concluded with a review of the impact of urbanization on health, illustrated by evidence from Cape Town (Chapter 18). Chetty argues that health is not just a medical outcome but a result of unequal living standards and access to care, themselves rooted in the structure of apartheid society. As in other aspects of life, fundamental causes rather than symptoms need to be addressed in the formation of a new social order.

13 *The absorptive capacity of the informal sector in the South African city*

J71 R11
Ō17
South Africa

C. M. ROGERSON

One of the most pressing policy matters concerning contemporary South African urbanization surrounds the 'absorptive capacity' of the cities. With the abandonment of influx control and adoption of orderly urbanization, critical importance attaches to the propensity of the country's urban centres for managing employment generation, income enhancement and the accommodation of increasing numbers of new migrants (Rogerson 1989a). The dramatic reduction in the capacity of the formal sector to create new income opportunities for work-seekers (from 73.6 per cent between 1965–1970 to a meagre 12.5 per cent for 1985–90) has directed attention to the income and job-creation potential of the informal economy. Against the backdrop of economic recession and increasing structural unemployment, during the 1980s the economies of South Africa's largest cities showed evidence of 'informalization' (Rogerson 1987, 1988a, 1988b; Rogerson and Hart 1989). More specifically, the process of informalization witnessed visible new growth across a range of street or pavement-centred activities (hawking, street barbers, taxi operations, herbalists, prostitution, begging), an expansion in an array of home-based enterprises (shebeens, spaza stores, hairdressers, backyard workshops) and a small rise in the category of newly 'formalized' ventures situated within fixed business premises (small-scale industry, liquor taverns, vendors). In the rapidly changing policy climate of the 1990s, amidst all the feverish debates about 'fundamental economic restructuring', 'democratizing economic growth' and new strategies of 'growth through redistribution' (African National Congress 1990; Gelb 1990; Kaplinsky 1990), sight must not be lost of the central economic role which small-scale enterprise and the informal sector continue to play in South Africa's cities.

Although precise statistics on the size of the country's informal economy are unavailable, the best estimates suggest that at the close of the 1980s the sector afforded a means of livelihood to at least 1.8 million and perhaps as many as 2.5 million black South Africans or roughly 30 per cent of the total labour force (Kirsten 1988). More significant, however, is the dynamic of a very rapid expansion of informal sector activities, a trend observed in virtually all recent South African research. For example, within the shack settlements girdling Greater Durban it has been estimated (conservatively) that the number of informal-sector enterprises grew from 62,000 to 72,400 over a three-year period during the late 1980s (May and Stavrou 1990: 44). Indeed, the momentum of new informal economic initiatives, especially in cities, has been on such a scale that it is no longer valid to characterize South Africa as a special region in Africa where the informal sector can be written off as 'negligible' and of little policy consequence. Rather, in smoothing the transition from the apartheid- to post-apartheid city, planners, policy-makers and development agencies must take cognizance of the importance and circumstances of the informal sector (Rogerson 1988a).

The objective in this chapter is to raise a number of critical questions concerning the absorptive capacity of the informal sector under late apartheid and beyond. Three sections of material are presented. First, the shifting policy context and recent growth of South Africa's urban informal sector are discussed. Against the background of the state's new focus on inward industrialization and economic deregulation, the second section reviews the developmental potential of production, construction and retailing activities as representing different key spheres of informal economic endeavour. Finally, in the wake of current and projected policy initiatives, the concluding section reflects on the present and future absorptive potential of the country's urban informal sector.

Reform and the re-formation of the urban informal economy

The architects and managers of the post-1948 South African city evinced scant tolerance for the enterprise of the urban informal sector. The development of the apartheid city was marked by a history of successful policies directed to suppressing informal sector activities whenever and wherever they surfaced. Illustratively, between 1940 and 1960 the state undertook a systematic purge to destroy a flourishing community of Black backyard furniture manufacturers on the Witwatersrand. A similar fate befell several communities of Black street traders who from 1930 began invasions of inner city 'white space' in order to secure a livelihood (Rogerson and Beavon 1985; Beavon and Rogerson 1986; Rogerson 1988c, 1990a). The most dramatic campaign was, perhaps, the 'war' declared by Johannesburg's city fathers during the 1950s against 2,000 women coffee-cart traders vending daily refreshments to Black industrial workers (Rogerson 1989b). Threats of police raids, harassment and fines undermined also the everyday existence of women brewers, the operators of shebeens or illicit township liquor outlets (Rogerson and Hart 1986, Rogerson 1990b) and the growing numbers of informal Black taxi enterprises (see Chapter 15). Finally, the state steadfastly continued to oppose the actions of the 'squatter builders' functioning on the margins of the apartheid city, albeit from the 1960s increasingly losing this particular battle against the informal sector (Mabin 1989).

The arsenal of repression built up by the apartheid state embodied such restrictive legislation as the Group Areas Act, harsh licensing, strict zoning regulations, official campaigns to encourage consumer boycotts by white customers of Black informal business and the formation of special police squads dedicated to the persecution of informal enterprise. Underpinning the state's policy was a mixture of factors, some distinctively local and others which parallel situations elsewhere in the developing world. In terms of the local ideological creation of Blacks as 'temporary sojourners' in designated 'white' urban space, the very existence of an informal sector was regarded as an affront to the managers of apartheid urbanization (Wellings and Sutcliffe 1984). The argument of temporary sojourners, for example, was rehearsed to rationalize the forced closure and excision to township areas of all Black small-scale industrialists still operating within white urban space during the 1950s. Additionally, in a manner characteristic of South Africa's racial obsessions, the operations of several Black-dominated informal sector niches were either construed hazardous to public health or stigmatized as a 'social evil', a threat to the existing political order. Binges of slum clearance and other periodic attacks on the illegal spaces within which informal entrepreneurship thrived therefore served to 'cleanse' the South African city of such imminent dangers.

As in many urban areas of the developing world, the informal sector in the apartheid city suffered also from conflicts with the ideology of developmentalism and planners' visions of a 'city beautiful' (Rogerson 1988c, 1990a). Stereotyped as unsightly, unsanitary or dangerous, the activities of common hawker, backyard artisan or shebeener were viewed as contrary to official images of what constituted a 'modern' South African city and thus vulnerable to actions of persecution (Rogerson and Hart 1986, 1989). In addition, public objections that stressed the threat of economic competition posed by informal sector enterprise to established formal businesses were a powerful force for the underdevelopment of communities of Black street traders and small-scale manufacturers. Equally, the competitive threat posed by women brewers and shebeeners to municipal-run liquor outlets and beer halls was a factor in accounting for the venom of repressive measures directed against their existence.

The anti-informal-sector programmes, which functioned in South Africa's cities from 1948 to the end of the 1970s, were an aspect of the state's broader strategies towards urban unemployment during a period when the bantustans were the critical pillar of state policy. With the crystallization of the 1980s reform initiatives, however, the policies of influx control coupled with industrial decentralization began to yield to programmes for orderly urbanization, economic deregulation and inward industrialization. The state's implicit acceptance of the inevitability and desirability of Black urbanization required a set of measures designed 'to maximise this process for stability rather than attempt to undermine it' (Morris and Padayachee 1989: 78). In this regard, the strategy of 'inward industrialization' tied to programmes of deregulation assumes special significance as a vehicle for resolving an urban crisis of escalating levels of Black unemployment by generating job opportunities in the sphere of small business and the informal sector (Hindson 1987; Zarenda 1989). Implicit within certain formulations of inward industrialization is an element of income redistribution in favour of the poorer classes; less clear is the way this redistribution will occur or the nature of the political base which is to initiate and sustain such a strategy.

At present, South Africa's inward industrialization strategy essentially is a labour-intensive urban employment creation policy that aims to supplement the traditional economic strategies of export promotion and import substitution, which are conceded as having failed to stimulate job creation (Gelb 1989; Hendler 1989). The policy is premised upon the notion that a rapidly urbanizing Black population will require commodities ranging from low-cost housing through to inexpensive clothing, furniture, basic foodstuffs and the like. Programmes for economic deregulation are designed to tap the growth potential of small business and the informal sector for the production of these goods and services (Zarenda 1989). Nonetheless, one observer points to a further important role for inward industrialization, to restore South Africa's ailing export competitiveness by potentially alleviating pressures on rising wages in urban areas (Lewis 1989). Although the exact directions of South African inward industrialization still remain undefined, certain parallels have been drawn with Colombia's 'development through urbanization' strategy of the 1970s using construction as a 'kickstart' or 'lead sector' for national development (Hendler 1989; Rogerson 1989c).

In terms of South Africa's new policy initiatives the ironical situation exists that 'activities previously seen as threatening to urban control are now presented as a means of resolving the urban crisis through creating employment' (Hindson 1987: 92). In the age of urban reform and inward industrialization the state reconceptualized communities such as informal builders, backyard producers, hawkers, pirate taxi drivers, shoe-shiners or shebeeners as vital sources of economic growth and employment creation to be unleashed by policies of deregulation (Wellings and Sutcliffe 1984; Rogerson 1987). Yet, such a viewpoint conveniently relieves the state of responsibility to provide welfare assistance to those such as the beggars, car-washers or garbage pickers, who endure bare survival existence in the 'dungeons' of the informal sector. Moreover, it allows the state to characterize elements that do not fit its stereotype image of entrepreneurship as a 'nuisance' to be harassed, prosecuted and banished (Rogerson and Hart 1989).

The shift in state policy towards greater tolerance and even selectively assisting the informal sector may be traced back to early phases of the apartheid reform period (Rogerson 1987). In 1984 a report of the influential National Manpower Commission recommended the preparation of 'a comprehensive policy to encourage small business enterprise' (South Africa 1984a: 1). This was followed up by a state commitment that 'measures at any government level that might restrict the development of the formal and informal small business sector should, where at all possible, be relaxed' (South Africa 1984b: 10). One step towards realizing this goal was taken in the 1986 White Paper on Urbanization which committed the state to making land available in or near township areas for light and service industries. Deregulation is, however, the watchword of inward industrialization strategists (Hendler 1989: 9); in terms of the 1986 Temporary Removal of Restrictions on Economic Activities Act, the State President may exempt certain industries or certain areas from any 'law, condition, limitation or obligation which he believes is unduly impeding the economic progress of people, competition or the creation of jobs' (Benjamin 1987: 374). By designating certain geographical areas as 'free zones', where existing wage, health or safety regulations do not apply, the Act creates a set of low wage, trade-union-free zones within the country's existing metropolitan areas (Rogerson 1990c). The emphasis upon deregulation as a chief tactic for stimulating inward industrialization was reiterated in a 1987 White Paper which stressed the imperative for 'encouraging entrepreneurship' and stated that 'the approach to regulation must therefore emphasise the promotion of economic activities and be less directed towards their control' (South Africa 1987: 13).

In the new reform climate of the 1980s and 1990s there has undoubtedly occurred a re-formation and restructuring of the urban informal economy in South Africa. Long-established and long-suppressed informal economic spheres, such as backyard production, street vending or informal transport, have experienced a phase of revival and resurgence out of the decline recorded during the repressive apartheid period. In addition, new legitimacy and even limited encouragement has been accorded to the operations of other urban activities, including small-scale builders, liquor selling and the running of spaza shops. Growth has been taking place in central city as well as township locales and in 'independent' forms of informal sector activity as well as through the fostering of closer linkages and dependent subcontracting networks with formal business enterprise. To probe further the progress and directions of the unfolding regime of inward industrialization and deregulation the discussion is narrowed to focus on three spheres of informal sector activity: small-scale industry, construction, and retailing.

The absorptive capacity of the informal sector

South African policy-makers' discovery of the informal sector precipitated sweeping and contradictory evaluations of its developmental potential. On the left the process of informalization via economic deregulation was condemned as a threat to the working class with calls made for alternative strategies to support the so-termed 'real informal sector', that is the promotion of cooperative enterprise (Jaffee 1990: Rogerson 1990d). On the opposite front the informal sector was celebrated as a reassertion of the entrepreneurial spirit with its achievements in job creation extolled as illustrating what can be achieved by the 'free market'. In assessing critically the absorptive capacity of the informal sector it is useful to distinguish between two differing evaluations of success. The first contrasts the availability of informal income opportunities with the absence of alternatives and stresses that, while formal work may be seen as preferable, given its scarcity informal sector enterprise is viewed as better than total destitution. The second argues that, under certain conditions, the process of informalization may have consequences superior to those 'of a model of development based on large-scale enterprises and a fully regulated labor market' (Portes *et al.* 1989: 301). Proponents of inward industrialization and accelerated informalization are suggesting that the urban informal sector in South Africa has an absorptive and developmental potential beyond bare survival and may rival successful 'informal economies of growth' such as central Italy or Hong Kong. To what extent are informal small-scale industry, construction and retailing in South Africa presently moving in this positive direction?

Small-scale industry The underdevelopment of the production sphere is one of the most distinctive facets of South Africa's informal economy. To some degree this situation is a historical product of apartheid planning with the state's crushing of Black industrial entrepreneurship in cities paralleled by efforts to rechannel it towards bantustans (Rogerson and da Silva 1988). Reformist initiatives during the 1980s reversed past policies of repression, instead seeking to rekindle and stimulate Black urban industrial enterprise. Yet, initially the impact of new policies was tempered by Group Areas legislation which geographically marginalized Black small-scale manufacturers to township locales. Under the leadership of the Small Business Development Corporation (SBDC) a network of township industrial parks, comprising small factory units or flats, was established to provide working space for new industrial operations as well as for expansion of existing backyard producers. The SBDC launched also a series of cluster schemes or industrial hives to function as incubators for fledgling 'micro-enterprise'. Upgrading of small-scale industrialists was facilitated by a package of loans and advisory services. In addition, the deregulation of many of these areas was interpreted as an important step towards reviving small-scale enterprise by creating conditions for the evolution of greater subcontracting networks, the gestation of the flexible firm and of flexible specialization in South African manufacturing. In order to galvanize further subcontracting relationships the SBDC initiated a special programme to foster linkages between large and small-scale industrial enterprise.

At the beginning of the 1990s the extent of Black small-scale manufacturing, albeit growing, was still limited in employment impact and the sphere of informal production remained marginal to the South African industrial system. For example, in June 1990 the three SBDC industrial parks in Soweto afforded a total of less than 1,000 job opportunities; nationally the scatter of almost 50 industrial parks was providing income opportunities for 7–8,000 workers. It is evident that the constraints on small-scale industry are not likely to be resolved simply by free market deregulation and current initiatives for inward industrialization. Problems include lack of access to capital and credit, lack of management skills, poor marketing information and little or no access to new technologies (Gelb and Miller 1988; Rogerson and da Silva 1988). Moreover, the economic space within which small-scale producers operate is severely narrowed by the expansion by formal large-scale production of relatively low-cost, mass produced and extensively advertised goods. Among a range of limitations on the expansion of small firm subcontracting and flexible specialization in South Africa are: the underdevelopment and shortage of artisanal, technical and managerial skills; concentrated distribution networks under the hegemony of large firms; and the flexibility of labour markets which allows firms seeking to lower their labour costs to decentralize plants to bantustans therefore making unnecessary the development of urban subcontracting networks.

Construction The expansion of residential construction for the provision of low-income housing is at the heart of South African debates concerning inward industrialization (Hendler 1989; Zarenda 1989). Growth in construction activity is viewed as a trigger for the development of a sub-economy, using labour-intensive production techniques, which regenerates itself on a base of petty entrepreneurs functioning as subcontractors to formal sector builders (Leiman and Krafchik 1990). The perceived advantages of construction as a catalyst to inward industrialization are several. Specifically, because construction serves internal demand it is not vulnerable to fluctuating world commodity prices; by using domestic raw materials it would not strain foreign exchange resources; and, as a labour intensive activity it could offer new jobs requiring little skill or education with a corresponding multiplier effect on small-scale industrialization (Hendler 1989). Overall, construction as a leading sector for 'development through accelerated urbanization' offers a promise of greatly expanded opportunities for small (mainly Black) entrepreneurs, developers, builders and subcontractors.

The absorptive capacity of petty construction in the current trajectory of inward industrialization is disclosed by Leiman and Krafchik's (1990) rich analysis of subcontracting in the Western Cape. The most striking finding was that, whilst inward industrialization requires upward mobility of petty construction enterprises, little evidence was discerned of successful movement from subcontracting to independent contracting enterprise. As underpinnings of this failure, several constraints were identified on the growth and profitability of subcontractors. First, severe limitations are placed upon the development of construction activity by the state's continuing restrictions on making available suitable land for sheltering the urban poor. Second, the effective demand for residential construction has been curtailed both by South Africa's monetary authorities' use of high interest rates in national economic management and of biases in state housing subsidies towards white households. Third, attention is drawn to the pervasive problems of small construction firms' lack of access to finance and limited managerial skills. Finally, the involutionary rather than the evolutionary development of small builders is linked to the wider distribution of economic power and the current system of tenders which accords considerable strength to large housing developers. Overall, therefore, a combination of internal policy contradictions surrounding inward industrialization and of structural features of the construction sector are reducing the likelihood that deregulation alone will spur the absorptive capacity of informal or petty construction activities.

Informal retailing The most noticeable impact of the state's economic deregulation initiatives upon the South African urban landscape is the burgeoning of a variety of forms of informal retail activities. In the central business districts and immediate environs of the largest metropolitan centres, the relaxation of licensing controls, the lifting of certain restrictive by-laws, the planning of special hawker zones and the approval of vending carts have contributed towards stimulating new hawker activity as well the rebirth of old forms of street-trading operations (Rogerson and Hart 1989). Outside South Africa's largest urban centres, deregulation has been applied unevenly with politically-conservative white local authorities clinging to repressive local by-laws and thwarting the advance of hawker activity in many of the country's intermediate-sized and smaller urban centres (Hart and Rogerson 1989a, 1989b). In township areas generally, however, a surge of hawker activities has been recorded alongside a proliferation of spaza shops, small township stores which are commonly operated from home or a garage (van Zuydam-Reynolds 1990). Ironically, a major stimulus to the expansion of spaza operations has been apartheid legislation and planning which historically restricted the development of formal retail business outlets within Black townships (Beavon and Rogerson 1990).

Retail activities comprise the largest element in South Africa's informal economy with national estimates of a total of 900,000 traders (Kirsten 1988). Nevertheless, the 1980s expansion of urban informal retailing is essentially a reflection of the absence of formal income opportunities in an economic environment of escalating unemployment levels and worker retrenchments. The great majority of South Africa's hawkers and spaza dealers function at bare survival levels with only limited prospects of income enhancement and a transition to formal retail outlet. Their growth horizons are darkened variously by inadequate access to credit and finance for expansion, lack of business management skills and competition and threats from the formal retail sector (Rogerson 1990a; van Zuydam-Reynolds 1990). Overall, the bulk of urban informal retailing is a struggle for sheer survival rather than a transition to autonomous growth; in terms of absorptive capacity, the potential is for a trajectory of continued involutionary rather than evolutionary expansion.

The informal sector beyond apartheid

It is evident that in the transition from the apartheid to post-apartheid city the process of informalization may offer both pitfalls and opportunities for labour absorption and economic growth. However, current state policy initiatives that favour deregulation and getting the state out of the economy appear to be creating conditions merely for an informal economy of survival rather than an informal economy of growth. In addressing this somewhat discouraging scene it would be irresponsible to neglect the opportunities of informalization, 'particularly if this neglect were merely to be founded on reaction to the current state agenda of reducing real wages through deregulation' (Kaplinsky 1990: 45). Post-apartheid policy debates on the informal sector must absorb the key lessons which emanate from the success-stories of informalization (see Portes *et al.* 1989). Most importantly, effective action to capture the absorptive capacity and income potential of a process of dynamic informalization will demand substantive state intervention in the economy to support small-scale, informal initiatives. A programme of 'growth through redistribution' in a new South Africa (African National Congress 1990; Gelb 1990) necessarily must incorporate innovative state responses to promote the developmental potential of the urban informal economy rather than return to earlier policies of rigid planning and control or deregulation.

Acknowledgements Thanks are extended to the Richard Ward Endowment Fund for financial support of the research.

References

African National Congress 1990. *Discussion Document on Economic Policy*. Johannesburg: African National Congress, Department of Economic Policy.

Beavon, K.S.O. and Rogerson, C.M. 1986. The council vs the common people: the case of street trading in Johannesburg. *Geoforum* 17: 201–16.

Beavon, K.S.O. and Rogerson, C.M. 1990. Temporary trading for temporary people: the making of hawking in Soweto. In D. Drakakis-Smith (ed.), *Economic Development and Urbanization in Developing Areas*, pp. 263–86. London: Routledge.

Benjamin, P. 1987. Labour law. In *Annual Survey of South African Law 1986*, pp. 374–94. Cape Town: Juta.

Gelb, S. 1989. The economic crisis in South Africa – the factors beyond. *Vierteljahresberichte* 118: 393–408.

Gelb, S. 1990. Democratising economic growth: alternative growth models for the future. *Transformation* 12: 25–41.

Gelb, S. and Miller, D. 1988. The spirit of small enterprise. *Work in Progress* 56/57: 66–70.

Hart, D.M. and Rogerson, C.M. 1989a. Towards accommodationist planning in South Africa's secondary centres: the case of hawker deregulation. *Development Southern Africa* 6: 161–72.

Hart, D.M. and Rogerson, C.M. 1989b. Hawkers in South Africa's small urban centres: planning and policy. *Development Southern Africa* 6: 295–310.

Hendler, P. 1989. *Politics on the Home Front*. Johannesburg: South African Institute of Race Relations.

Hindson, D. 1987. *Pass Laws and the Urban African Proletariat in South Africa*. Johannesburg: Ravan.

Jaffee, G. 1990. Worker co-operatives: their emergence, problems and potential. In M. Anstey (ed.), *Worker Participation: South African Options and Experiences*, pp. 191–212. Cape Town: Juta.

Kaplinsky, R. 1990. A policy agenda for post-apartheid South Africa. *Transformation* 12: 42–52.

Kirsten, M. 1988. A quantitative perspective on the informal sector in Southern Africa. *Development Southern Africa* 5: 251–7.

Leiman, A. and Krafchik, W. 1990. Inward industrialization and petty entrepreneurship: an evaluation of recent experience in the construction industry. Unpublished paper, University of Cape Town.

Lewis, D. 1989. Some fine new buzzwords, but whither do they lead? *Weekly Mail* (Johannesburg) 17–22 March: 14–15.

Mabin, A. 1989. Struggle for the city: urbanization and political strategies of the South African state. *Social Dynamics* 15: 1–28.

May, J. and Stavrou, S. 1990. Surviving in shantytown: Durban's hidden economy. *Indicator South Africa* 7 (2): 43–48.

Morris, M. and Padayachee, V. 1989. Hegemonic projects, accumulation strategies and state reform policy in South Africa. *Labour, Capital and Society* 22:65–109.

Portes, A., Castells, M. and Benton, L.A. 1989. Conclusion: the policy implications of informality. In A. Portes, M. Castells and L.A. Benton (eds), *The Informal Economy: Studies in Advanced and Less Developed Countries*, pp. 298–311. Baltimore: Johns Hopkins University Press.

Rogerson, C.M. 1987. The state and the informal sector: a case of separate development. In G. Moss and I. Obery (eds), *South African Review 4*, pp. 412–22. Johannesburg: Ravan.

Rogerson, C.M. 1988a. Late apartheid and the urban informal sector. In J. Suckling and L. White (eds), *After Apartheid: the Renewal of the South African Economy*, pp. 132–45. London: James Currey.
Rogerson, C.M. 1988b. Recession and the informal sector in South Africa. *Development Southern Africa* 5: 88–93.
Rogerson, C.M. 1988c. The underdevelopment of the informal sector: street hawking in Johannesburg, South Africa. *Urban Geography* 9: 549–67.
Rogerson, C.M. 1989a. Managing urban growth in South Africa: learning from the international experience. *South African Geographical Journal* 71: 129–33.
Rogerson, C.M. 1989b. From coffee-cart to industrial canteen: feeding Johannesburg's black workers, 1945–1962. In A. Mabin (ed.), *Organization and Economic Change: Southern African Studies Vol. 5*, pp. 168–98. Johannesburg: Ravan.
Rogerson, C.M. 1989c. Inward industrialization and orderly urbanization in South Africa: lessons from Colombia. *South African Geographical Journal* 71: 157–65.
Rogerson, C.M. 1990a. Informal sector retailing in the South African city: the case of Johannesburg. In A.M. Findlay, R. Paddison and J. Dawson (eds), *Retailing Environments in Developing Countries*, pp. 118–37. London: Routledge.
Rogerson, C.M. 1990b. Consumerism, the state and the informal sector: shebeens in South Africa's black townships. In D. Drakakis-Smith (ed.), *Economic Development and Urbanization in Developing Areas*, pp. 287–300. London: Routledge.
Rogerson, C.M. 1990c. Late apartheid, emergent black trade unionism and industrial change in South Africa. In G.J.R. Linge and D. Rich (eds), *The State and the Spatial Management of Industrial Change*, pp. 183–204. London: Routledge.
Rogerson, C.M. 1990d. 'People's factories': worker co-operatives in South Africa. *GeoJournal* 22: 285–92.
Rogerson, C.M. and Beavon, K.S.O. 1985. A tradition of repression: the street traders of Johannesburg. In R. Bromley (ed.), *Planning for Small Enterprises in Third World Cities*, pp. 233–45. Oxford: Pergamon.
Rogerson, C.M. and da Silva, M. 1988. Upgrading urban small-scale forms of production: the South African case. In G.J.R. Linge (ed.), *Peripheralization and Industrial Change*, pp. 199–213. Beckenham: Croom Helm.
Rogerson, C.M. and Hart, D.M. 1986. The survival of the 'informal sector': the shebeens of Black Johannesburg. *GeoJournal* 12: 153–66.
Rogerson, C.M. and Hart, D.M. 1989. The struggle for the streets: deregulation and hawking in South Africa's major urban areas. *Social Dynamics* 15: 29–45.
South Africa, Republic of 1984a. *Report of the National Manpower Commission on Investigations into the Small Business Sector in the Republic of South Africa, with Specific Reference to the Factors that might Retard the Growth and Development Thereof.* Pretoria: Government Printer.
South Africa, Republic of 1984b. *White Paper on a Strategy for the Creation of Employment Opportunities in the Republic of South Africa.* Pretoria: Government Printer.
South Africa, Republic of 1987. *White Paper on Privatization and Deregulation in the Republic of South Africa.* Pretoria: Government Printer.
van Zuydam-Reynolds, A. 1990. Shack shops: from spazas to plazas. *Indicator South Africa* 7(2): 64–6.
Wellings, P. and Sutcliffe, M. 1984. 'Developing' the urban informal sector in South Africa: the reformist paradigm and its fallacies. *Development and Change* 15: 517–50.
Zarenda, H. 1989. The rationale underlying the policy of inward industrialization in South Africa. *Development Southern Africa* 6: 409–20.

14 *Travelling under apartheid*

G. H. PIRIE

Apartheid required that urbanization was accompanied by the enforced segregation of people of different race. This necessitated a gigantic programme of spatial engineering in terms of which Blacks were allocated housing on the fringes of urban areas or in rural bantustans. In both instances regular, efficient and inexpensive public transport was imperative to ensure that the massive displacement of the workforce did not interrupt the smooth working of the economy. The extensive construction of commuter railways and roads, and the subsidization of commuter fares, were essential ingredients of this deliberately distorted form of urbanization.

The patterns of commuter travel that have been generated by apartheid in South Africa show marked racial differences. Unlike Black people whose residential location is circumscribed, whites enjoy a wide range of options and can arrange short work trips. The 2.5 million Black commuters who have a more rigid configuration of trip origins and destinations are least able to afford the flexibility that private cars offer and they rely heavily on public transport that follows fairly fixed routes, if not fixed schedules. By contrast, whites enjoy a high rate of car ownership and superior mobility. White commuters also generally use different routes because they live apart from Blacks. Until recently, whites who did patronize public transport did not have to mix with Black passengers. In places where spatial segregation was imperfect, and Blacks had to cross a designated white racial zone or group area for instance, racial segregation was enforced aboard public transport.

These broad features of transport and travel under apartheid are only some of its elements. They would be an adequate account if the passengers themselves were less than human. Try as they might, however, even the social engineers could not surmount the fact that Black commuters were not just units of unconscious freight. To its users, public transport is more than just uniform and passive mobility. Their everyday encounter with one of the most palpable creatures of apartheid is part and parcel of the experience of urbanization in contemporary South Africa.

Territorial segregation and travel

The geography of race in apartheid South Africa has created two major streams of Black commuting on trunk routes. One of these comprises the long-distance journeys between bantustan towns and metropolitan areas; these are equivalent to inter-urban travel. A second stream embraces the shorter trips between places of employment and Black dormitory townships which are located within cities. The two trip types blur inside the city as passengers join one another in onward journeys to dispersed work places.

Inter-urban travel In all the bantustans there are 'cross-frontier' commuters, but they are concentrated numerically in KwaZulu and Bophuthatswana (including Botshabelo in the detached Thaba'Nchu district) (Lemon 1987: 201–5). These territories are least remote from the major work magnets in the Durban–Pinetown and Pretoria–Witwatersrand complexes (and from Bloemfontein in the case of Botshabelo). In 1984 residents of these two bantustans accounted for 73 per cent of all 773,000 bantustan commuters who made 2.1 million daily journeys. Half these trips involved journeys averaging 20 km in length and 90 minutes door-to-door travelling time. The passenger market was split almost equally between train and bus modes. At the extreme, 20 per cent of the journeys averaged 45 km in length and 2 hours 20 minutes in duration. Buses carried two-thirds of the commuters on these longer journeys. To begin with, at least, planning and managing this trek occupied five committees in the Department of Bantu Administration and Development. A licence to transport commuters was highly prized: it was said to have more value even than a liquor licence (Burger 1970; Dippenaar 1972; Witulski 1986).

Apartheid artificially elongated the work journeys undertaken by the least mobile urban poor. Long-distance commuting occurred much sooner than it would have done under circumstances of organic urban growth and sprawl: it was premature and could not be funded out of the fare box. The public expense and sudden urgency of providing routeways and vehicles meant that other infrastructure was overlooked: bus shelters, benches, kerbs and pavements were as rare as transport schedules, tariff listings and public toilets. Interchanges between trunk and branch transport services were inadequately integrated and synchronized; terminals were sited at the edges of townships and on the fringes of downtown areas, necessitating long and sometimes hazardous walks at either end of a journey. Bantustan commuters spend approximately half their total journey time walking, waiting for and transferring between vehicles (Witulski 1986).

Lengthy bantustan commuting is inconvenient and it is also expensive and time-consuming. In extreme cases, transport pares disposable income (and standards of living) by as much as 20 per cent, and trims discretionary time budgets to as little as two hours daily (Witulski 1986). Travel in uncomfortable vehicles at odd hours and over long distances is also gruelling and debilitating. There is a point beyond which passengers cannot adjust any more to the rigours of the journey, and regularly arrive at their destinations exhausted. At factories and construction sites fatigue may undermine concentration and endanger the lives and limbs of manual labourers; productivity may be reduced, followed by the imposition of penalties and ultimately dismissal. On the home front, the price of long-distance commuting is minimal free time and energy for normal domestic and family activities.

The 20,000 people who commute daily by bus between the 'bus stop' bantustan of KwaNdebele and metropolitan Pretoria are among those whose lives have been reduced to wage-earning functionaries. They are a listless cargo of 'nightriders' who shuttle to and fro in badly ventilated, dimly lit vehicles. The shortest journeys are 100 km and last nearly two hours; the longest may reach 160 km and last over three hours. The first buses start operating before 03.00 h; the last engines are switched off at 21.30 h. The passengers pay to sit on hard seats which were designed for use on short hauls and on tarred city streets rather than on marathons across the rutted, dusty roads of the veld. On board the buses are a crowd of groggy, swaying bodies, some seated, some standing, others coiled around one another in the centre aisle. They are 'a congregation of specters, souls in purgatory' (Lelyveld 1986: 130; Goldblatt *et al.* 1989).

Intra-urban travel The number of intra-urban Black commuters far exceeds the number of bantustan commuters. Approximately 1.5 million people are transported daily to and from urban Black townships such as Soweto, Khayelitsha, Katlehong and Mamelodi (Witulski 1986; McCaul 1990). The detailed geography of travel reflects apartheid planning which maximized the need for public transport in impoverished communities while minimizing the possibilities for walking and cycling to work. Journeys are long, tiring and expensive, and female domestic servants in particular walk long distances between termini and their work (Preston-Whyte 1982). Low levels of car ownership persist in the Black townships, but the stake of buses, trains and taxis in the top end of the commuter market has shifted substantially in favour of shared minibus taxis (see next chapter): their superior speed and flexibility outweigh the higher fares which users must pay. Taxi passengers have to queue, and they squash together in the vehicles, but all of them do sit. Soft seats probably offer more comfort than the loud music which drivers play. Overloading and reckless driving means that travelling in taxis (nicknamed 'Zola Budds' and 'Mary Deckers' after the two Olympic track stars who collided with one another during a race) is not necessarily a safer option than travelling in trains or in the menacing buses which, when they were painted green, were likened to lethal snakes and were dubbed 'green mambas'. Inside the taxis, however, passengers are not liable to personal attack.

Travel in the lumbering and strictly functional ghetto trains is an unremitting struggle. As with mass commuting elsewhere, stampeding is common. When the tide of humanity on the platforms surges towards the trains, the exclamation 'Fudua! Fudua!' rings out as people shove bottoms-first through the carriage doors. Inside the carriages, limp bodies wedged in tight proximity are contorted into bizarre arrangements. Passengers stare blankly or doze in the stuffy and sweaty atmosphere. 'These are the hours to hell and back', wrote one African poet (Molusi, 1981). On late-night trains and on pay-days especially, train-friends keep alert for the pickpockets and bag-snatchers, youths who 'work' the trains instead of sitting behind a school desk or earning wages. For them, the crowd and the noise is a perfect screen (Siluma 1978; Tlali 1989).

Petty theft on the trains is trifling compared to the violent thuggery. Hooligans operating in organized gangs favour the unguarded third-class carriages containing the power switches which they tamper with to control the lights and doors. Before trains fully stop next to platforms, apprentices to the 'big bosses' of train crime leap aboard to take up positions at doorways and to book seats for their masters who target victims for robbery. In the moving trains terrified commuters watch helplessly and soundlessly while they are fleeced. They huddle in corners while the gangsters sprawl their card games provocatively over the floor. In the toilet there might be a brutal rape; 'train-pulling' means gang rape. Passengers who resist or speak out about the terror may be killed by the quick thrust of a sharpened bicycle spoke, or simply flung from the train. Reporting crime to the authorities brings swift revenge. At station after station the gangs swap coaches, preventing passengers from disembarking before they too are mugged (Setuke 1980; Themba 1985a, b; *Sowetan*, 10 December 1986).

Overcrowding, fear and daring on township trains have also merged in a more individualistic form, one which is not necessarily less dangerous. Young men have taken to travelling on the outside of commuter trains: they hang on to the roof gutters and window ledges, ride astride the carriage couplings, and scamper on the roofs. These practices probably originated when people ran to clamber aboard departing trains, or when they tried to evade fare payments. Gradually, the technique of riding the trains as if one were a member of the railway staff became a prank, and then was self-consciously developed into an expression of masculinity and athleticism, and a vital skill for train criminals. 'Staffriders', as they were called, became symbols of a new order. By spurning convention and ignoring regulations they epitomized freedom and revolt (Tshabangu 1978). The word 'staffrider' entered into the common lexicon and was adopted as the name for a new magazine of popular culture that was launched in 1978.

Crowding, violence and staffriding, as well as drinking and singing, are part of a distinctive culture on township trains. These trappings of Black travel help distinguish it from the sterile environment in which whites habitually travel on mass transport. The technology may be similar, but Black transport is more thoroughly public. Adding to the ritual of Black train travel is the element of public preaching and worship such as occurs on train services in the Witwatersrand and KwaZulu. In the latter case, passengers participated in religious ceremonies to protect them on their journey. Confronting the emotionally charged experience of train travel, 'the Zionist worked as a community exorcist to hold off public and impending danger' (Kiernan 1977: 220). Whether for the same reason, or for fellowship, commuters also preach on trains elsewhere. Friends and strangers congregate in two or three coaches to sing, simultaneously ringing bells, using the walls of the coach as a drum, stamping their feet and gyrating. The train ride is not only a means to an end, but an end in itself: trains become churches (Mofokeng 1987; *Contact*, 28 January 1981; *Weekly Mail*, 31 March to 7 April 1988).

As venues for a range of activities, Black commuter trains have also been used recently for political recruitment and reinforcement. In carriages on Witwatersrand trains, commuters gathered to express their political feelings and their solidarity. Because of their mobility, the trains are also used to unite workers at scattered work sites, and to roll strikes from one industrial area to another. In 1989 so-called 'train committees' (themselves including strikers) galvanized support for striking metalworkers during a national strike. Similarly, they were instrumental in winning support for a rent boycott in Tembisa, for the mass three-day protest in June 1988 against a new labour Bill, and for the boycott of a fish-and-chip shop which sold rotten food.

The carriages which are used for political purposes have acquired a name: 'Emzabalazweni'. It means 'in the struggle'. They were turned to this use during the States of Emergency when outdoor gatherings were banned. In a remarkable perversion, trains were used to undermine apartheid rather than simply serve its labour and segregation requirements. Inside the carriages trade union organizers met to discuss strategy (the 'shop stewards' working breakfast') and youth activists waved posters that impressed a political consciousness on commuters and kept insurrection alive. Speeches were made. Plays were performed, including 'Workers' Lament' and 'Women Stand up for your Rights'. Revolutionary songs reverberated to the accompaniment of swaying bodies and stamping feet. Carriages shook to the rhythm of the 'tovi-tovi' dancing, the thumping beat muffling the clatter of the wheels. While passengers clapped their hands and banged the sides of the coach with their fists, they chanted political slogans: 'Forward with the workers' struggle'; 'Down with informers'; 'People, remember June 16'; 'Be strong, workers, be strong'. Occasionally a less political note was struck: 'Down with the gang rapes of schoolgirls'. Borrowing a phrase first applied to buses, these train carriages were the 'moving cocoons' of apartheid in which a mass political culture was hatching.

The 'train rallies' occurred indoors (and therefore were not in contravention of Emergency regulations) and by their mobile nature they were difficult for security forces to monitor and control. Nevertheless, they were not beyond the long arm of the state; the Congress of South African Trade Unions claimed that in 1989 the police arrested about 450 people in connection with singing freedom songs on trains (Shubane 1988; *The Star*, 12 August 1989; *Weekly Mail*, 15 to 22 June 1989, 11 to 17 August 1989, 9 to 15 February 1990).

Train rallies are a new expression of the way in which public transport has long been used, both literally and figuratively, to mobilize political resistance in South Africa. The country has a long history of bus boycotts, and vehicles owned by the state especially have often been the target of attack. This politicization of transport by its users is in part a response to the way in which the apartheid state manipulated transport for its own political ends. For decades the state-operated trains and the subsidized buses were a daily reminder to Black people of their exclusion from white residential areas. They also measured the pulse of industrial life, reminding people of their inferior utilitarian status in urban South Africa. Public transport symbolized oppression and subservience (Kiernan 1977).

As public transport has played a crucial economic and political role it has been an obvious target for protesters. On occasion, buses and train carriages were immobilized by stoning them or setting them alight. Some were diverted by hijackers. Public transport was paralysed by threats against vehicles and drivers. Calls for stay-aways from work were made more effective by blocking entrances to bus depots, taxi ranks and railway stations. As happened when transport workers themselves went out on strike, soldiers and police patrolled these terminals in an effort to collect fares and protect both public property and the people who ventured to travel to work (Mathiane 1987). On countless occasions commuters have had to face great uncertainty about whether or not transport would be operating, and whether or not it would be safe to risk travelling. Sometimes taxis offered an alternative, but not during the periodic 'taxi wars' when passengers became embroiled in some fierce and fatal struggles between rival taxi operators (McCaul 1990). The manner in which transport may become a war zone is illustrated starkly in Pietermaritzburg where feuding Black groups have made travel a perpetual hazard: termini and streets are little more than battlefields (Kentridge 1990). During August 1990 the violence in Natal spilled over into the Witwatersrand townships. Early one morning during the mayhem and slaughter, about 300 Inkatha supporters armed with axes, spears and guns stormed a train which had stopped at Inhlazane railway station in Soweto. Nine startled commuters were killed, and more were injured. On another occasion, a hit-and-run attack by gunmen left a trail of blood at a taxi rank at Jeppe station. In the following month, more than a hundred Soweto commuters were hacked and stabbed, and 26 were fatally shot, or leapt to their death, when an unknown gang went on the rampage on a moving train near Johannesburg.

Social segregation and travel

Territorial segregation in apartheid South Africa reinforced the racially segmented passenger market that would in any event have emerged in response to discrepancies in the affordability of private transport. Within major cities, however, spatial segregation was never perfected to the point that Black and white commuters could avoid sharing rail lines and roadways in the same transport corridors. In an effort to prevent racial intermingling on vehicles and at vehicle stops along common routes, the apartheid government elaborated and extended the 50-year-old system of racially segregated public transport. Racial separation, racial exclusion and racial discrimination in buses, trains and taxis reached a pinnacle in the 1960s; throughout the apartheid period it has been a prominent thread in the travel experiences of South Africans.

The racial restrictions that were applied to public transport affected vehicles as well as platforms, toilets, benches, restaurants, ticket counters, waiting rooms, staircases and bridges. Signs on all these facilities helped to stoke racial consciousness and ingrain racial conceits. Many Blacks and whites found the overt racism repugnant, not least when it was sharpened by tactless and overzealous drivers, conductors, ticketing clerks, or fellow passengers. Verbal insults could escalate into ugly scuffles and violent ejection from vehicles. The racial restrictions were not only demeaning, they were also a great inconvenience. Travellers hurrying about their daily tasks would have to hunt down the racially correct entrance to railway stations, and then find the appropriate ticketing counter and staircase. Those needing to make a quick taxi trip would have to be certain that they summoned one which their pigmentation entitled them to hire. Bus passengers would have to queue at stops designated for their race group and could not simply board the first vehicle that appeared. The inconvenience was always greater for Blacks than it was for whites: apartheid on public transport never entailed complete duplication of all services.

The nuisance and indignity of racially restricted transport that was a way of life under apartheid began to ease slowly and selectively in the mid 1970s. The dismantling of apartheid on trains started on luxury intercity services in 1975 and was of no concern to Black commuters. Four years later, however, it was announced that Black commuters on the Witwatersrand who held first-class tickets could use 'white' coaches if no other accommodation was available, and if the train conductor approved. This localized concession was doubtless put into practice elsewhere by a few conductors who turned a blind eye to contraventions of seating codes. Desegregation advanced another step in 1985 when certain coaches, waiting rooms, restaurants and toilets were declared open for use by people of all races. A few of all these facilities were still reserved for use by whites exclusively. Suddenly, and without any fanfare, all suburban trains and all stations in the Cape peninsula were totally desegregated in June 1988. Despite dire predictions of inter-racial violence, the heavens did not fall in, and in the next month racially restrictive signs were removed from all other suburban trains and from train stations everywhere (Pirie 1989). From then on, train passengers were sorted according to economic means rather than skin colour. The envy and rage with which tightly jammed Black commuters gazed at white people seated comfortably in passing carriages (Tshabangu 1978; Matthews 1983; Tlali 1989) would subside or alter.

In keeping with the pattern in rail transport, the desegregation of road passenger transport proceeded on a trial basis. No dramatic court case or statutory revision ended bus and taxi apartheid except in the newly 'independent' bantustans. There, however, all public transport users were Black, and repealing institutionalized racism was merely token. The particular form which desegregation took in urban South Africa, and the pace at which it occurred, depended on political and other local circumstances in different centres. In the 1970s the government began relenting on strict transport apartheid, allowing local authorities to be guided by the wishes of their own constituencies. The desegregation that ensued was negligible. By 1986 only 16 per cent of the 12,406 buses (an estimated 75 per cent of the national fleet) operated by 53 organizations were licensed to carry passengers of all races.

Enforced racial segregation on buses in Cape Town was truly a product of the apartheid government. Alone among South African cities Cape Town had a tradition of racially mixed road vehicles. The privately owned bus company succumbed to apartheid in 1956, and largely for financial reasons was the first to allow it to slip away in 1979. In the following year bus apartheid was officially terminated in Port Elizabeth and East London where shrinking passenger loads made segregation increasingly unviable. In Durban desegregation was more protracted. The City was the first to request relaxation of apartheid, but annual petitions and appeals from 1975 onwards brought no relief until 1986. Then, for the first time, permission was granted for limited desegregation of some buses previously reserved for Blacks. After a trial period of a few months, restrictions on designated (and higher fare) 'white' buses were lifted. Codified racism on the buses has ended, but the persistence of a dual fare structure effectively maintains a racial division of travel. This is enhanced by the geographically divergent routeings of buses in race-specific passenger market areas.

Johannesburg was among the last of the major South African cities to desegregate its bus fleets. In a timid gesture, Chinese and Indian residents were admitted aboard 'white' buses as from 1980. Against the backdrop of several unpleasant incidents on the buses, and spiralling financial losses, services on three routes to the city's middle-class, politically liberal suburbs were desegregated on a trial basis in 1986. The offensive wording on the buses was removed, and racial prohibition ended, but brightly painted marker boards advertised which vehicles still offered service at the old higher and lower fares. Gradually apartheid was relaxed on other routes, bus drivers being given permission to admit people of any race at their discretion. In February 1990 the increasingly toothless racial restrictions on all buses in Johannesburg were formally cancelled (Pirie, 1990). Pretoria followed suit in July 1990. The vestiges of apartheid on public transport elsewhere disappeared finally by government decree in October 1990 with the repeal of the Separate Amenities Act. The confusion, inconvenience and resentment that was endemic to travelling in forcibly segregated vehicles ended. Township and bantustan commuting is wedded to less malleable settlement patterns and will remain race-specific; the redirection of public spending in a new political order may alleviate the expense, danger and disruption of that intense daily experience.

References

Burger, J. J. 1970. Vervoerstelsels as basis vir die toepassing van afsonderlike ontwikkeling. *Yearbook of the South African Bureau of Racial Affairs* 6: 55–66.

Dippenaar, M. 1972. Vervoer tussen blanke stede en Bantoetuislande. *Journal of Racial Affairs* 23: 11–15.

Goldblatt, D., Goldblatt, B. and van Niekerk, P. 1989. *The Transported*. New York: Aperture Foundation.

Kentridge, M. 1990. *An Unofficial War: Inkatha and the UDF in Pietermaritzburg*. Cape Town: David Philip.

Kiernan, J. P. 1977. Public transport and private risk: Zionism and the black commuter in South Africa. *Journal of Anthropological Research* 33: 214–26.

Lelyveld, J. 1986. *Mind your Shadow*. Johannesburg: Jonathan Ball.

Lemon, A. 1987. *Apartheid in Transition*. Aldershot: Gower.

Mathiane, N. 1987. Working on the railway. *Frontline* 7: 16–19.

Matthews, J. 1983. Whites only. In J. Matthews, *The Park and other stories*, pp. 51–60. Johannesburg: Ravan.

McCaul, C. 1990. *No Easy Ride*. Johannesburg: South African Institute of Race Relations.

Mofokeng, S. 1987. Train churches. *TriQuarterly* 69: 352–62.

Molusi, M. 1981. Black transportation. *Contrast* 13: 52–3.

Pirie, G. H. 1989. Dismantling railway apartheid in South Africa, 1975–88. *Journal of Contemporary African Studies* 8 (in press).

Pirie, G. H. 1990. Dismantling bus apartheid in South Africa, 1975–90. *Africa Insight* 20: 111–17.

Preston-Whyte, E. 1982. Segregation and interpersonal relationships: a case study

of domestic service in Durban. In D. M. Smith (ed.), *Living under apartheid*, pp. 164–82. London: George Allen & Unwin.
Setuke, B. 1980. Dumani. In M. Mutloatse (ed.), *Forced Landing*, pp. 58–68. Johannesburg: Ravan.
Shubane, K. 1988. Emzabalazweni! *South African Labour Bulletin* 13: 43–7.
Siluma, M. 1978. Naledi train. *Staffrider* 1(4): 2–4.
Themba, C. 1985a. The Dube train. In E. Patel (ed.), *The World of Can Themba*, pp. 33–9. Johannesburg: Ravan.
Themba, C. 1985b. Terror in the trains. In E. Patel (ed.), *The World of Can Themba*, pp. 111–15. Johannesburg: Ravan.
Tlali, M. 1989. Fud-u-u-a! In M. Tlali, *Footprints in the Quag*, pp. 27–42. Cape Town: David Philip.
Tshabangu, M. 1978. Thoughts in a train. *Staffrider* 1 (2): 27.
Witulski, U. 1986. Black commuters in South Africa. *Africa Insight* 16:10–20.

15 *Changing state policy and the Black taxi industry in Soweto*

J71
L92
South Africa

MESHACK M. KHOSA

Introduction

The Black taxi industry in South Africa has grown from a few hundred six-seater sedans in the late 1970s to over 80,000 ten- and sixteen-seater minibuses in the 1990s (Barolsky 1990; Khosa 1990; McCaul 1990). Today, the industry is a powerful force in the urban economy and estimates reveal that taxi owners buy over 800 million litres of petrol and purchase over 3.5 million tyres per annum. The Black taxi industry provides four motor manufacturing companies with a turnover of about R2 billion a year, capital investment of about R3 billion, and has created some 300,000 jobs (equivalent to 60 per cent of the entire gold-mining industry in South Africa) (Khosa 1990).

Despite the economic regulation, entrance control and geographic limitation on operational freedom, the Black taxi industry has grown dramatically and is acclaimed as one of the 'success stories' of Black small businesses in South Africa (Barolsky 1990). This chapter traces events which led to the Black taxi revolution in the 1980s with special reference to Soweto, the largest African township. The aim is to investigate the emergence of the Black taxi operation in the township, and provide an analysis of state policy. The theme that emerges is that the shift from repression to tolerance of the taxi industry in the past few years was concomitant with the advent of Black capital accumulation and economic empowerment.

Repression in the taxi industry, 1930–76

Before 1976, the policy of the apartheid government was to discourage Black trading and business in Black-designated townships by using various repressive strategies including the infamous 'one-man-one-business' policy whereby Black traders were limited to only one business, and they were precluded from forming companies and partnerships with the object of combining their resources in order to embark upon larger business ventures

(Beavon and Rogerson 1990). Not only were legalized Black taxi operators restricted to carrying five passengers, but they were also harassed by traffic officials and their vehicles confiscated for petty offences. Direct control on taxi licensing was (and still is) firmly in the hands of the white state apparatus. Black people were often sceptical about state legislation and perceived it as an imposition from above by the white state. This resulted in disregard for law and hence prosecutions. Discretionary powers to license or cancel taxi licences were at the mercy of Local Road Transport Boards (hereafter, LRTBs). However, the LRTBs would grant or cancel taxi licences depending on the information gathered from police and officials.

Both local and central governments used apartheid laws such as influx control to regulate issuing of taxi licences to Black people. Newcomers in cities who had no urban rights were refused a Daily Labourer's Permit (DLP) which was a key precondition for a taxi permit. Indeed, plenty of applications were set aside as soon as state officials obtained information that applicants were in urban areas 'illegally'. To get a taxi permit, any applicant had to be in urban areas legally, be a registered tenant, be in possession of a DLP, and have a good employment record. Moreover, to retain the favour of administrators, taxi operators had to avoid any 'subversive activity' such as encouraging bus boycotts.

The repression epoch saw the reign of sedan cars (such as Valiants and Chevrolets) as taxis. 'Taxilords' needed only a car to continue in business. Operating a taxi was not a capital-intensive activity and operators did not need book-keepers or accountants to manage their businesses. Whereas literate taxi operators took care of the administrative side of their business, others left such matters to able and experienced members of their family. Only a handful operated more than two taxis. Even if they had enough capital, the policy of the state was to grant only a single permit per applicant.

Faced with a myriad of obstacles to obtaining legal status, it was not surprising that many taxi operators resorted to 'pirate' operations. 'Pirate' taxis could be seen as a rejection of the entire transport policy imposed by an unrepresentative white minority regime. Nor was the gulf between legal and 'illegal' operators unbridgeable. Permit holders also duplicated their permits, thus joining the ranks of 'pirate' operators. Locally based taxi associations denied the state vital information (on the operation of 'pirates') which white officials needed to exert control over the taxi industry.

From repression to limited tolerance, post-1976

The student uprisings in Soweto in 1976 were a watershed in terms of Black business (Rogerson 1988). Prior to them, state policy was more repressive and thus curtailed opportunities for capital accumulation in the Black taxi industry. The changing material conditions were responsible for the shift in state policy from repression to tolerance of the taxi industry. Before 1977, Black taxis were not allowed to carry more than five passengers. But this changed with the enactment of the Road Transportation Act which redefined the taxi to eight passengers plus a driver, commonly known as an '8+1' vehicle.

As a result of the 'reform' policies which began in the late 1970s, the government established a commission of inquiry in 1980 with the aim of 'depoliticizing' transport in South Africa (South Africa 1983). The commission heard evidence mainly from two sides: on the one hand, bus operators whose market had been heavily eroded by the Black taxi industry, and on the other hand, there were fractions of capital which favoured deregulation. The recommendation of the commission sided heavily with the case presented by bus operators (South Africa 1983: F18 and F20). It was only after vociferous and popular protests waged by Black political organizations and sections of the private sector that the government withdrew the draft bill in 1985 based on the report of the Welgemoed Commission. In recent years, the government has adopted a more favourable attitude to the Black taxi industry (South Africa 1987).

From the 1940s to the late 1960s, there were unsuccessful attempts to forge unity amongst taxi operators, but this only came to fruition in the 1970s. The most significant milestone occurred on 23 October 1980 in Soweto. Representatives of more than 40 taxi local associations from various parts of South Africa gathered in Orlando (Soweto), to seek cooperation and to amalgamate into one united taxi front (hereafter, UTF). Delegates unanimously resolved to adopt what later became known as the pragmatic coordinated approach in the taxi industry. A seed which was planted at the Orlando convention germinated some seven months later.

In line with the October 1980 resolution, various regional taxi associations merged to form the Southern Africa Black Taxi Association (SABTA) in order to promote solidarity with colleagues in other geographical areas and to tackle problems as the UTF. The first president of SABTA, Jimmy Sojane, served until 1984 when James Ngcoya, the present president, was elected. SABTA membership rose from 20,000 in 1982 to over 50,000 in 1990. SABTA as presently constituted is an amalgamation of more than 400 local, 45 regional and five provincial taxi associations, each with its management structures. SABTA has begun to cast its net wider and recruit members in the frontline states of Botswana, Lesotho, Swaziland, Mozambique and Namibia. Whereas SABTA represents short-distance taxi operators, long-distance taxi operators are organized under the banner of the South African Long-Distance Taxi Association (SALDTA) (Khosa 1990).

Finance capital in the Black taxi industry

The various roles played by taxi associations, the state and finance capital were central in the process of capital accumulation in the taxi industry. Three forces were responsible for the revolutionary growth. First, the bargaining power of taxi organizations in mobilizing the involvement of big business in the taxi industry; secondly, the penetration of finance capital in the Black community; and thirdly, the shift in the apartheid state from previous ruthless and savage policies on Black trading in general, to promotion of the taxi industry in the 1980s in particular.

With the meteoric rise in the significance of the taxi industry, advocates of capitalism were quick to point out that capitalism was colour-blind and argued that taxi operators were participating in the process of capital accumulation. However, this was not without the active involvement of big business and finance capital which recognized in the taxi industry a new arena for investment. Major corporations which have entered deals with the Black taxi industry include Toyota, Volkswagen, Nissan and Unipart. SABTA has forged links with multinational companies (such as Shell) as evident in SABTA's apparent anti-sanctions campaigns (Khosa 1990).

Until very recently, taxi operators found it difficult to raise finance to purchase and insure their vehicles. Finance houses were unwilling to finance taxi operators because few had credit records or references, they kept no books of account, the finance houses were not aware of the profitability of taxi operation and there were cases of racial bias by bank managers granting hire-purchase credit. It was through the intervention of the organized taxi industry that a number of favourable financial deals were raised with major finance houses in South Africa. The shares of markets of those supplying finance to the taxi market are: Wesbank 61 per cent, Stannic 28 per cent, Santam 9 per cent and Nedfin 2 per cent (Barolsky 1990).

Major financial and banking institutions such as Wesbank which specializes in vehicle finance have entered the taxi industry and introduced concessions to SABTA members. According to a SABTA–Wesbank agreement, taxi operators can get vehicles on a deposit of 20 per cent, instead of the normal credit requirement of 50 per cent. This was facilitated by SABTA providing Wesbank with adequate securities to allow a relaxation by Wesbank of their normal credit requirements. SABTA acts as a default guarantor and its members contribute only 20 per cent of the value of vehicles to SABTA–Wesbank Foundation, and then Wesbank finances the purchasing and insuring of minibuses. By September 1987 Wesbank had negotiated about 2,500 deals and this rose to 4,000 deals by the end of 1989 (McCaul 1990).

Wesbank has now financed nearly R100 million worth of minibus taxis using an adaptation of a traditional African system of fund raising called 'stokvel'. Although dealers report that minibuses are often bought for cash, this does not mean that there are big reserves of liquidity among individual owners. Rather small collective schemes along ancient stokvel tradition club together informally.

Another phenomenon is that of whites informally financing Black taxis through titular owners. A white entrepreneur would buy a taxi and let it to a Black operator for a weekly or monthly rental. SABTA claims that up to 60 per cent of taxis in operation are white-owned. However, recent research suggests that the actual figure does not exceed 25 per cent as whites do not necessarily own the vehicles but are heavily involved in financing taxis (Barolsky 1990; McCaul 1990).

There are essentially four major suppliers of minibuses to the Black taxi industry: Toyota, Nissan, Delta (formerly General Motors) and Volkswagen. So great was the demand in November 1988 that dealers, for example, ordered 660 sixteen-seater Hi-Ace minibuses but Toyota could only supply 228 vehicles. Prices for minibus taxis range from R38,000 to about R50,000 before tax. Manufacturers are jealous of their business with taxi associations, who buy between 300 and 350 sixteen-seaters from them each month. Special SABTA versions of minibuses are on offer from the manufacturers; six minibus manufacturers have so far produced 23 different minibus models on the South African market (McCaul 1990).

The Black taxi operation has developed into a successful business, creating opportunities for the development of small-business skills and employment. Until recently, the Black taxi industry was a backyard operation with minimal financial backing. By 1989, the industry had captured the single largest share of the Black commuter market which resulted in a dramatic reduction of the number of passengers using buses and trains (McCaul 1990).

Taxi operators have become a major consumer of petrol, for example, SABTA members alone consume about 10 per cent of all retail petrol in South Africa. Shell, Total, Castrol South Africa and Cera Oil South Africa have become partners with the organized taxi industry. At present SABTA owns some 20 petrol stations in Black townships and their establishment costs between R65,000 and R90,000 each (Barolsky 1989). Other oil and accessory companies, including Trek, Unipart, Goodyear, and many other suppliers of spares have negotiated some form of discount and kickback with SABTA. The involvement of big finance capital in the industry is indicative that capital accumulation in the taxi industry has not only taken root, but is likely to be part of the post-apartheid economy.

The marriage between big business and the taxi industry has not, however, been a bed of roses. The growth in economic power of Black operators has its dark side in terms of often exploitative labour practices. As Black taxi operators increasingly secure fleets of minibuses, they become taxi managers and employ workers in the capacity of drivers, cleaners, mechanics, taxi marshals, taxi receptionists, and accountants to name a few. These employees are often forced to work extremely long hours (between 16 and 18 hours) for between R80 and R150 per week (Matiko 1989).

In response to numerous exploitative labour practices, taxi drivers formed their own taxi union (in the Vaal triangle) and affiliated to the Congress of South African Trade Unions (COSATU) in 1988. However, it is telling that SABTA and SALDTA have not yet responded positively to this new development. Not only have they refused to register taxi drivers with the National Manpower Commission, but they have also refused to recognize the union and threatened to fire unionized taxi drivers. In fear of employing unionized drivers, there has emerged a general tendency among taxi owners to recruit their labour force from rural areas. Often rural drivers would be coming to urban areas for the first time with no other visible means of existence and they depend on backyard shelters provided by their taxi masters.

In the morning, owners would give drivers a target figure to bring in the evening. In an attempt to raise the target amount set by taxi owners, taxi drivers overload passengers and break legal speed limits, which put the lives of their passengers at risk. Statistics show that collisions and injuries involving Black minibus taxis have increased in recent years: accidents increased from 28,636 in 1985 to 47,952 in 1988. The number of recorded deaths rose from 252 in 1985 to some 679 in the first six months of 1989. However, more people die in bus accidents per 100 million km travelled than in minibuses: the figure for buses in 1988 was 15.4 deaths per 100 million km; that for minibuses was 11.7; that for all vehicles was 5.8 deaths per 100 million km travelled (McCaul 1990).

Daily intakes after cost deductions (petrol, oil, etc.) vary from R100 to R200 for short distances (e.g. between Johannesburg and Soweto) and between R300 and R700 for long distances (e.g between Johannesburg and Durban). Out of an average of R1,000 generated per week, a short-distance taxi driver would be paid about R100. Of some R5,000 income generated per week about R200 is paid to a long-distance taxi driver.

Having given a broad overview of the taxi industry nationally, attention now turns to the emergence, operation and growth of the taxi industry in South Africa's largest township.

The making of Soweto

Soweto is an acronym for South Western Townships, construction of which began in the early 1930s some 20 km south-west of Johannesburg. The apartheid policies enshrined in the Natives (Urban Areas) Act of 1923 and the Group Areas Act of 1950 forced Black people to the fringes of urban areas. As the establishment of the Black segregated townships rested crucially on the provision of the transport facilities, the Government paid enormous bus and train fare subsidies to facilitate the spatial reorganization of urban areas (Khosa 1988). In Johannesburg, several Black settlement areas were demolished and Black workers were separated from their workplaces and residents were forced to settle in the south of the city where it became increasingly necessary to commute.

In response to critical housing shortages in Soweto, a number of townships were established over the years and what is now known as Greater Soweto (Soweto, Diepmeadow and Dobsonville) is spread over an area of 100 square km. Because of government policy regarding allocation of housing and restrictions on the mobility and choice, the spatial distribution of Soweto resident tended (until recently) to follow ethnic rather than economic-status lines. With the abolition of influx control in 1986 and subsequent construction of shacks on some 15 to 20 per cent of the sites in Soweto, the population density which was 100 per square hectare in 1980 (as opposed to Johannesburg's 23) is by now about 150 per square hectare. The population in Soweto is estimated at 2 million people with general characteristics of low income and very low car ownership per household, thus making public transport as a means of commuting essential. Statistical estimates reveal that about 85 per cent of Soweto commuters travel to work by public transport. At present, Soweto is served by two main railway lines and a network of roads which focus on Johannesburg.

The Black taxi industry in Soweto

Until very recently, public transport consistently reminded Black people of their lack of representation in the white Government, lack of share in the capitalist economy and their exclusion from urban areas except when selling their labour power in the apartheid city. The dangerous, overcrowded and uncomfortable character of the public transport network provided by the apartheid state played into the hands of the Black taxi industry which was able to provide a safe, quick and efficient mode of transport.

Unlike taxis in the western world, Black taxis operate on a shared-cost system, with commuters paying a predetermined fare for the particular route they are travelling irrespective of where they boarded. Although they charge higher fares than buses and trains, the Black taxis have become popular and fast means of transport for urban commuters.

The predecessors of the modern Black taxi operators were the horse-drawn cabs which emerged in Johannesburg at the end of the nineteenth century (Van Onselen 1982). The first motorized taxi operator for Black commuters was established in the late 1920s. The number of both legal and 'illegal' taxis has grown over the years in Soweto; now some 7,000 legal taxis operate in the township and one-third serves between Soweto and Johannesburg.

The supply of taxis in Soweto rose from 1.4 permits per thousand people in 1980 to 3.5 in 1990. Essentially, the Black taxis provide for work trips for the residents of Soweto. During weekends, taxis provide other trip purposes such as shopping, recreation and social visits. As there is no public regulation controlling fares, local taxi associations set their fixed fares for particular routes. There is no reduction of fares to passengers boarding or alighting between any two extreme points of taxi ranks. The most important person on taxi ranks is the queue marshal (employed by local taxi associations) who functions as a fair administrator to give taxi drivers equal business opportunity, especially with the turn in loading of passengers at the rank.

The taxi industry is far from being homogeneous; there are fleet owners or proprietors, and employee drivers. Over the years, legal constraints restricted Blacks from increasing their taxi permits, however, some taxi operators managed to obtain more than one taxi permit through the registration of taxis on different names, by transfer of the public carrier permit or by duplicating taxi permits.

Because of their special advantages of high speed, reliability, safety, comfort and their ability to drop passengers closer to places of work, the Black-owned minibus taxis have eroded the market control held by white-owned transport operators (Table 15.1).

Table 15.1 Comparative modal percentage split by transport modes of travel in Johannesburg, Soweto, and nationally.

Mode	Johannesburg 1989	Soweto 1990	Nationally 1987	Nationally 1989
Train	41	41	29	24
Taxis	20	29	9	30
Bus	23	19	39	27
Car and other	16	11	23	19
Total	100	100	100	100

Sources: Collated from Khosa 1990 and Morris 1990. National figures indicate a shift of commuters from cars and other modes (e.g. lift clubs, cycling, employers transport) to using minibuses.

The Black taxis can be categorized into two types, zonal and magisterial permits. Although zonal taxis operate between any point within Soweto, magisterial district taxis operate between Soweto and Johannesburg. Zonal taxis compete with rail and bus services by operating parallel or along the same routes.

Taxi operation in the township is primarily administered by local government (e.g Soweto City Council, Johannesburg Local Road Transport Board, Traffic Department and the South African Police). There is no legal numerical limitation on the entry into the market or monopolistic franchise to serve particular facilities or jurisdiction. According to the Road Transportation Act of 1977, the success of the applicant depends on the applicant's justification for providing the service in particular where the area was already served by existing transport facilities. The applicant has to prove that the transport facilities are not satisfactory, prospective service will be for the public interest, the applicant belongs to the same race (as applied under the Separate Amenities Act – repealed in October 1990) as the prospective patrons, and has the ability to provide satisfactory transport. The difficulty in securing permits has led to permit sale racketeering and in several cases a taxi permit has been sold for over R2,500.

Various estimates indicate that between 45 and 60 per cent of the taxis operate outside the legal parameters; they are known as 'pirate taxis' (Barolsky 1990; McCaul 1990). Although 'pirate' taxis convey passengers for reward without satisfying all the legal requirements of taxi regulation, they charge the same fares as legal taxis. 'Pirate' operation is common at month-end and weekends, partly because of high demand for taxi services since a greater number of Black people visit relatives and families in the homelands. Undoubtedly, 'pirates' are perceived by both the local state and legal taxi operators as undesirable and unfair competition respectively. Whereas local authorities lose revenues from licensing, legal taxi operators suffer severe losses because of the decline in volumes of passengers who are captured by the 'underground' taxi movement. This kind of 'mafia' operation seemed to be organized in a manner which has over the years puzzled both legal operators and the state. Even police who usually form special 'pirate' squads to curb 'pirate' taxis have hardly succeeded in eliminating this burgeoning and sophisticated operation.

To complicate things further, agreements are usually entered into between legal and 'pirate' operators whereby legal taxis are allowed to load first. For this favour, queue marshals would be given financial rewards. The majority of taxi associations have established 'taxi squads' to police 'foreign' vehicles in their area of operation. When on duty, squad members are compensated by taxi associations. If 'illegal' taxis are found, passengers would be forced to alight and drivers of the 'pirate' taxis escorted to the nearest police station where charges would be made. If a 'pirate' driver refuses to cooperate, the vehicle could be damaged and physical violence could even take place. Conflict frequently flares up between taxi associations operating in similar areas or on similar routes. These 'taxi wars' are often violent, and frequently lead to death and destruction of property.

Taxi associations are organized so that, after a taxi operator has obtained all the legal documents, the taxi driver must then apply for membership at one of the ranks. Taxi associations control their own routes, ranks and fares. Each association considers the application and if they feel that there are adequate taxis at that rank, the applicant would be turned down. The unfortunate taxi operator could seek membership from other ranks or routes for the operation, but if found operating along routes where membership application had been unsuccessful, would be liable to aggression.

Eighty-seven formal taxi ranks and a further 50 informal taxi ranks were identified in Soweto in February 1990. Baragwanath, White City and Merafe station taxi ranks are the busiest throughout the day. Because of the nature of the route system, passengers living in the western part of Soweto usually have to use two taxis to get into town. In a typical weekday, some 4,000 taxis with passengers arrive and 2,000 taxis loaded with passengers leave Baragwanath taxi rank (the largest in Soweto). A considerable number of passengers (about 7,000) arrive by taxis from various parts of Soweto and transfer to buses leaving for destinations in Johannesburg. A large number of passengers (nearly 8,000) also arrive by taxis from Johannesburg and Soweto for destinations in Soweto. About 4,000 taxi passengers work in or visit Baragwanath Hospital daily. At Baragwanath complex, over 100 hawkers and small shopkeepers have established their businesses as a result of these passenger movements.

The 1980s witnessed a meteoric rise in the number of taxis in Soweto, from 1,000 legal taxis in 1980 to 7,000 in 1990, and today the taxi industry has become a powerful force in the township economy. This phenomenal growth coincided with changes of government policy from stringent political and economic repression of Blacks to encouraging small-business development through deregulation and privatization policies.

Concluding remarks

The residential segregation of South Africa's urban landscape and the removal of Black townships to the urban fringes paradoxically re-created the conditions for the development of a dynamic taxi industry. Black taxis have come to play a pivotal role in the reproduction of the labour power in Soweto and Johannesburg. The late 1970s and the early 1980s witnessed a relentless effort of the state in harnessing the small Black business in order to foster the growth of a supportive Black petty bourgeoisie which might exert a stabilizing conservative influence in the face of the radical politicization in urban townships. Arguably, the state has become capital's crusading disciple in promoting the development of a free market economy, thus the Black taxi business has become a test case in Black participation in the capitalist economy. The recent state policy has unambiguously come to serve not only the dominant classes in society, but also to ensure that the present capitalist economy survives in the post-apartheid society.

References

Barolsky, J.B. 1989. *Black Taxi Case Study*. Parktown: Wits Business School.

Barolsky, J.B. 1990. Follow that taxi: success story of the informal sector. *Indicator South Africa* 7: 59–63.

Beavon, K.S.O. and Rogerson, C.M. 1990. Temporary trading for temporary people: the making of hawking in Soweto. In D. Drakakis-Smith (ed.), *Economic Development and Urbanization in Developing Areas*, 263–86. London: Routledge.

Khosa, M.M. 1988. 'Black' bus subsidies in 'white' South Africa, 1944–86. Masters thesis, University of the Witwatersrand, Johannesburg.

Khosa, M.M. 1990. The Black taxi revolution. In N. Nattrass and E. Arbington (eds), *The Political Economy of South Africa*, pp. 207–16. Cape Town: Oxford University Press.

Matiko, J. 1989. Taxis – conflict between drivers and owners, *South African Labour Bulletin* 14: 77–83.

McCaul, C. 1990. *No Easy Ride: The Rise of the Black Taxi Industry*. Johannesburg: South African Institute of Race Relations.

Morris, N. 1990. *Black Travel Today: Some Facts and Figures*. Pretoria: National Black Panel.

Rogerson, C.M. 1988. The underdevelopment of the informal sector: street hawking in Johannesburg, South Africa, *Urban Geography* 9: 549–67.

South Africa 1983. *Final Report of the Commission of Inquiry into Bus Passenger Transportation in the Republic of South Africa*. (Welgemoed Commission.) Pretoria: Government Printer.

South Africa 1987. *White Paper on National Transport Policy*. Pretoria: Government Printer.

Van Onselen, C. 1982. *Studies in the Social and Economic History of the Witwatersrand, 1886–1914, New Babylon and New Niniveh*. Johannesburg: Ravan.

16 *The Regional Services Council debacle in Durban*

UDESHTRA PILLAY

Even before the Regional Services Council Act, 1985 (Act 109 of 1985) became law on 31 July, 1985, it was confidently expected both at central government and local level that the first Regional Services Council to be established under the Act would be in Natal, more specifically in the Greater Durban Area. This was due to the high degree of acceptance of the principle of regionalization already reached amongst local authorities and other representative bodies in the area, including commerce and industry, and also because considerable progress had already been made towards the establishment of a regional authority for a metropolitan area long regarded as displaying a high degree of economic bonding and interdependence. As things have turned out however, this has not been the case.

(METROCOM memorandum accompanying letter to the State President, 19/6/87).

The historical development of regionalization of services initiatives in the Durban metropolitan area

The province of Natal has a population of some 6,301,000 people, of whom 5,049,900 are classified as African (more than 90 per cent being Zulu-speaking), 496,560 as white, 664,360 as Indian and 90,180 as coloured. Within this political entity, there are a number of subdivisions. The most important is between the KwaZulu Legislative Assembly and the Natal Provincial Administration. At the local level, divisions include magisterial boundaries and areas demarcated for local administration such as municipalities and Tribal or Regional Authorities as well as the racial and ethnic patterning of apartheid (see Fig. 16.1). The functional interdependence of the various politico-administrative components of Natal's metropolitan areas has been recognized by a variety of groups over a considerable period of time. A guide plan for the Durban Metropolitan Area (DMA), published by the Natal Town and Regional Planning Commission in 1973 argued that the future growth of Greater Durban (i.e. the DMA) was intimately bound up with that of its surrounding region and that a holistic approach in the formulation of a development policy was therefore required. It served as the catalyst for the development of the Durban Metropolitan Consultative Committee (METROCOM). METROCOM was formed in January 1977 at the instigation of the Natal Provincial Administration, for the purpose of achieving cooperation between all local authorities within the DMA and a coordinated approach to the economical and efficient planning. METROCOM was dominated by debates on the relative priorities of white local authorities in the DMA, as well as matters of inter-government liaison in the provision of services.

In November 1979, the Natal Provincial Administration requested the Consultative Committee to undertake the task of examining the concept of regionalization of services for the area. According to directives, the study was to include services to the adjacent parts of the KwaZulu 'self-governing homeland' and other areas of African residence under the control of Administration Boards. In early November 1980, following some preliminary investigative work, METROCOM directed that for the purpose of the investigation into the regionalization of services in the DMA, and as there can be no artificial boundaries in the supply of essential services, the area of jurisdiction of the body be the complete area as defined by: (a) the Durban Metropolitan Transport Advisory Board; and (b) the Natal Town and Regional Planning Commission in respect of the DMA. These boundaries extend from Illovo in the south to Umhlanga in the north to Hillcrest in the west (Fig. 16.1). A memorandum dealing with the proposed regionalization of services was subsequently submitted to the Provincial Secretary in November 1981, after exhaustive consideration of the matter.

However, it was not until late October 1984 that a comprehensive report on the financial implications of the regionalization of services in the DMA was released. The study, undertaken by Pim Goldby Management Consultants, was commissioned by METROCOM. It was the first such report in South Africa and served as a blueprint for other metropolitan areas estimating the cost of establishing certain municipal services on a regional basis. The area proposed for the regionalization of services in the report was the planning boundary as depicted in the Metropolitan Durban draft guide plan (identical to the 1980 METROCOM jurisdiction for regionalization of services in the DMA; Fig. 16.1).

According to the report, substantial savings would occur if minimum standards of services were provided by a regional authority. There was cautious optimism on the part of most of the participating local authorities in the DMA at the proposals. Many cited as extremely promising the fact that, unlike the many major planning moves affecting the lives of local

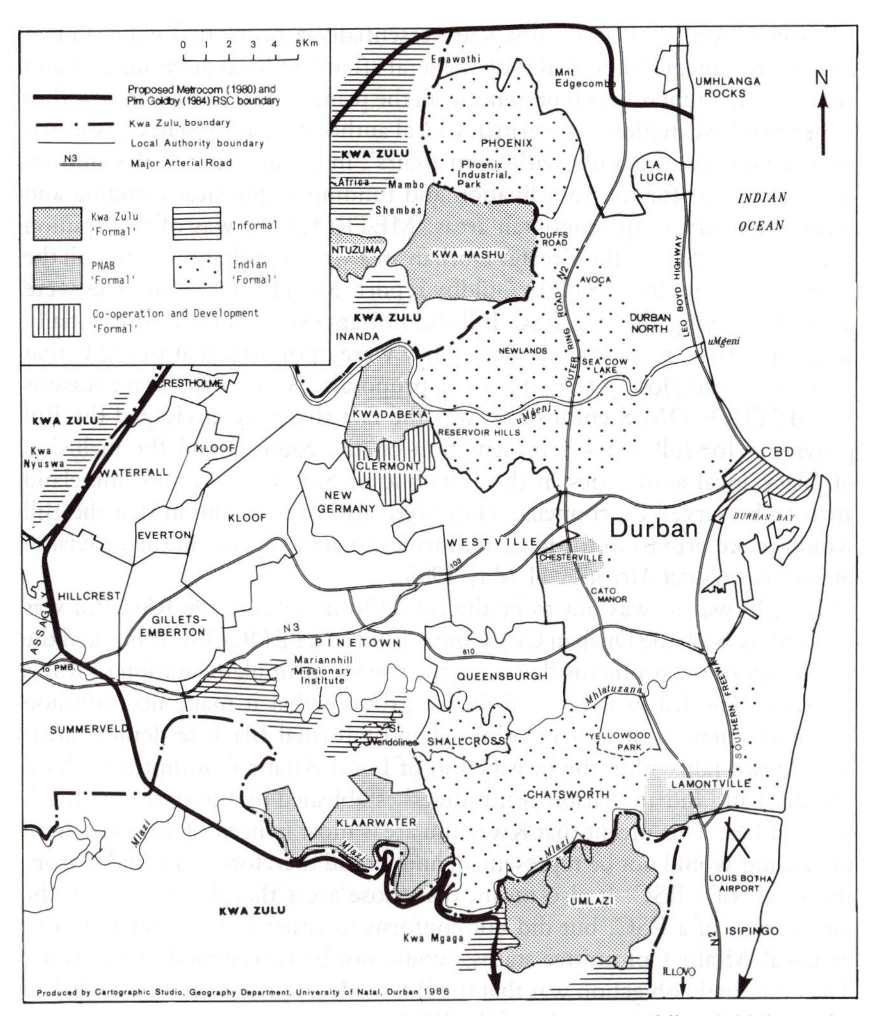

Figure 16.1 Proposed METROCOM (1980) and Pim Goldby (1984) service boundaries for the Natal/KwaZulu region, with other administrative boundaries.

people, the plan was not drawn up in remote national government offices in Pretoria, but by local people who best understood Durban's needs and who would be able to live with the results.

Introduction of the Regional Services Council (RSC) Bill—initial reaction

The government-drafted Regional Services Council Bill was introduced in Parliament in May 1985 to allow for the establishment of Regional

Services Councils (RSCs). RSCs are essentially a horizontal extension of local government. They deal with 'general affairs' of local governments and are made up of members representing all the primary local authorities within a specified RSC region. The primary local authorities, at present constituted on a racial basis, deal only with 'own affairs'. In broad terms, RSCs amount to mechanisms for the coordination and funding of physical planning and service delivery in metropolitan areas. METROCOM was of the opinion that the new RSC for the Greater Durban area would probably include all the areas covered by the 1984 Pim Goldby report. METROCOM sources were quoted as saying that the RSC Bill did not appear to deviate substantially from their basic recommendations. They were optimistic that the RSC that evolved would closely resemble their proposals. Among the main reasons for METROCOM'S enthusiasm was the fact that they envisaged the Bill providing for full Black representation on the councils and the inclusion of Black local authorities in the Republic of South Africa and homeland areas to be served by councils. They were also of the opinion that the Bill would make provision for local authorities to act as agents in the provision of services (*Natal Mercury*, 11 May 1985).

This, however, was not to be the case. On its release, the RSC Bill was rejected by both the Durban City Council and METROCOM. It marked the beginning of the problems that were to arise in the implementation of RSCs in Natal. The Bill was rejected on the grounds that it made no provision for representation of Indian, coloured and informal Black settlement areas that do not fall within the jurisdiction of Local Affairs Committees – local coloured and Indian 'liaison committees' established by the state – or Black local bodies. This meant that KwaZulu areas most in need of infrastructural upgrading would not be represented, and would therefore be excluded from the ambit of a RSC. It also meant that those areas that did fall within the jurisdiction of a RSC, but did not conform to either Black local authority or Local Affairs Committee status, would not be represented on the RSC. The other main objection was that the Bill made no provision for established corporations to act as agents of the RSC.

METROCOM made strong representations to the Parliamentary Standing Committee on constitutional affairs. In view of basic decisions made in 1981, METROCOM was totally opposed to the introduction of a RSC which would exclude areas in KwaZulu (*Natal Mercury*, 6 June 1985). Subsequently, in August 1985, METROCOM submitted a further memorandum to the Provincial Administration updating the original 1981 submissions in accordance with the RSC Act and highlighting the financial conclusions reached as a result of the 1984 report of the consultants. This memorandum was adopted by, among others, the KwaZulu cabinet.

It seems somewhat ironic that METROCOM should have become engaged in a confrontationist stance with central government on the issue of RSCs, given that METROCOM was instigated by the Natal Provincial

Administration, and provincial government in South Africa generally owes allegiance to central government. However, Natal provincial politics, up until recently, has never fallen under the hegemony of the ruling National Party. It would also seem to be the case that the central government put the brakes on METROCOM in fear of its own local government restructuring programme being overtaken by local initiatives (although it has been argued that METROCOM had played a significant role in actually shaping the RSC Bill). As the *Financial Mail* (26 October 1984) has pointed out, sentiment around Pretoria during the course of these debates was that METROCOM, a purely Natal initiative, was attempting to pre-empt central remodelling of the lower levels of government.

In the Durban case at least, initial government pronouncements of maximum devolution of power and minimum administrative control at the local level in its new ostensibly democratic system of local government seems to have been tainted from the very start. As Cameron (1988) has concluded, government reform (including local) was to take place solely within the parameters set by the government itself. There would be concessions, but it would not be dictated to.

Ambivalence, acceptance and opposition

Organized business and commerce Organized business and commerce in the DMA voiced their objections to RSCs, although not to regionalization of services initiatives in general. In fact, RSCs met their first stiff opposition throughout the country in the form of local business coalitions. Business groups in Durban hit out at the tax regulation implications of the Bill, arguing then as they continue to do now, that the method of funding RSCs – levying extra taxes on local commerce and industry – would increase inflation and put many people out of work. Spokesmen for the Durban Metropolitan Chamber of Commerce and the Natal Chamber of Industries argued that the taxes would further depress industry and commerce, lead to higher inflation, higher unemployment and more businesses going under (*Natal Mercury*, 10 May 1985). These sentiments were shared and endorsed by other business sectors and coalitions throughout the country.

There was sufficient ambivalence among ruling urban elites about RSCs in Natal by the mid 1980s to allow an initial acceptance of the structures by late 1985, at least among white local government officials and councillors (substantive amendments to the RSC Bill had much to do with this). The more conservative Indian politicians were similarly well disposed given their generally cliental relationship under reforms sponsored by the National Party (McCarthy 1988). Nationally as well, while the Bill was meeting with resistance on a wide front, it was likely to receive the necessary support from government's junior 'coalition' partners in the (coloured) House of Representatives and the (Indian) House of Delegates (*Cape Times*, 10 June 1985). On the other hand, some urban elites, given their generally negative disposition to any constitutional reforms sponsored by the National Party, began to favour other forms of local and regional government.

Populist responses to RSCs—the emergence of Inkatha as a key player It was around this time that the two major populist political groupings in Natal, the United Democratic Front (UDF) and Inkatha expressed dissatisfaction with RSCs. The UDF is a mass-based nation-wide extra-parliamentary opposition grouping formed in late 1983 in a 'Charterist' tradition and closely allied with the African National Congress (ANC). Inkatha, by contrast, is a mass political and cultural organization formed in KwaZulu in 1975 to increase participation in the homeland system, although its leaders have national aspirations as well. Chief Gatsha Buthelezi, who is both President of Inkatha and the Chief Minister of KwaZulu, has cleverly used his base as the leader of a traditional clan and a member of the Zulu royal family to build a strong political following. Members of the Mass Democratic Movement (MDM) in South Africa regard him as a government collaborator and political stooge. Despite their political differences, both organizations have rejected RSCs on the grounds that there has been, firstly, a lack of consultation with them in the formulation of the RSC Act, secondly, a bias towards the wealthy (largely white) areas in the allocation of voting powers and, thirdly, an effective entrenchment of racial segregation at the local authority level from which RSC representatives are to be nominated.

It is commonly held that Inkatha has been the chief protagonist in the RSC debacle in Durban. Friedman (1988), for instance, contends that the main reason RSCs have not been introduced in Natal is because Inkatha is opposed to them and it is they who control African local authorities in the area. It has also been argued that Inkatha's opposition to RSCs results primarily from the fact that they were at no time party to discussions which resulted in the formulation of the RSC Act. Indeed, this has always been a contentious point, and has, in no uncertain manner, contributed to the impasse in the region. However the reason why the KwaZulu government and Inkatha have for so long rejected RSCs, and delayed their implementation in the region beyond expectation, is more deeply rooted in the region's history and geography, as will be explained below. The point must be made at the outset, though, that Inkatha has never been hostile to the concept of regionalization of services in the Greater Durban area. As it has already been noted, the KwaZulu cabinet in fact adopted the 1985 METROCOM proposals.

The specificity (in spatial and historical terms) of the Natal/KwaZulu region has been the subject of much debate and a small, but significant, amount of scholarly work in recent years. It is due to the economically and demographically interwoven sub-regions of Natal and KwaZulu, coupled with a climate of opposition from powerful Natal-based interests at Pretoria's insistence on KwaZulu taking independence, that certain calls were heard, from 1977 onwards, for a special Natal dispensation in terms of local government structures. Inkatha was intimately involved in these calls given that the various elements of the pro-Inkatha elite are dependent on the KwaZulu bantustan, and their horizons – the rhetoric of their national political aspirations notwithstanding – rarely extend beyond it. Over the years, Inkatha has increasingly come to represent the linking of the regional representatives of an African middle class and the interests of monopoly capital that perceived a base for political reform residing in such an alliance.

Since the late 1970s, therefore, there have been at least two commissions and other less formal calls for a separate Natal/KwaZulu option regarding local government. These calls essentially boiled down to two recurrent demands. Firstly, they demanded that Pretoria devolve a measure of authority to Natal, giving 'moderate' forces in the region licence to pursue their own regional political solutions. Secondly, they demanded that white Natal and KwaZulu be allowed to collaborate, and set up joint political and administrative structures in the search for a politically credible, economically viable alternative to Pretoria's apartheid programme for Natal (Glaser 1986). There was also an implicit attempt in these calls to engage those propagating more radical sociopolitical models for South Africa in serious ideological competition.

Interestingly, the government-drafted RSC Bill was introduced in Parliament in May 1985 in the wake of initiatives by a growing reformist consensus (embracing the former opposition, the Progressive Federal Party (PFP), conservative coloured and Indian parties, Inkatha and prominent businessmen) in favour of a special Natal dispensation, a consensus which ultimately found expression as the 'KwaZulu/Natal Indaba' ('Indaba' is a Zulu term meaning a meeting or discussion). In fact, the term 'KwaZulu/Natal Indaba' was coined by erstwhile Progressive Federal Party leader Van Zyl Slabbert at the very same time as the RSC Bill was being introduced in Parliament (Cameron 1988). He used the term to describe the process of working towards a joint legislature for the Natal/KwaZulu area as the last phase of initiatives that had started with the Ulundi Accord in November 1984. The Ulundi Accord was signed between the KwaZulu government and the Natal Provincial Council (then dominated by the former NRP, a national white party with its base in Natal) to allow for direct negotiation and cooperation between Natal politicians and KwaZulu.

It seems that Inkatha (one of the major promoters of the Indaba) felt that its initiatives at consolidating the region administratively and politically, which involved years of hard work and cooperation, and which were finally to find expression in April 1986 as the KwaZulu/Natal Indaba, were now being pre-empted and superseded by government created RSCs (over which it had not even been consulted) scheduled to begin at around the same time. As KwaZulu Minister of Education, Culture and Local Government, Oscar Dhlomo has pointed out:

> We have coexisted in this region to a far greater degree than all the other provinces. One needs to look at the political developments in the region since the early 1980s. If you study the political developments, you would see conditions for closer cooperation between the regions and even in fact unity in the region. So it would not have been possible for the government to forcibly introduce any regional structure without direct consultation because that would be working against the tradition of cooperation in the region

He went on to add:

> . . . apart from attempting to pre-empt (our initiatives), the government was trying to stifle and kill off any trend towards co-operation by bringing in the whole thing [RSCs] using methods that we were totally opposed to in the region. Perhaps if it was only pre-empting, we would have tolerated it because we don't mind pre-empting if it is in line with what we ourselves are interested in. I think it was more the problem that the government was trying to override and destroy any initiatives that came from the people themselves and, in other words, trying to discourage consultation and cooperation across the political boundaries. (O. Dhlomo, personal communication, 24 November 1989)

There is a sense therefore, implicitly at least, that Inkatha (as well as other major sponsors of the Indaba) viewed the government's insistence on going ahead with RSCs as insensitive and ill-timed. There was therefore no way that Buthelezi was simply going to sit back and succumb to government pressure to participate in the establishment of these bodies.

While not attempting to reduce the Indaba initiative to a straightforward competition with RSCs, it must be pointed out that both initiatives address similar concerns. Interestingly though, McCarthy (1988) points out that both the timing of the Indaba (beginning at much the same time as the anticipated formation of the Durban RSC) and the alliance of the major promoters (PFP/NRP/Inkatha) appear to be more than coincidentally related to the RSC initiative in Natal. With hindsight, it will be recalled that it was in the period following Inkatha's initial rejection of RSCs that the KwaZulu/Natal Indaba emerged as an alternative concept of regional government contra RSCs, but not necessarily in direct competition with it.

In mid 1987, the government announced its firm rejection of the Indaba proposals (after hardly considering them) on the grounds that they made insufficient provision for the protection of group rights. This was a curious avenue of criticism given that RSCs themselves make no explicit provision for group right protection, but the Indaba proposals threatened the group-based distinction between 'general' and 'own' affairs on which the government's reform initiative depended. It set the scene for a scathing attack of the RSC system by Buthelezi. The fact that Inkatha was one of the major sponsors of the Indaba, coupled with the manner in which the proposals were summarily dismissed, and the way in which the new Provincial Executive constantly berated dissident groups – business, the KwaZulu Legislative Assembly and town councils – for their perceived reluctance in implementing RSCs, has had much to do with this. Fearing nothing to lose, but still guarding the credibility of the Indaba proposals cautiously, Inkatha was to adopt a totally uncompromising stand on the RSC issue.

RSCs were deemed part of a prescriptive politics that the government was trying to 'ram down people's throats' (*Natal Mercury*, 29 July 1987) and were 'introduced into legislation by a White Parliament that totally excludes us and they were being applied elsewhere by the hideous politics of prescription' (*Natal Mercury*, 3 August 1987). These were sentiments expressed by Buthelezi throughout addresses and press releases in the 1986–87 period.

Recent developments

By late 1987, RSCs had still not been implemented in the region. At around the same time, a Joint Executive Authority for Natal and KwaZulu was established through the introduction of the Joint Executive Authority Bill in Parliament. The Bill allows for services and planning to be undertaken by a joint body of KwaZulu and the now-appointed provincial government. The Joint Executive Authority arose out of the 1981 Good Hope Conference where the National Party addressed related problems of interregional coordination between homelands and provinces. The concept of 'multilateral development coordination', at an executive level, subsequently became the favoured mechanism for resolving such problems. Given its timing, a number of interpretations hold on the formation of the Joint Executive Authority. The first is that the Authority is the government's attempted compromise with the Indaba initiative. Another is that the National Party hopes to use the Joint Executive Authority to get KwaZulu to cooperate with RSCs (McCarthy 1988). On reflection, the latter interpretation seems to be a fairly accurate one; very recent developments seem to suggest that the Joint Executive Authority has played a significant role in resolving the RSC impasse in the region.

In early 1988, the Natal Provincial Administration Special Committee established for the purpose of formulating boundary proposals (urban regions) identified four regions as a basis for considering the establishment of RSCs in Natal: the Durban Metropolitan region, the Pietermaritzburg region, the Upper South Coast region and the Lower South Coast region (NPA Special Committee Report, 1988). All KwaZulu areas were excluded from suggested boundaries. This arose from a legal opinion furnished to the Provincial Secretary by the senior state law adviser attached to the Natal Provincial Administration. It argued that because the National States Constitution Act of 1971 (under which a homeland is declared a self-governing or independent territory) precludes it, no areas of KwaZulu can be included within the domains of these RSCs without the KwaZulu Legislative Assembly's permission.

Both METROCOM and the KwaZulu Legislative Assembly rejected these recommendations. METROCOM reiterated and endorsed earlier sentiments that the exclusion of those portions of KwaZulu which are inextricably part of what is a cohesive and integrated economic community will have serious implications for the provision of services on a regional basis in terms of the Act. This would defeat long-term objectives. Buthelezi urged Blacks to have nothing to do with RSCs, warning them that 'RSCs will burn all the fingers that touch them.' (*Daily News*, 12 January 1989).

The Durban City Council and the Durban Metropolitan Chamber of Commerce also expressed concern at KwaZulu's exclusion. The Chamber of Commerce argued that RSCs in Natal should be deferred until KwaZulu participated as it was immoral that KwaZulu's citizens should contribute financially, yet receive no benefits (*Daily News*, 21 November 1988).

In May 1989 the Natal Provincial Administration received the recommendations of the Demarcation Board (which can, at the request of a Provincial Administrator, hold an enquiry and advise him on the desirability or otherwise of demarcating, altering or withdrawing the demarcation of the area of jurisdiction of a local authority). Peter Miller of Natal Provincial Administration, subsequently announced that RSCs would begin operating in Natal by October (*Natal Mercury*, 9 May 1989). The final boundaries for the RSCs were very similar to the Natal Provincial Administration Special Committee recommendations. All KwaZulu areas were excluded. Miller's announcement, especially in the light of the formation of the Joint Committee of the South African and KwaZulu governments in late 1988, took many by surprise. According to the Natal leader of the National Party Stoffel Botha, this Committee was set up to 'identify and address obstacles impeding the process of negotiation between South Africa and KwaZulu' (*Daily News*, 27 February 1989). The sudden urgency to go ahead with RSCs therefore seemed anomalous at the time.

The announcement did not meet with much enthusiasm. Even the National Party chairman of the Durban City Council management committee voiced disapproval (he favoured a Joint Executive Authority system for the region). Buthelezi attacked Miller on the grounds that the Joint Executive Authority was at the time finalizing an alternative constitutional structure for the regionalization of services. If acceptable to central government, it would enable KwaZulu to cooperate. Notwithstanding the disapproval voiced at the prospect of having RSCs operational in Natal by October, the Natal Provincial Administration remained adamant. Four years after the RSC Act had been passed, three and a half years after a RSC should have become operational in Durban (if METROCOM had anything to do with it), and two years after their implementation in the rest of the country, it seemed as if RSCs were finally to become operational in the Natal region. Even the chairmen were appointed and the consultants commissioned.

However, almost as one had come to expect in the business of setting up a RSC in the region, events yet again took another turn. The Provincial Executive announced, in early June 1989, the likely scrapping of RSCs in favour of a system that would include KwaZulu. In what was reported as a major breakthrough, 'an ad hoc committee of the Joint Executive Authority has reached agreement on a draft Bill which would make it possible for municipal services to be provided on a joint, coordinated and regionalized basis in Natal and KwaZulu' (*Natal Mercury*, 3 June 1989). Importantly, this announcement came shortly after a meeting between Chief Minister Buthelezi and leader of the National Party (although not yet State President) F. W. de Klerk under the auspices of the Joint Committee. It was subsequently learned that the RSC issue was extensively debated at that meeting (O. Dhlomo, personal communication, 30 November 1989).

Joint Services Boards are now to replace RSCs in the Natal/KwaZulu region in terms of the Natal and KwaZulu Joint Services Bill published in Parliament in April 1990. The provision of all essential services in the region will now fall under joint Natal and KwaZulu control (*Daily News*, 19 April 1990). Their brief will be, as a matter of priority, to establish, maintain and improve infrastructural services in areas of greatest need. The Bill brings the Natal Provincial Administration and the KwaZulu government closer together administratively, as members of the Joint Services Boards will be drawn from local government bodies in both areas. It also gives an increased importance to the Joint Executive Authorities which will now be responsible for defining the regions, establishing the Joint Services Boards and administering them (ibid).

Conclusion

An explanation of the impasse that has been created over the establishment of RSCs in the Natal/KwaZulu region invites an examination of a complex set of relations and developments, political and economic, within and beyond the Natal/KwaZulu region. Two main reasons have been advanced for the delay in the establishment of RSCs in Natal. Firstly, there was a great deal of innovation from key actors in the DMA, particularly local state agents and organized business and commerce, spanning almost ten years from the late 1970s, towards regionalizing services. These initiatives, which had pre-empted government thinking on the matter, were eventually superseded by government-created RSCs. This situation gave rise to a profound change in central–local relations which was to impact considerably on the process of establishing RSCs in Natal. Secondly, the regional level (the different state forms, the distinctiveness of Natal/KwaZulu, political developments and the accompanying changing political dynamic) inserted itself between the local and central state levels (during attempts to implement RSCs in the DMA) in a way which had not been the case in any of the other major metropolitan regions in South Africa. It was a significant development, to which the ensuing debacle bears testimony.

The point has often been made (e.g. in Chapter 2 of this volume) that there is a new appreciation of local politics in South Africa, and that from all over the political spectrum comes the suggestion: in whatever direction future political developments may go, sub-national institutions seem destined to play a significant role. This being the case, future research documenting contemporary struggles over the spatial redelimitation of state powers in South Africa would do well to take a lesson from the Natal experience if the parameters (in political, economic and spatial terms) of these sub-national institutions are to be clearly defined, and the required theorization and conceptualization of this research problematic is to be sound.

References

Cameron, R. 1988. The institutional parameters of local government restructuring in South Africa, in C. Heymans and G. Totermeyer (eds), *Government By The People*. Johannesburg: Juta.

Friedman, S. 1988. Implementing the RSC concept. *Reality*: special double issue on Local Government.

Glaser, D. 1986. Behind the Indaba: the making of the KwaNatal option. *Transformation* 2: 4–29.

McCarthy, J.J. 1988. The last metropolis: RSC stalemate, Indaba checkmate? *Indicator South Africa* 5(4): 45–8.

METROCOM Memoranda, Minutes and Letters on Regionalization of Services in the Durban Metropolitan Area, November 1981 to January 1988.

METROCOM 1984. *Investigation into Regionalization of Services*. Durban: Pim Goldby Management Services.

Natal Town and Regional Planning Commission 1974. *Metropolitan Draft Guide Plan*. Pietermaritzburg: Natal Town and Regional Planning Commission.

Pillay, U. 1989. A critical appraisal of Regional Services Councils with a preliminary insight into the RSC issue in Natal. Working Paper, Community Research Unit, Durban.

17 *Tourism and development needs in the Durban region*

L83 R12

LINDA J. GRANT &
JOHN BUTLER-ADAM

South Africa

Introduction

In the eyes of white explorers and settlers and their descendants, the area around Durban has been synonymous with travel and tourism for a period which began some time before the origins of the city itself. Early visitors to the area recorded in their diaries comments on the beauty of the sun-washed coast, where a subtropical forest of banana palms, cotton-woods and coastal bush swept down to broad sandy beaches, clear lagoons and a sparkling sea. The development of the small colonial outpost of D'Urban in the mid nineteenth century served, in the long run, to heighten the attraction of the area by offering the possibility of a tourist experience which combined the natural beauties and opportunities of the coast with the advantages of a growing urban area.

Since then, and in particular since the 1920s, Durban has become South Africa's major tourist and recreation destination, expressing in its tourist areas the broader political, economic, social and cultural events and changes through which the country at large has passed. Yet while tourism is a major source of employment and income for the city and its region, there remains much which the industry needs to achieve if it is to offer holiday and related opportunities to more South Africans, and if it is to play a more constructive role in the economic and social development of the region. It is to the issues of tourism (as a major element of the regional economy) and development (as a major regional need) that this chapter is devoted.

We start by describing the socio-economic circumstances of the Durban Functional Region. In the context of this information, we offer an introduction to tourism, development and Durban, followed by a brief perspective on the historical growth of tourism in the region and its apartheid and structural character. In the remaining sections, tourism is evaluated as an agent for development, and some suggestions are offered for the future planning and application of tourism in the Durban Functional Region.

The socio-economic context

Durban is generally acknowledged to be amongst the fastest growing cities in the world. In 1970, its population was about 1 million; in 1980 the figure had doubled to 2 million and the population is now estimated to be of the order of 3.4 million. By 2000 it will have reached 4.5 million, largely as a result of internal population increases rather than in-migration from rural areas. This pattern of rapid growth has given to the Functional Region much of its present socio-economic character – a character which is of considerable relevance to the relationships between tourism and urban development.

Perhaps the most important facet of the population is not its mere size but its age structure. In 1990, 44 per cent of the residents of the Functional Region were under 18 years old and over 50 per cent were under 20 years old. This youthful structure has considerable implications for the region's needs in the areas of economic growth, education, employment, housing and recreational opportunities for the next two generations – needs which will be compounded upon starting positions which are already far from being satisfactory or even adequate. If there were no other reason, the age structure of the population of the region would serve on its own to indicate the critical need for one of the region's major income-earners to play an essential role in regional development.

There are, however, clear circumstances which demand a stronger relationship between tourism and development. In the area of education, classroom ratios and per capita expenditures on the education of black children serve to exacerbate other educational hurdles, with the result that failure rates amongst those who write school-leaving and tertiary institution entrance examinations are excessively high. In addition, about 2 million of the 2.3 million African residents of the Durban Functional Region have less than a Standard Ten level of education.

Along with very poor educational opportunities, the rate of population growth, in relation to the presently low rate of economic growth (about 1 per cent as opposed to the needed 5 per cent), means that unemployment is already high and that it will continue to grow in the future. Some researchers have placed the unemployment rate in the Durban Functional Region as high as 50 per cent of the employable population (which excludes the people under 18 years of age and who have yet to enter the labour force).

Of the present population, more than 75 per cent live in townships or in 'informal' areas. Two points relevant to this chapter need to be made about living conditions in such areas. The first is that there is a substantial shortage of housing so that occupancy rates in existing dwellings, whether formal or informal, are very high. The second is that within both township and informal areas facilities for recreation and tourism are almost non-existent, despite the intention of the policy of grand apartheid to provide 'separate but equal' facilities for all people 'in their own areas'. The implications are that housing, like education, is a high priority in the development of the region, and that domestic living conditions and environments at both household and neighbourhood levels are of a very low order. Not only direct development in these areas, but also the provision of tourism and recreation opportunities throughout the Durban Functional Region are a high priority.

Major economic and social problems and differences characterize the Durban Functional Region, and these have distinct spatial forms created by the laws of petty and grand apartheid as well as by structural processes at work within the South African economy. The notion of two cities in one has been invoked (not altogether accurately) to describe the situation because apartheid laws have served to give some truth to that metaphor.

In the case of Durban, it has been observed that a 'third city' might be added – the tourist city which stretches along the beachfront and into the city centre and on to some suburban shopping centres, and which boasts an international character of considerable wealth and comfort. This 'third city' is used by the residents of the 'first city' and at times, amidst scenes of low- or high-profile conflict, by the residents of the 'second city'. Although it is not an altogether useful metaphor for the nature of the Durban Functional Region, the 'three cities' notion at least serves to highlight problems of contrast and conflict – problems which are at present heightened, but which could contrariwise actively be lessened, by tourism development in the region.

The circumstances described here provide an insight into some of the more pressing issues which face Durban. Many of the conditions outlined are the product of apartheid legislation of various kinds, but many are equally the product of class-related considerations which have become part of the race–class structure of the South African space-economy and space-polity. In order to address the structural problems and inequities and those of their manifestations which have been set out above, the region's greatest need is for the coordinated planning of development strategies and projects. It is in this context that the role of tourism takes on special significance.

Tourism, development and Durban

The complexity of everyday life in the Durban Functional Region makes for difficult planning and development. To consider the phenomenon of tourism in the context of such a contradictory and fragmented landscape is, many would argue, to deal in trivia and merely to compound the problems facing planners. Tourism is, after all, closely associated with 'having a good time' and so a temptation exists to consider it a lesser social issue. Instead, we believe that tourism (and recreation) are *essential* aspects of society. Any meaningful discussion of them involves consideration of a wide range of fundamental social, economic and political issues, including those outlined earlier.

Such sensitivity and insight are crucial, not least because tourism is the second largest industry in Durban. It is very much a 'consumer activity' (Butler-Adam 1988) with the driving force of incentive being for both purveyors of tourism and tourists to 'make a profit at all costs'. Little respect for the resources and needs of the vast majority of the local inhabitants is shown, resulting in tourism being labelled a potential trap – or 'tainted honey' (Britton and Clarke 1987). Planning for tourism development therefore requires both rigorous *analysis* and acute *understanding*. Only strategic planning based on such an approach will ensure that tourism and recreation in the Durban Functional Region are *appropriately* developed. This in turn will facilitate the role which tourism and recreation can play in unravelling the complexities of the region. Their role is crucial, for example, in aiding racial interaction in a society long perverted by the policies of apartheid. And such appropriate development is vital, particularly in the current process of social reorganization.

To assist in providing an analysis and understanding of the role of tourism in Durban, both now and as an agent for development in the future, the remainder of this chapter sets out the historical reality of tourism in the South African context; the problems and ambiguities that almost always come up when pursuing tourism as an agent for development in Third World countries; and an alternate, yet sustainable (De Kadt 1990), role which tourism can play in the Durban Functional Region's future development.

Understanding tourism in the South African Context

'Disposable income, occupation, taste, patterns of association and other factors of class culture are the factors which shape the tourism and recreation landscape in most countries in the world.' (Butler-Adam 1988: 1). These variables, however, have been considerably (although not completely) influenced in the broad South African situation, by the spatial and social patterns created by the legislation and apparatus of apartheid.

A major result of these constraints, and in particular of legislation such as the Separate Amenities Act (abolished in October 1990) and Group Areas Act (abolished in 1991), has been that there has, until very recently, been almost no black tourism at all (Silva and Butler-Adam 1988). The Reservation of the Separate Amenities Act of 1953, for example, empowered local authorities and all persons 'in charge of public premises with the right to reserve these for the exclusive use of persons belonging to a particular race or class' (Silva and Butler-Adam 1988: 16). Since virtually all persons in positions of power were (and frequently still are) white, the best beaches, and most hotels, restaurants and cinemas were reserved exclusively for use by whites. This exclusivity was compounded by the Group Areas Act which precluded people of different races from residing in the same area. Thus the Durban beachfront, despite its being a tourist attraction, was also a white Group Area. Hence, it was virtually impossible for persons who were not white to find accommodation in the hotels or holiday flats along the beachfront.

The Job Reservation clauses of the Industrial Conciliation Act meant, furthermore, that blacks could not rise above certain categories of industrial employment. This Act had profound long-term economic implications, precluding the development of black managers, entrepreneurs and other skilled workers. 'To be black, then, meant to be working class and hence unable to purchase much by way of tourism opportunities' (Butler-Adam 1988: 4). Apartheid policies, of course, marginalized those potential black tourists who had managed – against the odds – to make a decent living.

For whites, the opposite circumstances historically hold in most cases: they enjoy a wide and even distribution of facilities and opportunities, receive preferential treatment and their privileged economic and political status affords them access to virtually all tourist attractions.

Recent structural changes in the country have, however, altered the patterns of black domestic tourism. Domestic tourism for black South Africans now falls into two categories (Butler-Adam 1988: 5). The first is a small but growing middle-class component which displays preferences and patterns very similar to white tourism patterns. This is not surprising and does not contradict our earlier claims that tourism patterns in South Africa have been determined by enforced racial segregation. Instead, it serves to illuminate the tragic social catastrophe which apartheid ideology has created, and should encourage planners and developers to plan for the future on the basis that 'tourism is a matter of class culture and not ethnicity' (Butler-Adam, 1988).

The second category of black domestic tourism consists of a much larger working-class component of sporadic tourism, such as is evidenced on holiday weekends on Durban's beaches (Butler-Adam 1988). Whilst tourism growth is perceived as fastest in this very low-income group – because of the large numbers – it does not mean that this group is dominating the market. The African holiday-maker and spender is very much a middle- and upper-income person, especially seen in the context of the low average earnings in the African community.

In spite of this, lower-income tourists must be catered for in planning, particularly in Durban. There are a large number of lower-income people in the area immediately surrounding Durban (the majority without any local recreational facilities), and an even greater number from the Transvaal for whom Durban is the closest and most accessible holiday destination. The 'beach', clearly, is an inexpensive tourist attraction and appeals to the lower-income tourist who for so long has been denied access. If Durban is to preserve its tourist image and avoid the severe overcrowding already experienced on its newly racially integrated beaches, lower-income tourists must be provided for. Tourism in Durban thus requires sensitive strategies and sustainable policies, if it is to serve the mass of the people.

Evaluating tourism as an agent for development in Durban

As the second largest industry in the city, tourism plays a crucial part in the economy of Durban. Together with related industries, it provides '190,000 persons in the Durban area with their daily bread' (Durban Publicity Association, personal communication 1990). Worth over R3.5 billion per annum in terms of buying and spending power, direct income generated by Durban's tourism industry is assessed as in excess of R840 million (ibid). These, and other claims, such as the '365-day holiday season' and '32,000 bed nights per annum' (ibid.) which the city boasts, make tourism a top priority constituent amongst the city's decision-makers, planners and advisers.

Thus the recent, and very real, threat of Durban losing its competitive edge to the Western Cape as a tourist playground has initiated extensive promotion campaigns and launched powerful planning groups. An example of the latter is TANK (Tourism Association for Natal and KwaZulu). Also, local business coalitions such as the Durban Metropolitan Chamber of Commerce have initiated promotion campaigns such as the current 'Be The Best We Can Be Campaign'.

In the face, too, of tourism in South Africa being on the brink of new boom times, thanks to the combination of an improved political climate and a weak Rand exchange rate (*The Star*, December 1989), city authorities and entrepreneurs have gone all out to 'enhance the image of the tourist in the eyes of all Durban's citizens' (Durban Metropolitan Chamber of Commerce, personal communication, 1990). Citizens are encouraged to recognize that 'tourism affects the business prospects of the entire city, regardless of industry'.

Durban's tourist industry then, plays a significant role in the city's dynamics. Its contribution is not, however, without controversy and complexity. The most blatant and challenging complexity lies in the simultaneous concentration of First and Third World characteristics inherent in Durban. Embedded, for example, in Durban's beachfront are on the one hand the 'crown jewels' (Soja 1986) of advanced industrial capitalism, and on the other hand the crude elements of a Third World city. The former are exhibited in the postmodern landscape, developed proudly in the recent and ongoing urban renewal programme. The latter, not exhibited of course, can be found in the 'cheap, culturally splintered, occupationally manipulable, Third World labour pool' (Soja 1986: 267), ever tangible within the city, on the pavements and on Durban's immediate periphery.

Many from this potential labour pool are trapped in a survival economy (De Kadt 1990), where *basic needs satisfaction* is their organizing principle. Dependent on the dominant market economy (ibid.), of which the tourism industry is an integral part, they seek employment as waiters, chars, ticket collectors or machine operators in the amusement park, as ice-cream and sweet vendors; others, mostly women, sell cheap curio kitsch and some expensive ethnic art. Their struggle for survival stoops insignificantly in the background of the 'fantastical place – a kind of papier-mâché confection' (Naipaul, cited in Britton 1987: 172) which Durban's tourist landscape is.

Thus, whilst the tourism industry in the Durban Functional Region can boast about the amount of income and employment it generates for the city, there are a number of inherent ambiguities and criticisms. These can be divided into three central themes. The first concerns the question of who benefits (De Kadt 1990), the second that of who participates in decision-making, and the third concerns the cultural impact of the industry.

In regard to who benefits, it would appear that those most in need – black South Africans – are left only with very modest employment opportunities. The severe housing crisis (amongst other things) suggests that the tourist industry's profits are not invested in projects such as low-income housing which would benefit the poor. Instead, those least in need – the local business elite and large capital investors, such as Sun International – benefit. This has facilitated their emergence as a powerful group where 'values of being' have been crowded out by 'values of having' (De Kadt, 1990). This reinforces their social powers, already entrenched by a perverted political system, and so they have become the all-powerful decision-making body for development.

Thus, and this brings up our second theme: for the majority of the local inhabitants in Durban, growth in the tourism industry implies less power in economic and political decision-making. This has been evident in the large-scale, government-backed beachfront renewal and development programme. Here, the local political and economic elite and, indirectly, the state have all received credit for 'transforming' and 'improving' Durban. It appears not to matter that there has been a lack of community consultation and consensus.

Whilst we hesitate to pass judgement on the aesthetics of the major remodelling which has taken place, its 'language' does not communicate 'human-centredness' (De Kadt 1990), nor cultural sensitivity. These are the values which should be given to *development*. The current overemphasis on economic progress in tourism at the cost of cultural processes must be challenged. This bias has given birth to another relationship of dominance and subordination: this time, through the sociocultural effects of tourism. This brings us to our final theme: the cultural impact of the tourism industry.

The fact that it has always been those who are better off – the political and economic elite – who get the most out of tourism, has enabled them to penetrate and dominate the tourism market. This is, of course, intrinsic to market relations, which automatically and necessarily benefits those with the most effective demand. The poor on the other hand, whose lifestyle and culture is also less like that of tourists, and who gain fewer benefits from tourism, often lose their desire to maintain an identity separate from the dominant culture reinforced by tourism (De Kadt 1990). Eventually, the main stimuli for cultural development come from the 'outside' which often destroys local initiative and self-reliance. This is already evident amongst Zulu women – famous for their beadwork and basketware – who have resorted to importing goods made in Taiwan. This dominance is also likely to generate 'the erosion of local character, and spread and growth of what Relph aptly terms "placelessness"' (Hudson, cited in Britton and Clarke 1987: 55). The irony and tragedy, of course, is that the loss of culture is destroying the ethnic image, the very source of the tourist attraction.

What emerges in these three themes is that the kind of tourism currently being pursued in Durban is promoting development which is not sustainable. Bruntland, quoted in De Kadt (1990), defines sustainable development as that which 'meets the goals of the present without compromising the ability of future generations to meet their own needs' (ibid: 3). Exploiting and 'homogenising' subordinate cultures is one such example of development which is *not* sustainable. The failure to encourage community participation in development planning, which might otherwise have succeeded in promoting the provision of alternate, cheaper accommodation for lower-income tourists, for example, is another. The absence of equity in the provision of facilities and of sustainability on the agenda of government policy are yet other examples.

The prospects for tourism as an agent for more equitable and sustainable development seem gloomy. However, developers would do well to turn the situation around and interpret it positively. Unlike many other developing nations, who cannot promote distributive justice because there is no cake and therefore nothing to distribute (De Kadt 1990), South Africa and Durban have some cake. We can afford, therefore, to pursue tourism development which will ensure that tourism advances the sociopolitical and economic needs of the local inhabitants in a spatially equitable way. It is possible too, to cater for the needs of tourists from all income groups. The section immediately to follow attempts to map the road to equitable and sustainable tourism development.

Making tourism development sustainable: some suggestions for Durban

The sustainable development policies which we wish to introduce to the tourism industry in Durban include the following: the investment of tourism profits into projects providing recreational, educational and other essential services, desperately needed in Durban's townships; shared decision-making in planning; cultural sensitivity; a reduction in the scale of organization; and an economics which favours stability rather than growth and material consumption. Policies of this nature are intended to bring development to Durban, as the following illustrations attempt to indicate.

Emphasis on the necessity of distributing some of the profit generated by tourism into projects aimed at meeting basic human needs will, for example, hopefully begin to alleviate the poverty in the Durban Functional Region. Community participation in planning and cultural sensitivity will promote equitable growth in the tourism industry, and see to it that the needs of lower-income tourists, for example, are appropriately met. Thus far the tourism industry has catered primarily for white, middle-class holiday-makers.

With reference to the issue of scale, a scaling-down in virtually all aspects of the tourism industry – in production, consumption, government involvement and administration – will facilitate the growth of small-scale enterprises. Unlike the 'massive integrated operations' (De Kadt 1990: 14) of multinationals or national capital, these are more likely to accommodate subordinate cultures and encourage local initiative. An alternate economic policy is envisaged 'to move away from the simple maximization of flows (growth) to paying much more attention to increasing efficiency and maintaining substance' (ibid.). This would require a reorientation on the part of planners to give preference to ecological and 'human' measures, so that account could be taken of the different effects (positive, neutral or negative) on the environment and on society.

It is easy enough to identify these areas of concern and recognize their importance. It will be difficult however, to convince those powerful institutions and individuals who are currently benefiting from existing arrangements of the need for alternate policies. They are unlikely to agree to policies which may compromise their profits. Similarly, the poor, 'whom the transistor radio and bicycle have wrenched out of their isolation, do not want to be told to discard their aspirations as consumers' (De Kadt 1990: 15). The fragmented institutional structure of the Durban Functional Region is also likely to make policy implementation very difficult (see Ch. 16). While it is compelling to recognize these constraints, it is equally important not to exaggerate them. Our concluding comments, therefore, will focus on the 'key players', who, we suggest, can facilitate the implementation of sustainable tourism policies in Durban.

The first is the state. Clearly, certain aspects of state and public administration are obstacles to more desirable development. In other respects, an appropriate incentive structure for sustainable tourism must come from government if it is to succeed. It will help, for example, if we can encourage the current negotiating process towards a new constitution to put sustainability and equity high on their agenda of government policy. This way, the activities of private capital and multinationals – often insensitive to local social and ecological issues – might be effectively prevented. Such problems are more likely to be addressed if the state insists that tourism planning be an integral part of development planning in general (De Kadt 1990). Shared decision-making, with community participation, must of course be the overriding principle.

The process of building the appropriate institutions needed for policy implementation in tourism is crucial and will provide key players in the industry. The Durban tourism industry will do well to take lessons, for example, from Gunn (1988), who suggests that tourism should be 'slow-paced and indigenous'. Many of the tourism ills of the past, he argues, 'can be traced to too much too fast and inappropriateness of development' (ibid: 273). Small, local groups could respond with much greater agility to changes in the market and hopefully facilitate the creation of networks of mutual support. Such networks would encourage community involvement in tourism development – a crucial consideration – particularly in view of the fragmented spatial and sociocultural character of Durban.

The road ahead is a long one. The difficulties lie not with distance, however, but with direction. This chapter has tried to demonstrate the importance of tourism development embarking on a journey in the direction of sustainable and equitable development. To do this there must be a *shared* and democratic spatial, social and political vision of the future of Durban and hence coordinated planning. Only this way can we ensure that tourism development *does not* contribute to uneven development in the Durban Functional Region.

Acknowledgements

We would like to acknowledge, with gratitude, the help which Karen Kohler and Win Venter provided in the preparation and checking of this chapter.

References

Britton, S. 1987. Tourism in small developing countries – development issues and recreation needs. In Britton and Clarke 1987, pp. 167–87.

Britton, S. and Clarke, W. (eds) 1987. *Ambiguous Alternative – Tourism in Developing Countries*. Suva, Fiji: University of the South Pacific.

Butler-Adam, J. F. 1988. *Recreation and Tourism in Post-Apartheid South Africa*. Durban: University of Durban–Westville.

Crush, J. and Wellings, P. 1987. Forbidden fruit and the export of vice – tourism in Lesotho and Swaziland. In Britton and Clarke 1987, pp. 91–112.
De Kadt, E. 1990. Making the alternative sustainable: lessons from development for tourism. Paper delivered at the International Academy for the Study of Tourism, Zakopane, Poland, August 1989.
Gunn, C. A. 1988. *Tourism Planning*. New York: Taylor & Francis.
Hudson, B. 1987. Tourism and landscape in Jamaica and Grenada. In Britton and Clarke, pp. 46–60
Silva, P. and Butler-Adam, J. F. 1988. *Terrae Incognitae: A Journey into the Unknown Lands of Domestic Tourism in South Africa*. Pretoria: South Africa Tourism Board.
Soja, E. W. 1986, Taking Los Angeles apart: some fragments of a critical human geography. *Environment and Planning. D: Society and Space* 4, 255–272.

Unpublished surveys and reports from the Tourism and Recreation Research Unit of the Institute for Social and Economic Research in the University of Durban–Westville provided useful empirical and conceptual reference material.

18 *Urbanization and health: evidence from Cape Town*

K. S. CHETTY

J71 R11
I18
South Africa

Introduction

South Africa is a country marked by extreme inequalities between the black and white population groups, the rich and the poor, and urban, peri-urban and rural communities. The state's policy of apartheid has been a source of social conflict and struggle for many years. We now find South Africa in a period of transition with the possibility of a negotiated settlement. The state has instituted a number of reforms, including reforms in the health sector.

In the face of a rapidly changing South Africa, where racism will no longer be a legalized institution, it is necessary to go beyond the emotive, racist aspects of South African society, to understand the social context of inequalities. Racism has been used to maintain class inequalities, and South Africa has been conspicuous by the extreme racial character of its class relations. This is fundamental in understanding the social, political and economic determinants of health in South Africa. Health is an outcome of politically determined structural economic and social changes. It is not a medical outcome, but rather a social outcome (Navarro 1984). An analysis, therefore, in South Africa of specific interventions without addressing the structural changes would merely be addressing the symptoms of the deficiencies in the present health system, without addressing the causes.

This chapter gives an overview of the health system in South Africa. An analysis is done of the health problems of the white, coloured and African population in a defined area in Cape Town to illustrate the impact that apartheid and urbanization have had on the health status of these groups. The distribution of hospitals and clinics in Cape Town is discussed, and the question as to whether the recent desegregation of hospitals will address the health problems of blacks will be examined. The Cape Town metropolitan area has a number of health authorities responsible for different areas and functions. This chapter will focus on the twenty four health regions under the Cape Town City Council.

These health regions exclude the major African townships of Khayelitsha,

Crossroads and Nyanga, which fall under the Western Cape Regional Services Council. The major metropolitan areas of Bellville and Parow also are under the control of the Western Cape. Therefore the statistics of these areas are not included. However, some statistics of Khayelitsha will be discussed for comparison.

The estimated population for the area in 1987 was 1,212,166, of which 50 per cent were coloured, 26 per cent African, 23 per cent white and 1 per cent Indian (Statistics for the Indian population group will not be presented because of the small numbers). There are only two African townships (Langa and Guguletu) included in this area, with an estimated population of 223,500, representing 19 per cent of the population. The rest of the African population is distributed throughout Cape Town, and includes mainly domestic workers living in their employers' homes (Medical Officer of Health, Cape Town City Council 1987). However, the population estimates should be treated with extreme caution as they are a marked underestimation.

In order to analyse the effects of the social, economic and apartheid policies on people's health and health care, the use of racial categories cannot be avoided. Where the terms African, Indian, coloured and white are used, they refer only to the legal category and this does not imply legitimacy of the system. Black is used to refer to Africans, Indians and coloureds collectively.

Overview of the health system in South Africa

Differences in the health status and allocation of health resources reflect the extreme inequalities between blacks and whites, urban, peri-urban and rural communities, the rich and the poor. The South African health policy is instrumental in furthering the social, political and economic system which has laid the foundations for the development of these inequalities. The essential features of the health system described below bear testimony to these arguments.

Fragmentation The present health system is a fragmentation of 14 different Departments of Health: the creation of the homelands has resulted in ten Departments of Health; the 1983 Tricameral Constitution created three separate parliaments, for whites, coloureds, and Indians, each with their own Department of Health for 'own affairs' and one Department of Health for 'general affairs'.

There has also been a proliferation of other authorities responsible for different aspects of health care. These forms of fragmentation have resulted in a system which is not cost effective, lacks coordination and affects the quality of care that is delivered.

Racial segregation One of the most obvious features has been the racial segregation of health facilities. Reports of overcrowded, understaffed black hospitals, with underused, overstaffed white hospitals have been confirmed by the Minister of Health in her 1990 budget speech. She also declared that hospital beds will be open to all. This is in fact laudable, and has been widely acclaimed both nationally and internationally. However, desegregation must not be seen in isolation, as racist laws can be abolished while at the same time control of access can be maintained through other strategies.

Primary health care The South African health authorities lack a coherent primary health care strategy. The state has for a number of years accepted that greater emphasis must be placed on primary health care, but as yet there is no evidence that this is being actively applied. Health is not just a medical outcome, but is an outcome of politically determined structural economic and social change. Unless the root causes of deprivation are addressed, the health status of communities will not change significantly. It is clear that this approach is lacking in the present policies.

Privatization of health services A fundamental characteristic is the current trend towards increased privatization of health care, which promotes an increase in the inequalities in health. One of the main problems with the private sector is that it is mainly accessible to whites and a minority of blacks who can afford private medical care. The private sector also undermines and drains resources from the public sector, has an urban bias, and focuses on curative hospital-based care rather than community-based primary care (Price 1987).

Mcintyre and Dorrington (1990) have shown that in 1987 South Africa spent 5.8 per cent of its Gross National Product on health care. Of this, 44 per cent was spent in the private sector which cares for perhaps 20 per cent of the population. The remaining 56 per cent was spent on the care of that 80 per cent of the population dependent on the public sector.

There is no doubt that privatization with its market-based approach increases the inaccessibility of health services and health care for those who need it most. Increasing privatization would only increase the inequalities between the rich and the poor.

The above briefly describes the deficiencies in the health system in South Africa. It is important to keep this picture in mind to fully understand the impact that urbanization has on health.

Urbanization and health

South African state policy and legislation has been aimed at reducing the rate of urbanization, but the situation is that the state has failed to prevent urbanization, which is rapidly accelerating. Van der Merwe (1988) showed that in 1985 approximately 56 per cent of the total population lived in urban areas. Unlike the white, coloured and Indian population groups, Africans have a relatively low level, but an imminent high rate of urbanization.

There is no doubt that this rapid urbanization will have an impact on health. While the positive aspects of urbanization such as improved employment and education opportunities have applied to some urban dwellers, the majority of rural dwellers who move into urban areas in the hope of finding employment and shelter soon find that this is extremely difficult. Some of the options open to these rural dwellers are to create or live in informal settlements. The Urban Foundation (1988) estimated the growth in shacks at 60 per cent a year since 1985.

Riley *et al.* (1984) stated that, based on the 1980 census, there was an estimated minimum backlog of 46,000 houses for the Cape metropolitan area. They also showed that 58.9 per cent of all coloured households in the region are overcrowded, with 48.8 per cent falling below the minimum income level.

These figures represent the situation in the established townships. The picture for the informal settlements is even more depressing. Hewat *et al.* (1984) found that in Crossroads, a squatter area in Cape Town, there was an average of 2.7 persons per room, 3.3 per sleeping room, and 12.0 per toilet.

The importance of housing cannot be overemphasized. Both overcrowding and the quality of housing have been associated with certain diseases. Both overcrowding and pollution of the internal environment predispose towards respiratory disease, and more especially tuberculosis. The incidence of diarrhoeal disease has been positively correlated with sanitation, water supply and sewage. Overcrowding and poor housing conditions are also associated with morbidity and mortality, skin diseases and infectious diseases (Lipschitz 1984).

Housing, of course, cannot be seen in isolation from other socio-economic determinants of ill health. In the same study Lipschitz stated that in areas characterized by poor housing, a number of adverse socio-economic conditions existed, and that modification of the physical environment can only have a maximum effect on health if accompanied by other socio-economic improvements.

De Beer (1984) stated that 'the picture of life as a member of the African working class is a bleak one. It is a reality of begging for jobs and food, of alcoholism and violence, of abandoned children and abandoned hope. It is a reality of precisely those conditions which cause psychological and physical stress, and result in widespread ill health.'

The picture for whites, including the white working class, is completely different. This is reflected in the different health problems experienced by whites and blacks, and is related directly to the social, political and economic status of the two groups.

Three broad groups of health problems that arise in urban areas have been described by WHO/UNICEF. The first relates to health problems associated with poverty. The second group relates to the impact of industrialization and conditions of the urban environment. The third is a result of social and political instability in certain urban areas. This chapter will look at some aspects within these categories.

Health problems related to poverty

In this section the health problems have been divided into mortality trends – what people die from, and morbidity trends – what diseases people suffer from.

Infant mortality rates The infant mortality rate (IMR) is the number of deaths of infants up to the age of one year per 1000 live births in a specified year. The IMR is a sensitive indicator, not only of infant health, but of the whole health of the population in general. Further, it is a useful index of the quality of health care and of socio-economic status, and can therefore provide valuable information on the impact of urbanization on health.

Pick *et al.* (1989b) recently showed that the IMRs for the coloured and African populations in Cape Town were highest in the low income groups. Africans in the lower income group had a greater chance of dying before the age of one year than coloureds in the same income group. The difference between the IMR for whites in the high income group and coloureds and Africans in the low income group is even more startling, with rates of 7.3 for whites (high income group) and 27.0 and 39.2 for coloureds and Africans respectively (low income group). However, the rates for middle income groups of coloureds and whites are similar (Table 18.1). It is important to note that the IMR is usually underestimated by up to 30 per cent (Yach 1990), and that the problem may in fact be worse.

Table 18.1 Infant mortality rates per 1,000 live births by income and population in Cape Town, 1982.

Income group	White	Coloured	African
High	7.3	–	–
Middle	9.7	12.4	–
Low	–	27.0	39.2

Source: Pick *et al.* 1989b.

The IMRs given by the Cape Town City Council per 1000 live births for 1987 is 4.9 for whites, 16.4 for coloured and 30.5 for Africans. These figures do not include Khayelitsha which has an IMR of 50.0 (Rip *et al.* 1988).

The IMR has two components: the postneonatal mortality rate which is a more sensitive indicator of social, economic and environmental factors, and the neonatal mortality rate, which reflects maternal health and care during pregnancy. In a study on perinatal mortality in metropolitan Cape Town, it was shown that those areas with low socio-economic status exhibited high postneonatal mortality rates (that is, deaths from 28 days to 1 day short of 12 months per 1,000 live births) and areas with high socio-economic status were characterized by low postneonatal mortality rates (Rip *et al.* 1986).

Gastroenteritis and pneumonia remain principal causes of infant death in the African and coloured populations (Medical Officer of Health, Cape Town City Council 1988), which are directly attributed to the effect of bad sanitation, poor nutrition, inadequate water supplies, poor housing, overcrowding and poverty.

It is alarming to note that measles is still one of the principal causes of infant mortality in Africans. Measles is a preventable disease and has been eradicated from many parts of the world by extensive immunization. Yach *et al.* (1991) have shown that the opportunity for measles immunization is being missed in a high proportion of children passing through curative facilities in the Western Cape. The fact that this preventable disease is a cause of infant mortality is a sad reflection of the South African health system.

Child mortality rates The child mortality rate is the number of deaths between the ages of 1 and 4 years in a given year. Childhood mortality is a good index to assess environmental conditions adversely affecting child health. Again these relate to socio-economic factors such as nutrition, sanitation and housing.

Pick *et al.* (1989a) have calculated child mortality rates in Cape Town by race and income group (Table 18.2). The childhood mortality of Africans is ten times higher than that of whites in Cape Town, and the lower income category is nine times higher than that of the higher income group. This shows clearly that blacks in low income categories have the worst health status.

Mortality in the general population The mortality rates given by the Cape Town City Council for the general population are crude rates and therefore not easy to interpret because of differences in age distribution. However, these crude rates show that the causes of death are a reflection of the socio-economic status of the different race groups. Whites die mainly from ischaemic heart disease, other heart disease and cancer, consequences of their destructive lifestyles and overindulgence. Blacks die from homicides, TB and pneumonia, consequences of their social instability and economic status. It is interesting to note that ischaemic heart disease is the third biggest killer in the coloured community, and rates for cerebrovascular disease, other forms of heart disease and chronic obstructive pulmonary disease are very similar to those of whites. The coloured community mortality patterns therefore show a classical picture of deaths associated with poverty and social instability, and also deaths associated with urbanization. With increasing urbanization these problems could be exacerbated, more especially in the African community who have not reached the same level of urbanization as coloureds.

Table 18.2 Child mortality in Cape Town per 1,000, 1982.

By race		By income category	
White	4.8	High	2.75
Coloured	15.2	Middle	9.75
African	47.2	Low	23.50
Overall	17.5		

Source: Pick *et al.* 1989a.

Tuberculosis (TB) According to the Medical Officer of Health for the Cape Town City Council (1988), 'TB remains the greatest single communicable disease in Cape Town reflecting the effects of general socio-economic deprivation.' The Medical Officer of Health for the Western Cape Regional Services Council (1989) stated that 'The epidemic may have reached its peak in the coloured population. Cases from the African population are however increasing in association with rapid urbanization. The Western Cape continues to have one of the highest TB incidence rates in the country and even in the world.'

Tuberculosis is endemic among black South Africans, but this was not always the case. The disease was virtually unknown in South Africa until brought there by infected individuals from Great Britain and other countries of Europe who came to the country to benefit from the agreeable climate. It was given immense impetus by living and working conditions in the mines (World Health Organization 1983).

There are two important factors in increasing the incidence of TB. Firstly the standard of nutrition is an important element in determining the level of resistance, and underfed families are therefore especially at risk. Secondly the level of exposure to the tubercle bacillus is also of major significance. Overcrowded conditions at home and at work increases the likelihood of contracting the disease. The poor are therefore especially at risk (Doyal 1979).

Yach (1988) has stated that an immediate improvement in the quality and quantity of housing will certainly improve the long-term prospects for eradicating TB, but in the presence of such a large infected pool, an immediate reduction in the rate can only be achieved by directly attacking this pool (the seedbed of TB), using a chemoprophylaxis programme.

However, tuberculosis will remain a problem for the poor, unless there is improvement in their socio-economic status. Yach also states that acceptance of the role of non-medical interventions lies at the heart of a future successful TB control programme for the region.

Health problems associated with an urban environment

Homicide Homicide is a reflection of social instability and social deprivation in any country. Unemployment, poor housing, lack of recreational facilities and powerlessness are some of the factors that contribute to this instability. In Cape Town, this is reflected in the mortality due to homicide which ranks first as a cause of deaths in Africans and sixth as a cause for coloureds.

A survey of crime for 1988 shows that homicide is highest in the African townships of Guguletu and Khayelitsha with 187 and 170 cases reported respectively, and lowest in the elite white areas of Camps Bay and Rondebosch with 1 and 0 cases respectively (Hansard 1989). Khayelitsha, however, does not fall under the Cape Town City Council. Therefore the mortality rates given previously reflect only those deaths in Guguletu and Langa. There has in fact been an increase in the incidence of homicide in 1988. In a recent international survey it was recorded that Cape Town has the highest murder rate of the world's 100 largest metropolitan areas with an annual rate of 64.7 murders per 100,000 people (*Argus*, 19 November 1990).

Smoking Smoking and the health problems related to smoking is another feature of urbanization. Yach (1989) stated that the rapidly urbanizing areas of South Africa are providing growing opportunities for the marketing of products such as tobacco, which are hazardous to health. The impact of tobacco marketing in South Africa has resulted in 60 per cent of men and 40 per cent of boys (by the age of 15 years) becoming smokers in Cape Town townships. Table 18.3 shows the mortality from lung cancer for males in Cape Town per 100,000 population between 1983–1987.

Table 18.3 Lung cancer mortality by race, 1983–7.

	1983	1984	1985	1986	1987
White	57.8	51.6	53.1	53.1	36.3
Coloured	45.2	44.8	40.0	50.2	53.7
African	22.2	18.3	21.3	22.9	11.5

Source: Medical Officer of Health, Cape Town City Council 1988: 97.

Lung cancer is directly related to smoking, and is a greater problem in the established urban communities, shown by the higher rates in whites. In the same paper Yach also stated that by the year 2000 and beyond, considerable cigarette-related early deaths will occur among the African and coloured populations.

Other problems of urbanization such as motor-vehicle accidents, occupational diseases, sexually transmitted diseases, psychiatric problems, child and women abuse are also major problems in an urban environment, but will not be discussed.

Health facilities

An analysis of the distribution of hospital beds per 1,000 population in Cape Town for the health district under the Cape Town City Council shows that there are 9.7 white beds per 1,000 whites and 6.6 black beds per 1,000 blacks. Whites have 1.5 times more beds than blacks. These figures include the long-stay psychiatric and TB beds. If these beds are excluded, the ratios would be 5.1 white beds and 3.2 black beds per 1,000 population. With the desegregation of hospitals the ratio of beds per 1,000 of the total population is 7.3 if long-stay beds are included, and 3.7 if long-stay beds are excluded.

It is obvious that with segregation, blacks in Cape Town were disadvantaged by the distribution of facilities. With desegregation, the ratios improve for blacks. However this does not necessarily improve the situation for blacks. If hospitals are categorized into referral and general hospitals there is a disproportionately high number of referral beds in the Cape compared to general beds (Chetty 1990). Furthermore, 65 per cent of the general beds are owned by private for-profit hospitals. A very small proportion of blacks can afford this private care, and therefore the majority are disadvantaged by financial inaccessibility.

If a similar analysis is done for clinics, the following figures would apply for population to clinic ratios in 1987: 14,838 whites per clinic, 31,008 blacks per clinic, overall total, 24,738 population per clinic.

The World Health Organization recommends 1 clinic per 10,000 people. The distribution of clinics falls well below this recommendation. Clinics, health centres, health workers and general practitioners form part of the primary care level, and act as a first point of entry into the health system. Greater emphasis should therefore be placed on this level, which is more accessible to people and more cost-effective.

Spatial distribution of facilities Analysis of total numbers alone, however, is insufficient as there are many other factors that affect equity in the provision of health care. One of these factors is geographical access. Desegregating all hospitals does not lead to immediate accessibility. Access depends on the location of the facilities, the distance people have to travel, and the availability of transport and communication.

Table 18.4 Number of hospitals according to type and location, 1987.

Type Of hospital	White suburbs	Coloured suburbs
Referral or general hospitals	7	1
Other hospitals, e.g. maternity, infectious, orthopaedic	5	1
Psychiatric hospitals	5	1
Private hospitals	19	3

Table 18.4 shows the number of the different types of hospitals according to location by racial area in 1987. There are no hospitals in the African areas, except two 'day hospitals' which only see outpatients. Most of the hospitals are located in white areas and therefore are geographically inaccessible to the majority of blacks. What is also striking is the large number of hospitals in the private sector, financially inaccessible to blacks.

A similar study done by Rip and Hunter (1990) analysing the participation of mothers in the Peninsula (Cape Town) Maternal and Neonatal Service shows that race and space are co-terminous. They state that 'because all five maternity hospitals are clustered in white areas near the city centre, referrals from mixed (coloured) and black residential areas are disadvantaged by distance . . . The inequitable geographical location of facilities in relation to residential areas is a direct result of the institutionalized structures of apartheid and the Group Areas Act . . . The study provides a classical example of how apartheid can influence health through spatial location.'

Conclusion

There is no doubt that a marked disparity exists between the health status and health care delivery for black and white populations in Cape Town. A similar picture exists for other urban areas of South Africa. Whites suffer from diseases of wealth whereas blacks suffer from diseases of poverty. With the rapid rate of urbanization the health problems will increase, more especially for blacks. An increase in urbanization without any changes in the social, political and economic status of blacks will deepen the poverty crisis facing them. The result might be an increase in poverty-related diseases, as well as an increase in diseases created by the urban environment. But at the same time, while this chapter has used one urban area as its focus, the impoverished rural areas must not be forgotten. The health care delivery and health status of people in rural areas is even more dismal than in the urban areas. It is because of the very conditions that exist in rural areas that people are forced to migrate. Typical responses in a study done in Crossroads (Hewatt *et al.* 1984) were 'But what choice is there?' 'Of course it is overcrowded in Crossroads. But if we stayed in the Transkei we would starve.'

What is clear is that there is no infrastructure in urban areas to support the rapid rate of urbanization. Thus far, there is no evidence of the political will of the South African government to create these infrastructures or improve the conditions for urban blacks.

The health problems of the majority of South Africans can only be addressed by a drastic restructuring of South African society. Until such time any interventions will be addressing the symptoms and not the root cause of the problem, and will prove to be ineffective.

References

Chetty, K.S. 1990. The provision of hospitals and clinics in South Africa: some steps to redress the problems. *Critical Health* 31/31 (double edition): 21–3.

De Beer, C. 1984. *The South African Disease. Apartheid Health and Health Services.* Johannesburg: South African Research Service.

Doyal, L. 1979. *The Political Economy of Health.* London: Pluto.

Hansard 1989. Crime 1988. Quoted in *South African Barometer* 3(17): 268–9.

Hewat, G., Lee, T., Nyakaza, N., Olver, C. and Tyeko, B. 1984. An exploratory study of overcrowding and its relationship to health at Old Crossroads. Second Carnegie Inquiry into Poverty and Development in Southern Africa, Conference Paper No. 14. Cape Town: Southern Africa Labour and Development Research Unit.

Lipschitz, M. 1984. Housing and health. Second Carnegie Inquiry into Poverty and Development in Southern Africa, Conference Paper No. 164. Cape Town: Southern Africa Labour and Development Research Unit.

Mcintyre, D.E. and Dorrington, R.E. 1990. Trends in the distribution of South African health care expenditure. *South African Medical Journal* 78(3): 125–8.

Medical Officer of Health, Cape Town City Council 1988. *Annual Report of the Medical Officer of Health. City of Cape Town 1987.* Cape Town.

Medical Officer of Health, Western Cape Regional Services Council 1989. *Annual Report of the Health Department of the Western Cape Regional Services Council 1988.* Cape Town.

Navarro, V. 1984. A critique of the ideological and political position of the Brandt Report and the Alma Ata Declaration. *International Journal Of Health Services* 14(2): 159–78.

Pick, W.M., Bourne, D.E. and Sayed, A.R. 1989a. Child mortality in Cape Town. Paper presented at Paediatrics Priorities Conference in Rustenberg, 1989.

Pick, W.M., Bourne, D.E. and Sayed, A.R. 1989b. Mortality and low birth weight – a new look. Paper presented at Epidemiology Conference, 1989.

Price, M. 1987. Health care beyond apartheid. *Critical Health*, Dissertation No. 8, March 1987.

Riley, N., Schuman, C., Romanovsky, P. and Gentle, R. 1984. Spatial variations in the levels of living in the Cape metropolitan area. Second Carnegie Inquiry into Poverty and Development in Southern Africa, Conference Paper No. 13. Cape Town: Southern Africa Labour and Development Research Unit.

Rip, M.R, Keen, C.S. and Kibel, M.A. 1986. A medical geography of perinatal mortality in metropolitan Cape Town. *South African Medical Journal* 70: 399–403.

Rip, M.R, Woods, D.L, Keen, C.S. and Van Coeverden De Groot, H.A. 1988. Perinatal health in the peri-urban township of Khayelitsha, Cape Town. *South African Medical Journal* 74: 633–4.

Rip, M.R. and Hunter, J.M. 1990. The community perinatal health care system of urban Cape Town, South Africa – Part 2: geographical patterns. *Social Science and Medicine* 30: 119–30.

Urban Foundation 1988. Cited in *Race Relations Survey 1987/88*. Johannesburg: South African Institute of Race Relations.

Van der Merwe, I.J. (1988). A geographical profile of the South African population as a basis for epidemiological cancer research. *South African Medical Journal* 74: 513–8.

Venter, E.H. 1990. Introductory speech: budget vote national health and population development. *Hansard*, cols. 9383–94.

World Health Organization 1983. *Apartheid and Health*. Geneva: World Health Organization.

Yach, D. 1988. Tuberculosis in the Western Cape Health Region of South Africa. *Social Science and Medicine* 27: 683–9.

Yach, D. 1989. Urban marketing to promote disease or health. *Critical Health* 28: 37–43.

Yach, D. 1990. Personal communication.

Yach, D., Metcalf, C. and Lachman, P. 1991. Missed opportunities for measles immunization in selected Western Cape Hospitals. *South African Medical Journal* (forthcoming).

PART FIVE

Towards a post-apartheid city

This final part of the book looks ahead to the post-apartheid city. Beavon sets the scene (Chapter 19): in reviewing possible changes against current uncertainties, he provides a reminder that the development of a post-apartheid city will be greatly constrained by what already exists – in the physical structures created by apartheid. Dewar follows (Chapter 20) with proposals designed to address the peculiarities of the cities formed under apartheid. He stresses the importance of urban management, in moving towards a new kind of city the nature of which is in urgent need of definition. The specific problem of an appropriate housing policy for post-apartheid South Africa is considered by Corbett (Chapter 21). Using the Durban region as an example, he points to the enormity of the Black/African housing problem and the need for policies which will reach everyone even if standards are less than ideal.

The next two contributions deal with change as it is actually taking place. Steinberg, van Zyl and Bond (Chapter 22) place the (re)development of Johannesburg's central business district in the context of the broader economic crisis besetting South Africa, and requiring a changing spatial strategy on the part of capital. A consequence is the restructuring of inner-city space driven by property speculation, threatening a gentrification process which would enable well-to-do blacks to settle there. The Tomasellis (Chapter 23) describe a process of integration in the Westville district of Durban, as Indian families have moved into a previously white residential area. They examine the role of a local residents' association set up to facilitate integration, a strategy which may be required well after the repeal of the Group Areas Act.

The remaining two chapters bring quite different perspectives to bear on the post-apartheid city. Robinson (Chapter 24) provides some historical reflections based on investigations of early township planning designed to provide housing for Blacks in a controlled environment. She suggests that aspects of the pre-existing discourse and practice of planning may be carried over into post-apartheid South Africa, shaping its cities as reflections of new power relations and threatening newly-won freedoms. Finally (Chapter 25), Davies examines the possible relevance of the experience of Harare, capital of Zimbabwe, to the transformation of formerly white residential areas. A tendency to concentrate in outer areas is evident, but otherwise Blacks distributed themselves fairly widely according to their purchasing power and prevailing land values, without conflict or much disturbance to established pattern of socio-economic differentiation.

19 *The post-apartheid city: hopes, possibilities, and harsh realities*

K. S. O. BEAVON

Introduction

An attempt to sketch the character of urban places representing the hopes and possibilities of a nation for some date in the future is a difficult task at the best of times. Such a venture must take account of the harsh realities reflected in the stability or otherwise of the political, social, and economic conditions that prevail and have prevailed for a significant period. Furthermore it must be noted that there is as yet no universally accepted urban policy for the 'new' South Africa.

Given the increasing volatility of the political situation in South Africa, particularly since the declaration of the State of Emergency in 1985, reliable predictions about the 'late-apartheid and post-apartheid South African city' have become increasingly difficult to make beyond the very short term. By way of an example, an attempt three years ago to sketch the essentials of the future South African city (Beavon 1991) was initially predicated on the cosy hope that the Progressive Federal Party would emerge as the 'government in waiting' after the 1987 general election – de facto an election of a white parliament. In fact the paper had to be written against a very different background; one that was coloured by the fact that the white right wing was on the move and that the government under P. W. Botha would bow to pressure from that quarter. Consequently rather gloomy predictions were made of violent outcomes of the inevitable interfacing in urban areas of different racial groups. Since 1987 there has been another election, frequently styled the 'last white election', further gains by the right wing, and a change of President in what was effectively a bloodless coup. Then on 2 February 1990 came the unbanning of the African National Congress (ANC), and a host of other political groups and organizations. The release of Nelson Mandela followed, as did serious attempts by both the ANC and the national government for negotiations directed at the creation of a non-racial democratic 'new' South Africa with a promise that the Group Areas Acts will be repealed in 1991.

Against the heady climate of the 'February days' more optimistic predictions about well organized, well managed urban places were more easily made and believed but the sun soon set on that rosy day. Once again a series of violent and tragic events have given some cause to be pessimistic in posing any scenarios for the post-apartheid city. More particularly, as various political factions and interest groups have begun jockeying for power so the tensions between them, aggravated by decades of apartheid administration, have erupted into violent confrontations at the grass-roots level. Included are the massacres of August 1990 which saw 500 killed in 'Black on Black' attacks on the Witwatersrand in one week. The meddlesome actions of white vigilantes and death squads also enter the equation. Furthermore there has been the endless killing in what has been styled 'the War in Natal', all of which is only part of the brutal background that complicates any attempt to present here a firm prediction about a post-apartheid South African city which in any event must to an extent be subjective.

The emergence of the apartheid city

Aspects of the evolution of the South African urban system and the character of the apartheid city are covered in other chapters. Readers might also refer to Fair and Browett (1976) for a discussion on the urbanization process, and to Beavon (1982) and McCarthy and Smit (1984) for references on the early impact of segregation and apartheid in urban areas. For a discussion of some of the problems of housing in Black areas in the post-1976 era the reader is referred to Mabin and Parnell (1983) and Hendler and Parnell (1987). The resultant geography of the South African city is succinctly captured in Figure 19.1.

The contemporary situation

Recent surveys undertaken by and for the Urban Foundation in Johannesburg have revealed the stark statistics that now confront urbanists, urban planners, developers, and any future government trying to cope with urbanization. The population of South Africa – including the 'independent states' of Transkei, Ciskei, Venda and Bophuthatswana – was 29.1 million in 1980 and it will double by 2010. Whereas in 1985 there were 20.7 million, or 63 per cent of the total South African population, resident in and on the edges of urban areas, by 2010 that will have increased to 43.7 million or 73 per cent. The rate of urban population growth will be higher than that for the population as a whole. The main component of the expected increase will be the Black population which is anticipated to increase from 13 million in 1985 to 33 million a mere generation from now. Most of the population

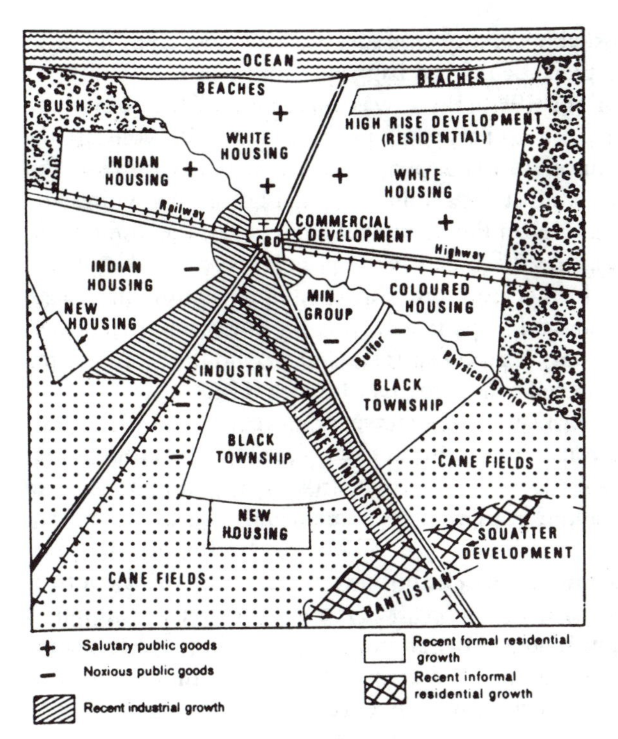

Figure 19.1 A graphic model of the apartheid city of South Africa (after Davies 1976 and McCarthy 1986).

growth will be concentrated in and about the existing metropolitan areas, some of which already straddle homeland borders. In 1985 population in those metropolitan areas was 8.7 million and by 2010 the same areas will have 23.6 million people: an increase of 270 per cent! The percentage of Black people there will increase from the 35 per cent of 1985 to just on 50 per cent twenty years from now. More graphically the total Black population increase in the metropolitan areas will run at approximately 600,000 per annum of which about one-third will be in-migrants. At present between 25 and 40 per cent of the total economically active Black population is formally unemployed and figures of 8 million unemployed by the turn of the century are now being mentioned (Urban Foundation 1990b). The mind boggles at the enormity of the task ahead, not only in terms of the erection of houses and shelters but especially in terms of job creation. All this must take place against the backdrop of an economy severely weakened by sanctions, and at a time when former European and American trading partners are increasingly looking towards investment in the newly emerging free markets of Eastern Europe. At the same time the ANC has yet to come up with an economic policy that will attract back lost investment. This will compound the problems facing the cities and almost certainly confuse those attempting to find workable solutions or survival strategies over the next several years whatever their ideological baggage.

At the beginning of the 1980s there was one 'formal' (i.e. brick and mortar) house for every 3.5 white people in the country. By contrast the ratio of formal units amongst Black people was 1:43 (Cosser 1990). Put another way there were almost 1.29 million formal houses for whites and only 486,000 for Blacks (Urban Foundation 1990a). Given the differentials in growth rates between the Black and white populations, and the general impoverishment of the former group in respect to opportunities in general and access to finance and land in particular, it is not surprising that shacks have become an increasingly typical feature in and around the metropolitan areas and the major and even minor towns of South Africa.

In 1989 the Pretoria–Witwatersrand–Vereeniging region, the PWV contained 412,000 formal houses in the Black townships but that number was exceeded by the 422,000 shacks in the backyards of formal units (Urban Foundation 1990a). In addition there were some 635,000 shacks on vacant land ranging from the old golf course in Soweto to land nominally open for industrial and agricultural use. The housing shortage for Blacks outside of the homelands is put at between 808,000 and 850,000 units (Tomlinson 1990; Urban Foundation 1990b). Overall more than 7 million urban people nation-wide live in shacks of one kind or another with 2.5 million of those located in the inner zones of the PWV (Fig. 19.2).

Towards the future

The reality of the problems are of course much more complicated than can be demonstrated here (Bernstein and Nel 1990; Swilling 1990; Tomlinson 1990). Given that the essence of the problem about the urban future of the South African city revolves around the racial prejudice and discrimination of the past and the allegiances and alliances of the present, then such problems can most likely be expected to appear in that component of the city which is least integrated at the present time and where the concept of 'own turf' has long been entrenched, that is, in the residential areas. Against the background of the unequal division of formal housing and the rapid expansion of squatter areas it is doubly appropriate to concentrate attention here on the residential component of the city and to examine those facets that reflect most sharply the changing urban geography of the South African city. Given a political settlement the prospects for peaceful coexistence of all South Africans could be high. Consequently the aspects of urban living to be addressed here include a consideration of the future in the zones of interface already styled the 'grey areas', and in the adjacent residential zones including both the almost exclusively white areas and Black townships. A brief resume is also provided of what the future might hold for those who

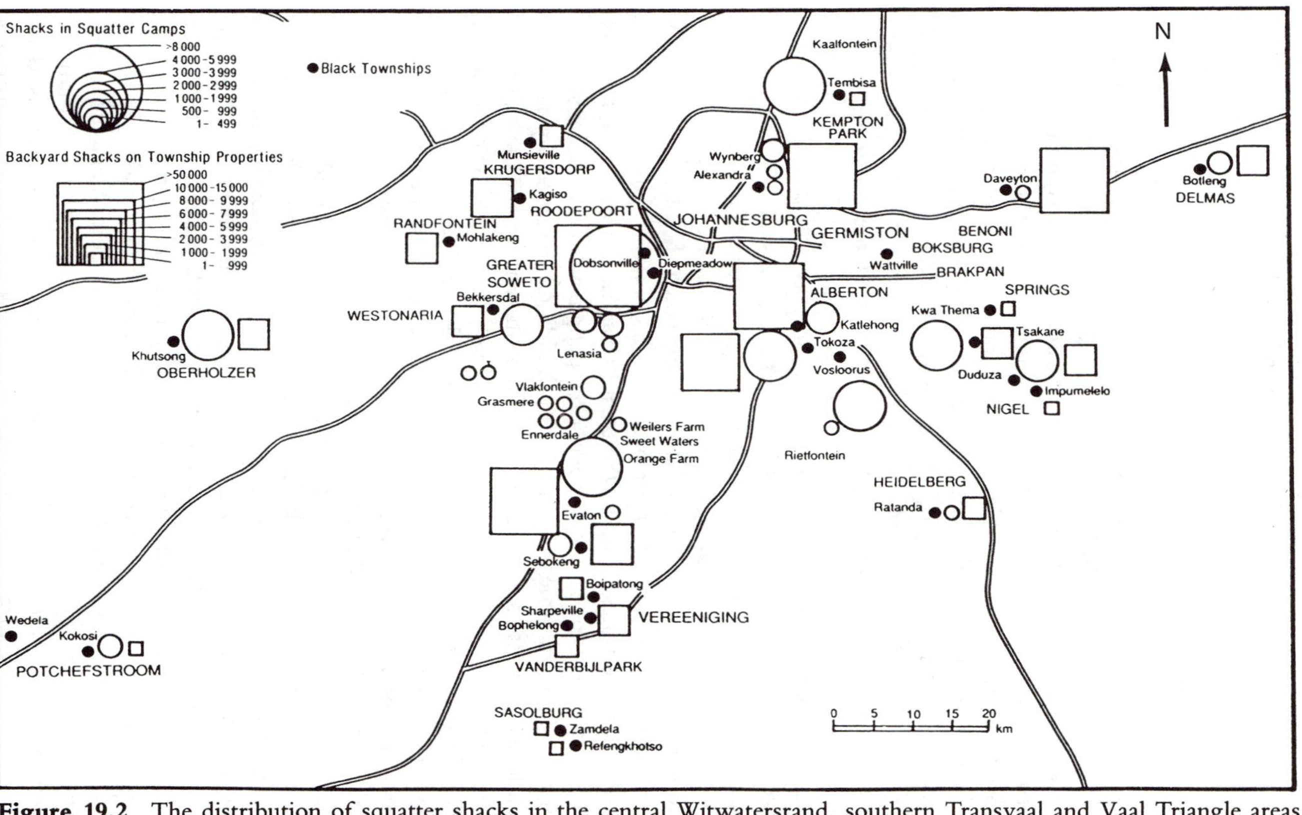

Figure 19.2 The distribution of squatter shacks in the central Witwatersrand, southern Transvaal and Vaal Triangle areas (based on data from the Urban Foundation, Johannesburg).

are increasingly being forced into shacks or squatter camps. In the discussion below no consideration is given to the possibility of an urban future that embraces the unpopular concept of 'free settlement areas' (Elder 1990).

Inner-city 'grey areas': the new slums

In the South African lexicon 'grey areas' are those white parts of the apartheid city where people other than whites have taken up illegal residence. The phenomenon of 'greying' first became apparent in a significant manner during the mid 1980s. It reflected the chronic housing shortage created by racial divisions of residential land even for those who were wealthy enough to buy, build, or rent substantial homes but who were precluded from the market as a whole merely on the grounds of race (Hendler and Parnell 1987). The increased demand for residential space close to the central business district was also promoted by trends in the centre of the city where business houses were increasingly offering employment to black people for what had formerly been white-only jobs. Although grey areas have been identified in many of the largest cities the comments that follow here are related specifically to Johannesburg (see also Ch. 22).

Four categories of greying can be identified that reflect both the economic status of the actors, the zone of the city in which they live, and the type of accommodation they own or occupy. The range is from affluent professionals and business people who own houses or apartments in relatively genteel middle-zone suburbs, to poorer people who are subtenants in a portion of an apartment in inner-city buildings managed by latter-day slumlords. It is that particular faction of tenants who are the most exploited of grey area residents – be they black or white people – and who face a grim future in the inner-city areas even with the Group Areas Act repealed. To understand that claim, and the predictions that follow, some attention to the current plight of residents in the inner-city grey areas is essential (de Coning *et al.* 1986; Fick *et al.* 1988; Pickard-Cambridge 1988).

Under the Group Areas legislation black people are illegal residents in white areas. Consequently black tenants in the crumbling inner-city apartments have no legal standing when they wish to take a stand against exploitative landlords and their rent collectors: and indeed could become victims of the racial laws before their complaints against owners could be heard. Certainly in the inner areas of Johannesburg numerous cases have been reported of rack-renting, evictions, and scandalous lack of attention to essential maintenance on the side of the owners and agents. The illegal residents have turned to social-area activists for assistance in trying to obtain recompense. Notable among these has been Actstop, a non-racial organization formed to act against all evictions. It would appear, however, that repeal of the Group Areas Act in itself will not do much to ease the plight of the poorest inner-city tenants. The basic premises on which illegals (because of race) might embark on legal action against slumlords, for illegal evictions and for undermaintained buildings, would be stronger. Nevertheless inner-city tenants will still be residents in a part of city which, even in a non-racial democracy, is sited on land that is very likely to appreciate considerably. With appreciating land being the real stake for the owners the decline in the quality of the apartment stock, even after a repeal of Group Areas legislation, could continue over a protracted period; possibly for longer than it takes to change both the constitution and government and until a clear indication of an urban policy in the 'new', South Africa is forthcoming. Furthermore building developments that will eventually follow on existing sites are likely to be commensurate with the value of the land. So the destruction of old, poor, but cheap residential stock could be followed by erection of new and more expensive apartment stock. Consequently hardship for the current, and predominantly black, poor in the inner-city slums will not be solved by the simple repeal of the Group Areas Acts. Such action will in all probability make the existing higher-rent apartment areas, and those that will emerge in time, attractive to both black and white yuppies, and even couples in the third stage of the urban life-cycle. The problem of assisting and housing the then former inner-city poor will remain. Although a non-racial character will form part of the future for the high-rise inner-city areas getting to that future is going to be traumatic for the poorest residents.

The white suburbs

Generally within any present South African town or city the average price of residential properties is considerably higher than in their associated black townships. Consequently the lifting of the Group Areas regulations will have little effect on the townscapes in the formerly all white areas unless such areas have already been abandoned. That said, for several years small numbers of so-called 'non-whites' have been buying into all-white suburbs in a variety of major urban places. Depending upon the character (and wealth) of the suburb concerned there has been little if any public mention of the changes taking place and the process appears to be peaceful and unproblematic: regarded as a favourable portent for continued peaceful change. In the same vein the recent decisions by parent bodies to open 96 formerly all-white state schools to all racial groups has also been a portent of better things to come. There is of course another side to the story and it comes in those areas where the dominant group is very conservative. Under those circumstances there is no 'invasion' by 'non-whites'. Alternatively violent and unpleasant incidents are associated with the moves or attempted moves. What is true of suburbs in large towns is also true of towns outside of the metropolitan regions where resistance to change and acceptance of mixed racial areas is vehemently opposed by whites.

It can be expected that the lifting of the Group Areas legislation will see the emergence of a new set of dominant-group rules and procedures. Such restrictive measures have already made their appearance in a number of 'platteland' (country) towns controlled by conservative local authorities. There the repeal of the separate amenities legislation was met almost simultaneously by demands for exorbitant deposits from those non-residents wishing to make use of the public library, and an insistence that public swimming pools can only be used by season-ticket holders the cost of which was also increased. On the positive side and within many of the major cities, which is where most of the black urban population is located, access to and use of central and major public and private facilities has been operating reasonably smoothly for some time without recourse to any deterrents. Given the high unemployment levels amongst black people in the country areas the recent charges for access to public services can only be regarded as discriminatory. Black populations will almost certainly retaliate with their most effective weapon, namely the boycott of white-owned businesses. That has proved to be very effective in the past and will no doubt do so again. The possibility of violent confrontations cannot be ruled out. Until such time as de Klerk's 'reform government', or a new government, make discrimination an offence the retaliation of whites who wish to resist change is likely to increase tensions in many white areas. The potential danger areas will continue for some time to be those suburbs close to the central zones of the city where white working-class people live. Although some of the successful greying has taken place in such zones, notably in Johannesburg's Mayfair, the outward movement of conservative whites will result in tighter and more resistant white enclaves. It is in those areas and not in the leafy suburbs of white affluence that the real unknowns lie and many white areas will remain white. For a whole set of different reasons most of the current black areas will remain black.

The Black townships

Ironically it is in and about the townships that some of the major problems for a non-racial democratic South Africa lie. I hasten to add that it is the actions of the white ruling elites at a variety of employment and administrative levels who are ultimately responsible for what now appears on the ground. A major component of the problem is that the townships were declared 'independent' municipal areas several years ago. The effect of that window-dressing decision has been not only to sever links with some of the supportive white municipalities but to place a heavy financial burden on the Black residents – many of whom earn low and very low salaries if employed at all. The real rateable base of the functionally integrated but administratively divided city remains the central business district in the white zone. Furthermore most if not all of the high-value and rateable industrial land also lies in the white zone. So, the Blacks sell their labour in the white zone which in turn benefits from the rateable base so created and the bulk of the funds raised are used to serve the white residents. Such alienation was further aggravated by the introduction of the three-chamber parliament that excluded only Blacks from a say in central government. As a form of protest residents in many townships joined the rent boycotts that have run since 1985 – effectively a refusal to pay rent for state-owned houses and a refusal to pay for electricity, water, refuse removal and other services. Without an income from rates the legal Black local authorities have been unable to maintain essential services let alone improve the quality of residential amenity in the townships. Although a settlement of the long-running rent dispute appears in sight the ravages it has caused will take a long time to heal.

So it is in the former 'locations' that the worst of some 70 years of segregation and apartheid is most visible. The repeal of all apartheid legislation, a new and popularly acceptable constitution, and the granting of a universal adult suffrage will not speedily eliminate the townships. For at least several decades into the future their monotonous and truly grey townscapes will haunt the First World component of the nearby cities, a situation reinforced by the fact that apartheid has lowered the residential value of land adjacent to Black townships.

As attempts are made to alleviate the chronic housing shortage for Blacks the danger continues to exist that the only available land in large supply for the low-cost housing needed by millions of Black people will be found in the vicinity of existing townships. Unfortunately making use of such land for low-cost public and private housing will in a sense entrench the existing framework of segregation. Evidence that such a pattern was developing in the central Witwatersrand is strong. In the mid-1980s following the repeal of the (Black) influx control legislation and in anticipation of 'reform' many speculative housing companies began erecting apparently good quality housing for Black people able to afford a housing bond. In an effort to reduce the total cost of a unit as low as possible the construction interests readily made use of the marginal land notwithstanding the incompatibility of the quality of housing. Even so many families lived in shacks on their new plots of land while the core unit of their house was constructed. Upon completion they would move into the formal house and move squatters into the vacated shack while the rest of the house was built. The backyard squatters would in effect through their rent assist the new landlord to pay off the bond on the formal house. What emerged was clusters of the new European-style housing in close proximity to some of the worst of the old state-built mass housing for Blacks. Likewise much of the up-market housing was literally within a stone's throw of squatter camps which mushroomed as people drifted in from the homelands and out of hiding in the existing townships.

The new Black housing bubble burst when the current wave of unrest erupted on the Witwatersrand in mid 1990. Substantial numbers of European houses were destroyed in the battles which raged between varying alliances of various factions of squatters, township residents, and migrant workers living in the infamous hostels. As building labourers found themselves threatened, as pillage of building materials increased, so the provision of new housing units ceased. At the same time the building societies began to repossess properties as owners affected by the unrest fell behind with their bond repayments. At the time of writing the problem of how to confront the housing shortage amongst Blacks hangs like a shadow over the veld. Some compromise will have to be made so that those who cannot afford housing but who hold regular jobs become the responsibility of the state. At the same time a way will have to be found to protect property from random arson attacks and thereby assure the providers of houses and the mortgagees that scarce money is not simply being torched. In theory a Black bourgeoisie could be poised to leave the townships for the current white suburbs with the Group Areas Act repealed thereby vacating some units which could be filled in a general shuffle and at the same time contribute towards the emergence of a more noticeable racial mix in the white suburbs. In reality the number of such moves will be small. Furthermore ties to community and race consciousness may well dictate otherwise and in a sense further entrench the great and long-running divide even in a country liberated from statutory apartheid. The prospect of the scrapping of municipal boundaries drawn on racial lines does however means that in future the tax base of the functional city can be used for all the people.

The squatters' areas

Although the number of squatters, in reality homeless but not always jobless people, has increased since the lifting of the influx controls the numbers of in-migrants from the homelands is considerably less than earlier anticipated (Crankshaw and Hart 1990). That said, however, the surprising statistic that has emerged is that most of the squatters are in effect refugees from the formal townships where they had been hidden in crowded houses and backyard shacks in the days when the right to live even in a Black township was restricted by the infamous Section 10 of the Natives (Urban Areas) Consolidation Act. It is anticipated that the percentage of migrants from the homelands will rise thereby increasing the pressure to find a viable policy for squatter settlements (Beavon 1989).

Following trends which have been observed in other Third World regions squatters have sought out land adjacent to existing impoverished communities and secluded spots beyond the eyes of officials even if such spots are in the interstices of the white urban fabric. Land invasions have only recently begun to occur. The response of the state has been to simply destroy the shacks. Such bulldozing tactics have done nothing to alleviate the plight of the victims and have aggravated the strained relations between blacks and whites over residential land. The cry from the squatters is for security of tenure above all else. Currently there are only a very small number of official site-and-service schemes which are being tried as a basis for a wider solution. Suggestions have been made by the Urban Foundation and others that, in accordance with a change of position by the World Bank, one-off capital subsidies should be used to create a more stable situation for many of the homeless (Beavon 1989). Currently the situation is in a state of flux. The only certainties in the matter are that the numbers of homeless and jobless people will increase in the future: a squatterscape will certainly be part of the new South Africa.

Conclusion

The presentation here has cut to only a few essentials in what is a much more complex situation. Certainly the prospects for solutions hinge as much on finding a way to re-stimulate the economy as it does on the rapid construction of first homes for millions of people. At the base of the problem are the vestiges of formal apartheid and racism. Looming large is the need to create millions of jobs in the next 10 to 20 years. In that respect the longer economic sanctions are in place the less likely is any solution at any time irrespective of constitutional change.

References

Beavon, K. S. O. 1982. Black townships in South Africa: terra incognita for urban geographers. *South African Geographical Journal* 64:3–20.

Beavon, K. S. O. 1989. Mexico City and colonias populares: hints for a South African squatter policy. *South African Geographical Journal* 71:142–56.

Beavon, K. S. O. 1991. To be or not to be: views of the South African city in the future. In D. Drakakis-Smith (ed.), *The Keele Symposium on Southern Africa* (in press). London: Routledge.

Bernstein, A. and Nel, M. 1990. Urban policy: the fiction of conspiracy. *Urban Forum* 1:111–14.

Cosser, E. (ed.) 1990. *Homes for all: the challenge of homelessness.* Johannesburg: Christian Research Education & Information for Democracy.

Crankshaw, O. and Hart, T. 1990. The roots of homelessness: causes of squatting

in the Vlakfontein settlement south of Johannesburg. *South African Geographical Journal* 72:65–70.
Davies, R. J. 1976. *Of Cities and Societies: a Geographer's Viewpoint*. Inaugural Lecture New Series, 38. Rondebosch: University of Cape Town.
de Coning, C., Fick, J. and Olivier, N. 1986. *Residential Settlement Patterns: a Pilot Study of Socio-political Perceptions in Grey Areas of Johannesburg*. Johannesburg: Department of Development Studies, Rand Afrikaans University.
Elder, G. 1990. The grey dawn of South African racial residential integration. *GeoJournal* 22:261–6.
Fair, T. J. D. and Browett, J. G. 1976. The urbanization process in South Africa. In D. T. Herbert and R. J. Johnston (eds), *Geography and the Urban Environment: Progress in Research and Applications. Vol. 2*, pp. 259–94. Chichester: Wiley.
Fick, J., de Coning, C. and Olivier, N. 1988. *Ethnicity and Residential Patterning in a Divided Society: a Case Study of Mayfair in Johannesburg*. Johannesburg: Department of Development Studies, Rand Afrikaans University.
Hendler, R. and Parnell, S. 1987. Land and finance under the new housing dispensation. In Southern African Research Service (ed.), *South African Review 4*, pp. 423–32. Johannesburg: Ravan.
Horrell, M. (ed.) 1978. *Laws affecting Race Relations in South Africa: to the end of 1976*. Johannesburg: South African Institute of Race Relations.
Mabin, A. S. and Parnell, S. 1983. Recommodification and working-class home ownership: new directions for South African cities? *South African Geographical Journal* 65:148–66.
McCarthy, J. J. 1986. Problems of planning for urbanization and development in South Africa; the case of Natal's coastal margins. *Geoforum* 17:267–88.
McCarthy, J. J. and Smit, D. P. 1984. *South African City: Theory in Analysis and Planning*. Cape Town: Juta.
Pickard-Cambridge, C. 1988. *The Greying of Johannesburg*. Johannesburg: South African Institute of Race Relations.
Swilling, M. 1990. Deracialized urbanization: a critique of the new urban strategies and some policy alternatives from a democratic perspective. *Urban Forum* 1:15–38.
Tomlinson, R. 1990. *Urbanization in Post-Apartheid South Africa*. London: Unwin Hyman.
Urban Foundation 1990a. *Policies for a New Urban Future, Urban Debate 2010: 1 Population Trends*. Johannesburg: The Urban Foundation.
Urban Foundation 1990b. *Policies for a New Urban Future, Urban Debate 2010: 2 Policy Overview, the urban challenge*. Johannesburg: The Urban Foundation.

20 *Urbanization and the South African city: a manifesto for change*

DAVID DEWAR

Introduction

Urbanization ranks as one of the most significant and far-reaching dynamics currently affecting South African society. No one really knows how big the major cities are or precisely how fast they are growing, but several features of the urbanization process are clear. First, most urban growth is occurring in and around the four major metropolitan areas: Pretoria–Witwatersrand–Vereeniging; Durban; Cape Town; and Port Elizabeth–Uitenhage. Second, although in-migration is occurring apace, most urban growth is the result of natural increase: the urban explosion is irreversible and will continue for a long time to come. Third, the highest rates of growth are amongst the poorest people: the dominant demographic tendencies are faster, younger, and poorer. Fourth, accompanying this dynamic of growth are high and increasing levels of poverty, inequality and unemployment, with a large and increasing proportion of people seeking survival in the informal economic sector: the socio-economic profile within South African cities corresponds increasingly with those of other Third World cities.

Surprisingly, despite the unprecedented rate and scale of the urbanization process, there is little deliberation about where the process is heading. Searching questions are being asked, and debates initiated, about the political and economic future of the country, but few relate to its urban future, particularly in relation to the larger cities, which will affect, fundamentally and directly, the lives of well over half the country's population. Particularly, there is no critical examination of the objectives which urban management should seek to promote or of the formal and structural characteristics which best advance those objectives. This chapter seeks to initiate that debate.

Current urban characteristics

The starting point lies in a brief evaluation of current formal and structural characteristics which are resulting from the urbanization process and management responses to it.

At the level of the urban whole, three spatial patterns, more than any others, characterize all South African cities. Significantly, they can be found in a great many other Third World cities as well. The first is explosive low density *sprawl*, the direction of which is largely non-managed. Three processes primarily determine the pattern of growth. One is speculative sprawl, involving higher-income people seeking to privatize amenity. Developers target places of beauty, frequently destroying, in that process, the very qualities that made the site attractive in the first place. Another is the crisis-driven search for land by authorities, in order to undertake low-cost housing schemes. Historically, the main factors informing the choice of land have simply been ease of acquisition and a desire for racial separation. A third process is illegal squatting by people who either cannot find a place in legally designated housing areas or who seek locations closer to their places of work. The primary factor affecting the location of these settlements is avoidance of harassment.

The low-density nature of development is fuelled primarily by the application, in urban management, of entrenched anti-city values of suburbia which promote the single-storey house on a large plot as the image of 'good' urban living. This model of urban development is based on the assumption that every family will have access to one or more motor vehicles: concerns about the free flow of traffic outweigh all other management considerations. Very high space standards for community facilities form another important dimension of the model and contribute to the loose character of the urban fabric.

The second pattern is *fragmentation*. The grain of the cities is coarse, primarily since development occurs in relatively discrete pockets or cells, frequently bounded by freeways or buffers of open space. The primary reasons for this are: the relatively unquestioned management belief in the introverted 'neighbourhood unit' or 'urban village' concept in which housing areas focus inwards on community facilities imbedded at their geographical centres, in terms of more formal housing developments; the tendency to undertake new housing developments as large entities on discrete, consolidated sites; and the dominant locational need to avoid harassment in the case of squatter settlements.

An inevitable corollary of this cellular pattern is a simplified movement hierarchy. In essence, these pockets of development operate in relative isolation. They are linked primarily by freeways and other limited access forms of movement, which bring few benefits in an urban structural sense. They emphasize the importance of a limited number of points only and smaller businesses and public facilities are frequently excluded from these via the land market. At lower levels, the emphasis is almost entirely on local routes within the cells. Frequently, these are so specific in their form that they, too, emphasize a very limited number of points. There are very few intermediate routes between these levels.

The third pattern is *separation*: land uses, urban elements, races and income groups are all separated to the greatest degree possible. In particular, the separation of places of work and residence is deeply entrenched in the philosophy of urban management. The dominant urban land-use pattern resembles a series of relatively homogeneous 'blobs' of different uses, tied together by high-speed transport routes. Because of the historical legacy of apartheid, which displaced poorer people to the urban periphery, and because most new growth is occurring amongst the poorest people and there is limited filtration of families through the housing stock, increasing numbers of poor people are living on the urban edges, further and further away from urban opportunities.

While there is little overt planning, in the sense of consciously seeking to achieve positive social economic and environmental outcomes, at the level of the city as a whole, there is a great deal of control at more local scales. The cities are growing primarily through the rapid expansion of largely monofunctional housing areas. In the case of lower income areas the prevailing suburban philosophy translates into a housing-estate mentality: the emphasis is still upon the single-storey house in the single plot, but plots are far smaller. Structurally, these housing estates reflect the conventional planning wisdoms which were imported from Europe and the United States of America: introverted neighbourhood units; the superblock; arithmetically organized community facilities; and everything scaled to the car. Whereas previously the emphasis was on industrialized mass housing schemes, planned down to the last detail, the emphasis now is on mass site-and-service schemes, which are also fully planned and which incorporate only essential social facilities such as schools and clinics. In essence, these are almost precisely the same as the older mass housing schemes: there is simply less 'house'.

The consequences of these characteristics, in combination, on the lives of the majority – the urban poor – are savage (Dewar *et al.* 1990a, 1990b). The sprawling, fragmented urban system generates an enormous amount of movement, but fails to create the preconditions for viable, efficient and widely accessible public transportation systems to emerge. The costs of this movement to urban dwellers, in terms of time and money, are becoming increasingly intolerable: the structural system is aggravating significantly the major development issues of poverty, unemployment and inequality. It is economically inefficient, inflationary, and mitigates against economic growth. The large distances and low densities ensure that distributional costs form an inordinately high proportion of total costs in the cost structures of most businesses. Further, the system differentially affects very small businesses. The market concentration necessary to generate vibrant local economies does not exist and the limited number of points of high accessibility, in combination with the spatially extensive market catchments, ensure that only large economic units can really flourish: the physical structure promotes economic centralization and monopolization. It fails to generate high levels of social and commercial services: indeed the costs of providing the sprawling system even with adequate utility services are becoming prohibitive. It wastes society's scarce resources such as land, energy and finance. It is resulting in extensive environmental destruction and pollution. Finally, the potential of the housing process to generate economic development and achieve a wider circulation of income via inward industrialization is not being realized.

Experientially, these dormitory housing areas constitute desperately poor living environments. They are highly inconvenient, particularly for those huge numbers of people without motor cars. Because of their fragmented, introverted nature, individual housing areas, in effect, need to be self-sufficient in terms of social and economic infrastructure: there is little reinforcement or interaction between areas. In reality, because of low thresholds, service levels are extremely low. Indeed, for many, the average day is dominated by survival activities such as the search for fuel and water and desperate attempts to generate a meagre income.

Being poor in these areas is an expensive business. Transport costs, although heavily subsidized in the case of public transportation, bite deeply and commodity prices are generally higher than in the more wealthy areas.

Rather than being arcadian retreats, the townships have very little growing in them. The public spaces are inhospitable, dangerous, and frequently serve as dumping grounds for rubbish. Atmospheric pollution is severe. Large amounts of residual land, awaiting new or expanded facilities that in all probability will never materialize, destroy all sense of human scale and there is no tradition of making positive urban spaces.

Finally, because of the high degree of planning and control, the settlements are inevitably sterile, monotonous, and boring, despite attempts to provide variety through 'design techniques' such as convoluted road configurations.

It is clear that current urban management practices are inappropriate and are failing to guide the urbanization process positively. A new management philosophy, based on the realities of context, is required if significant and accelerating improvements to the living conditions of the urban majority are to be achieved over time.

An appropriate philosophy

A new more equitable direction requires that the urban problem be rethought from first principles. It begins with the question, why do people come to cities in the first place? In terms of the prevailing urban management philosophy, the central urban issue is housing: indeed 'urban development' and 'housing' are seen as being synonymous and the individual dwelling unit is viewed as the basic building-block of urban settlements. However, in reality people do not come to cities to find housing. They come in order to experience the economic, social, cultural and recreational opportunities and facilities which can potentially be generated through the physical agglomeration of large numbers of people. Urbanization demands, and indeed is predicated on, increasing levels of specialization and diversity and these qualities, in turn, can exist only in the presence of large thresholds of support. The more the city generates economic, social and commercial opportunities and facilities, therefore, the better it must be adjudged to perform. Of particular importance in South Africa, given the contextual realities of rapid growth and increasing poverty and unemployment, is the need to generate opportunities for small-scale, self-generated and often informal economic activity. Significantly, the ability of urban systems to generate opportunities of this kind is not related simply to their size: it is profoundly affected by their form and structure.

The second central issue is that of equitable and easy access. It is little use generating opportunities if they are accessible only to a limited number of people. The clear implication of the need for access is that as far as possible people should be able to carry out most necessary daily activities on foot.

Obviously, this mode of movement does not cover all situations. The essence of urbanity is that with increasing agglomeration, higher-order activities, opportunities, and facilities can be supported. If poorer people are to gain access to these, however, a viable and efficient public transportation system is a prerequisite. The reality is that a large proportion of future residents of South African cities will not be able to afford personalized motor transportation. The technology of movement must be accessible to all.

In short, the central non-negotiable management task must be to generate qualities of 'city' as opposed to those of suburbia. The central assumptions of the low-density, sprawling suburban model of development, which controls almost all urban development in South Africa – unrestricted mobility for all based on car ownership, and the belief that the entire range of a family's needs can be met within the individual house – are simply wrong. They are middle-class perceptions which fail to describe the real conditions faced by the majority of people. They fly in the face of the economic realities of the time and, as has been shown, they impose massive costs, the impact of which is most devastatingly felt by the poor, on urban dwellers.

Appropriate urban environments are dense, complex and richly mixed in terms of uses. When there is a pronounced overlap of compatible activities, different activities reinforce each other and convenience is maximized. Further, positive environments are integrated: different parts reinforce each other and individual parts and elements are not entirely dependent upon their own resources to satisfy the full range of the needs of their inhabitants. Promoting complexity, therefore, is a central management issue but complexity cannot be designed. It results from freedom of action. An appropriate management plan, therefore, does not seek to determine everything. It is partial: it provides a framework which creates opportunities to which individuals can respond, exercising ingenuity to further their own self-interests. It is a plan of process.

While the importance of generating qualities of urbanity has been emphasized, it is equally important to maintain the relationship between built and unbuilt environments. The environment of society is a continuum of urban, rural and primeval landscapes: each enriches the others and access to all is a basic entitlement of people.

The philosophy applied

Application of this philosophy to South African cities suggests a number of specific actions which need to be incorporated into urban management practices.

Establish and maintain the relationship between non-urban and urban land It is vital to determine not only where urban development should go but where it should not go. The dominant idea in this determination should be to create a constant, enduring relationship – a fixed edge – between urban, agricultural and primeval land.

Contact with nature and agricultural landscapes provides opportunities for necessary escape from the intensity of urban living and represents the most fundamental form of recreation and opportunity for urban dwellers. Exposure to the totality of environment allows people to be a part of the place in which they live: it is the basic platform of regional identity.

The importance of this contact has long been recognized in the tradition of managing urban environments: almost all plans and town-planning schemes have defined areas of 'open space' in the form of parks and so on. Important as these may be, to a large degree this kind of provision is simply an overt recognition of the loss of, and a poor remedial substitute for, genuine contact with primeval and productive rural land. Further, the cost of maintaining these artificial substitutes is increasingly beyond the means of many city administrations. Rather than being socially beneficial, therefore, these spaces frequently become unsightly, unpleasant and dangerous barriers to be circumvented.

While the significance of this contact between man and land supersedes the use of the land, there are important economic reasons for maintaining a close relationship between urban and agricultural land, particularly in contexts which are characterized by high levels of poverty and unemployment. Primary sector activity must be seen as an important, though frequently supplementary, source of both income and nutrition: it is an important urban land use. A stable relationship between a dense local urban market and agricultural land stimulates intensive small-scale agricultural activity. Conversely, a sprawling, ever-moving urban edge ensures that only very large producers can serve the urban population, via centralized distribution points. This type of activity should not only be promoted on the edges of urban concentrations but also on publicly owned residual land within the urban fabric.

Compact the city by imploding growth A precondition for the achievement of high-performance urban environments is to compact the form of the city. There are several reasons for this.

First, it is essential in order to maximize the generative capacity of urban systems. The more compact the local market, the greater the range and diversity of potential economic opportunities which present themselves to all inhabitants. This is particularly important in the case of small economic enterprises. Second, levels of social and commercial services are much higher, and convenience and equity of access to them is much greater, in more compact systems. The more compact the system, the greater the range of services that can be experienced easily on foot. Similarly, in more compact systems, it is far more possible to initiate viable forms of public transportation than it is in diffuse sprawling systems. Third, unit costs of social and other services to the consumer tend to be lower in more compact systems, since levels of support per facility are higher. Finally, the services themselves tend to be less vulnerable to demographic change in more compact systems, since they are less tied to the fortunes of any one group of people.

The implication of this is that, as far as is possible, new growth should be strategically and sensitively imploded within the boundaries of the existing cities, as opposed to the urban edge being constantly pushed further outward. New growth should be seen as a resource to improve the performance of existing areas. There is an enormous amount of residual land within South African cities to implement policies of this kind.

Promote a more integrated urban form and more complex levels of order A central factor affecting urban performance is the degree to which urban activities and land uses within the city are integrated. Integrated urban systems, as opposed to ones characterized by separation, are more highly generative, in that they create more opportunities to which people can respond; they are more convenient and equitable, in the sense that people have exposure to a wider range of facilities and activities; and they are more efficient, in the sense that they make better use of infrastructure.

Despite the wealth of precedent to support this, the South African city is characterized by separation, fragmentation, and a simplified hierarchy of movement routes and points of high accessibility. In the face of this, there is a strong case to be made for developing intermediate levels of routes tying local areas together; designing these as spatial systems, so that they can accommodate and enrich a variety of activities, as opposed simply to being unifunctional roads; reinforcing them where possible through the insertion of fairly high density housing; and allowing more intensive activities, which are dependent upon public support, to respond to these. Inevitably, over time, these activity routes would result in more integrated and diverse environments and a more decentralized pattern of commercial and small-scale industrial activity. Over time, therefore, these actions, in combination with a gradual compacting, would increase accessibility (as opposed to mobility) by reducing enforced movement. This in turn, will result in a far more energy-efficient city and in decreased atmospheric pollution.

Redefine essential infrastructure Currently, almost all public investment in urban infrastructure is channelled into utility services, houses and 'essential' social services such as schools and health facilities. Almost everything else is regarded as a luxury. The problem of gearing up South African cities to accommodate the urban poor in a facilitating way demands that other actions assume prominence on the urban agenda.

One set of actions relates to site improvement and resource creation: using nature, in a managed and multifunctional way, as a source of renewable resources which the poor can use in the struggle to meet basic needs.

There are several forms which this can take. One, already discussed, is small-scale agriculture. A second is large-scale planting programmes – including the creation of urban woodlots – to improve the environment of low-income areas, to provide protection from the wind, to create supplementary sources of energy and building materials, to establish evapo-transpiration traps for grey water generated by low-cost, water-efficient sewerage systems, and to establish places of escape and recreation. On-going programmes of this kind would not only bring large local benefit but would be sympathetic to a growing international campaign to combat atmospheric destruction. A third form is the creation of water bodies that can be used as urban structuring elements, as places of recreation, as supplementary sources of protein via fish farming, and so on.

A second, vitally important and currently neglected, element of essential urban infrastructure is urban public spaces. Positively made and celebrated public spaces (green spaces, squares and other urban spaces, and so on) form the primary social infrastructure of successful urban environments. They are the places through which people experience the city and engage in its collective life. While being important for all, the role of public spaces in the lives of the urban poor is critical. When people are poor, the full range of a family's needs cannot be met through the individual dwelling unit which represents the locus of one family's, by definition, limited resources. The public spaces, however, can represent the foci of an entire community's energies and resources. If properly made and celebrated, they enable poverty to be tolerated with some dignity, since poverty does not become a badge, identifying particular individuals who happen to be worse off than others. They are the places where most social experiences are played out and they act, operatively, as extensions to the private dwelling unit: they are the places where people meet, children play, lovers court, teenagers read and study when the house is overcrowded, and so on. The way in which they are made profoundly affects the performance and the enjoyment of these activities. When the public spaces are rich social spaces, the entire environment is positive, regardless of the quality of individual buildings. Conversely, when the public spaces are unformed and poorly scaled, the entire environment is sterile, again regardless of the quality of individual buildings.

A third element of essential urban infrastructure is the public provision of economic infrastructure which allows people access to opportunities to trade and manufacture at viable locations with minimal overheads. There are several actions (both physical and a-physical) which can stimulate informal and other forms of small-scale economic activity. Examples include the promotion of street markets and other forms of both permanent and periodic informal trading; the provision of kiosks and other forms of small, cheap manufacturing space; alternative forms of service provision for periodic users (such as metered electricity and water); the removal of unnecessary restrictions and regulations; facilitating access to credit; assisting collective transport arrangements; and so on. Actions of this kind can increase the 'manoeuvring space' of the poor significantly.

A fourth set of actions relates to improved information and communication. Access to information is vital in the lives of the urban poor, particularly those newly urbanized. In South African cities, large numbers of urbanizing (particularly African) people are virtually excluded from conventional sources of information and knowledge. They do not speak either of the 'official' languages and they know few people: indeed the search for information determines the pattern of urban growth. Where new migrants initially live is determined primarily by the distribution of people they know. For these people, in particular, the city is a strange and hostile place.

There are several ways in which information diffusion can be increased. One, already discussed, is the creation of positive public spaces and other places of gathering. Another is the promotion of collective television points – or mini-cinemas – from which publicly generated information (relating to available jobs, sources of aid, and so on), educational material, and popular entertainment programmes can be screened. A third way is the creation of reception centres: places, centrally located in terms of public transportation, where people can get legal, building and other forms of advice; where public meetings can be held; places which double as building caches supplying building materials at wholesale prices; wholesale suppliers of wood and so on.

Finally, it is necessary to move away from the current situation whereby individual elements of urban infrastructure are associated almost exclusively with individual activities. The multifunctional use of urban elements and urban spaces is environmentally and economically essential. Roads, for example, should be seen (and made to operate) not simply as specialist movement channels, but as open spaces with important social functions, as potential market areas, and so on. Similarly, dimensions of 'school', such as playing fields, libraries and halls, should be seen as resources for the after-hours use of entire communities; community halls can double as play centres, crèches, adult education centres, or day-care health facilities. It is in the re-examination of the space and time budgets of urban environments that the greatest potential for cost savings and improved environmental performance lies.

The stimulation of more complex processes of urban management The processes through which urban development (and particularly housing delivery) take place are currently massively oversimplified. There is an overwhelming, almost exclusive, emphasis placed upon single forms of building delivery in the production of environments. This is particularly true of housing development for low-income communities. For years, policy-makers argued that the most efficient and cost-effective form of delivery was through centralized, high-technology mass housing systems, and the entire housing system was geared to this. Almost overnight, in 1983, the situation changed and self-help housing was promoted to the exclusion of all other systems.

The debate is a sterile one. The issue has little to do directly with technologies of building construction. It centres far more on the need to stimulate more complex *processes* of urban development, so that the widest possible range of agents – individuals, community groups, small developers, larger developers, local authorities, utility companies, employer agencies and so on – can enter the delivery system.

There are two main reasons why this is important. The first is ensuring that capital invested in the construction process circulates as widely as possible and particularly that it reaches the poorer sectors of society. At present that capital finds its way back into a very limited number of (larger) pockets and its spin-off effects are very limited. The second is environmental. The quality of environments which result from processes controlled by a very limited number of agents (be they mass housing schemes or large, authority-controlled, self-help schemes) is sterile and monotonous. The diversity, complexity and spontaneity that results from the ingenuity of many people being applied cannot be simulated through design.

The keys to stimulating a more complex process and in particular, ensuring a role for smaller agents, are two-fold. The first lies in a more complex form of land release, with a range of different-sized land parcels being made available to different agents. This will inevitably result in the use of a wide range of construction processes and materials, since different agents are better able to handle different technologies and materials. The increased range of choices, in turn, increases the degree to which people can manoeuvre around their own needs and resource limitations.

A vitally important requirement of the allocation process is that control of allocation remains a public responsibility. Currently authorities are entrusting the allocation of large tracts of land to a limited number of 'community leaders'. This in turn is fuelling the growing power of 'warlords' and a violent struggle for land. In effect, the allocation system is promoting a low-key but rapidly escalating civil war.

The second key lies in the stimulation of forms of necessary institutional back-up which are genuinely accessible to poorer and smaller agents. One is streamlined and simplified administrative and legal procedures. Another is a grass-roots financial institutional capability, in the form of savings and credit unions or housing cooperatives. These should be capable not only of mobilizing local savings, but ought to be integrated into a national system of housing finance which can effectively receive and channel externally available finance. At present the poor are almost entirely excluded from the private housing finance system and even state funds earmarked for low income housing are failing to reach that target. A third is the rapid training of large networks of small builders skilled in operating in the low cost housing market and the use of state contracts to stimulate the small builder sector. The use of public works programmes to provide bulk infrastructure is another possibility which requires careful investigation.

Conclusion

Whether the urbanization process is developmentally positive or whether it creates conditions of appalling and perpetuating misery depends almost entirely on how it is managed. Indeed, the economic and political options which are actually possible in post-apartheid South Africa will depend in no small measure on how successfully the urban management process is carried out.

There are currently no cohesive ideas about the form of city South Africa should be moving towards. The urban problem is interpreted almost entirely as the provision of shelter: as a consequence, kilometres of housing areas are emerging but few qualities or advantages of 'city'. The emerging urban structure and form is exploitative. It does not facilitate life: it massively retards it. It does not help people to improve their material well-being: it makes them poor. It is not sympathetic to nature: it destroys.

The successful management of urbanization towards the emergence of environmentally enriching, life-sustaining settlements requires innovative actions on a scale commensurate with the scale of the problem. The nature of the urbanization process is substantially different, in form and in scale, from anything experienced in the past. It is therefore necessary, as a matter of urgency, to redefine the rules of the urban game.

References

Dewar, D., Watson, V., Bossios, A., and Dewar, N. 1990a. *An Overview of Development Problems in the Western Cape*. Urban Problems Research Unit Working Paper No. 40. Cape Town: University of Cape Town.

Dewar, D., Watson, V., Bossios, A., and Dewar, N. 1990b. *The Structure and Form of Metropolitan Cape Town: its Origins, Influences and Performance*. Urban Problems Research Unit Working Paper No. 42. Cape Town: University of Cape Town.

21 *Post-apartheid housing policy*

R 31
R 50
South Africa

PETER CORBETT

Introduction

Post-apartheid housing policy-makers in South Africa will inherit two major problems: serious housing deficiencies for the 7 million people in informal settlements, and spatial distortions resulting from the Group Areas Act. Effective solutions must consider the total housing environment including services, facilities, amenities, and local government which can deliver them. Equally, government decisions about the amount of public funding available for housing programmes must be compatible with targets appropriate to needs and other priorities.

A new constitutional and fiscal structure in South Africa will affect both central and local government and, consequently, housing policy. Already the urgent need to address socio-economic problems such as housing has been acknowledged by the government. Substantial funds outside the normal state budget have been allocated, with special attention being given to Natal where some of the worst socio-economic conditions exist accompanied by widespread political and criminal violence. It has also published models for non-racial, democratic local government for discussion. Finally the Group Areas Act has been rescinded.

Housing policy may, therefore, be starting to assume its post-apartheid form. This chapter pulls together the various aspects of what, in the broad sense, it ought to encompass. The focus is on Durban, where some of the worst housing conditions are, though the conclusions are applicable in any metropolitan area in South Africa.

Current housing policy in South Africa

Housing policy for low-income white, coloured and Indian families, is an ‘own affair’ under the tricameral parliamentary system. Housing policy for Black South Africans is administered as a ‘general affair’ or by the ‘governments’ of ‘self-governing’ or ‘independent’ homelands.

Tricameral policy has been to provide subsidized housing of relatively high formal standards in well serviced suburban townships. It has provided a high proportion of the housing stock for coloureds and Indians (in Durban about 50 per cent, of which about one-third are rented, the rest privately owned). Criticisms of this policy are: its initial impetus derived from Group Areas Act removals in the 1960s; outlay is too high for many despite subsidies; there is still a large waiting list. Many families double up or rent outbuildings. Nevertheless these housing problems are relatively minor.

Housing policy for Black families also, initially, created large townships of mass-produced houses on the urban periphery (for those allowed in under influx control). Servicing standards and housing quality are lower, and provision of amenities less than in other areas. Most dwellings are rented (with subsidies) but some have been sold cheaply to the occupants, ostensibly to encourage home ownership. From the late 1970s, this township construction programme virtually ceased while influx control began to break down prior to its formal removal after 1986.

Consequently the 1980s has seen the growth of massive informal settlements, with much debate on appropriate housing policy. The concept of assisted self-help has been accepted and a few schemes have been implemented, but these have made little impact on housing problems. In the Durban Functional Region (DFR) less than 5 per cent of nearly 2 million people in informal areas live in assisted self-help schemes although about half of the total DFR population live in such areas (Table 21.1), in houses many of which are structurally adequate but lack the most basic services and amenities (e.g. potable water, electricity, schools, roads, police stations, post offices). These settlements tend to cluster around formal townships for access to services and amenities. Another 10 per cent of people live in backyard shacks in formal townships where the occupancy rate of houses is also high. Invasion of vacant land nearer the Durban central business district is growing.

An estimate of current housing needs is that half of informal dwellings need replacement because of structural inadequacies or to make room for roads and amenities. Removing backyard shacks and involuntary doubling up by families would add to that demand (about 160,000 new houses). Existing policy involving self-help schemes and upgrading could meet these needs but the necessary political, fiscal, and administrative actions are lacking.

In the DFR formal housing programmes for Indian, coloured and Black families have nearly stopped. Major initiatives to upgrade informal settlements are still in the planning stage. The Republic KwaZulu Development Project, which has identified 65,000 sites for new assisted self-help schemes, is making little progress in acquiring land (see below). Only the Urban Foundation has completed a number of small demonstration assisted self-help schemes and is currently embarking on a small in situ upgrading programme.

Lack of funds is not the reason for lack of progress in housing in the DFR. The Republic KwaZulu Development Project has arranged loans of R1.2 billion from the Development Bank of Southern Africa with the assurance of state budgetary allocations to pay the ensuing interest charges. The state-funded Independent Investment Trust was recently set up with R2 billion initial capital of which the DFR could expect a major proportion. 'Project Natalia', newly created to address the socio-economic problems of Natal, can also expect access to government funds. In 1991 the Joint Services Board for the DFR will commence operations with an estimated annual R150 million available for infrastructural development in deprived areas. Wealthier local authorities in the DFR are also willing to play a minor role.

The urgent need to move from planning to acting is shown by the projections in Table 21.1. A further 250,000 dwellings will be needed in addition to the backlog. Without the necessary action they will virtually all be in unplanned informal settlements and land invasion will be endemic.

Policy issues affecting housing

Several policy issues directly affect the overall housing environment. The delivery process alone (see below) can only partially satisfy the need for adequate housing in the broad sense. This section addresses issues not strictly included in housing policy, but which need to be planned. Insufficient provision of schools, police stations, bus shelters and other amenities in many housing schemes illustrates the importance of emphasizing these issues.

Urbanization strategies Natal is becoming rapidly urbanized (currently about 55 per cent). The projected annual population growth rate for the DFR is between 4 and 5 per cent most of which will consist of low income Black families (half rural–urban migration) (Table 21.1). Only Aids could make these estimates inaccurate if current trends continue.

The 1990 economically active population of the DFR is about 1.2 million. Approximately half are employed in the recorded (formal) sector, 20 per cent informally and the remainder are unemployed. Economic and employment growth lag behind population growth with increasing unemployment and falling real incomes. Per capita Black incomes are about one-sixth of non-Black incomes and educational levels show similar disparities. Therefore, urbanization strategies for the DFR must provide for rapid population growth of mainly low-income families with limited ability to pay for housing, services and amenities. Unemployment and poverty will create instability and stagnation if policies are inadequate.

South Africa has as yet no coherent urbanization philosophy (see previous chapter). Future policy must incorporate the following if post-apartheid governments are to achieve appropriate urban development:

Table 21.1 Estimated housing stock and population in the DFR, 1990.

Group/type	*Dwellings*	*Persons*	*Projected,* 2000
Black			
formal townships[a]	100,000	800,000	
townships shacks	25,000	100,000	
informal areas	175,000	1,750,000	
hostels and servants		100,000	
		2,750,000	4,500,000
Indian			
townships	55,000	325,000	
private sector	50,000	250,000	
outbuildings		25,000	
		600,000	700,000
Coloured	12,000	70,000	87,000
White	130,000	350,000	381,000
Total	547,000	3,770,000	5,668,000

[a] Includes a relatively small number of private sector developments.

(a) acceptance of demographic trends as irreversible, and recognition of metropolitan growth as an opportunity for, not a threat to development; existing decentralization or deconcentration policies, therefore, to be discontinued;
(b) coordinated, participative planning processes to replace fragmented, top-down planning; incorporation of economic growth objectives in urban planning;
(c) removal of barriers to efficient use of land and labour;
(d) development and maintenance of democratic, legitimate, viable, and effective metropolitan and local government structures, within a framework of supportive central government policies towards urbanization and development;
(e) mass provision of affordable shelter, amenities and transportation to poorer families; an affordable welfare net to cater for the unemployed and the indigent;
(f) creation for all urban residents of a feeling of permanence and safety.

The smooth functioning of cities is widely recognized as the key to economic growth in developing countries. South Africa has lost generations of opportunity but can now gain from international experience in the last 25 years thus avoiding costly mistakes. Existing infrastructure and management skills can give South Africa a flying start in the post-apartheid era.

Land use The main obstacle to effective housing policy in the DFR is the inadequate supply of land. A related land use problem is separation of employment (mainly centralized) from low-income housing (mainly peripheral). The major causes are: the Group Areas Act; tribal land ownership; reluctance of formal agriculture (mainly the sugar industry) to release land for housing; and town-planning attitudes.

At a gross density of 20 dwellings per hectare there is need by the year 2000 for a further 20,000 hectares of land in the DFR. Possible sources are: land in tribal ownership; commercial agricultural land; other vacant urban land.

To reduce commuting costs requires high density, central housing but also business development on the periphery. Apartheid land-use patterns cannot quickly be removed, but planning for a more compact central city with high-density activity corridors can minimise the residual costs.

Removal of apartheid land-use restrictions An important restriction on land use was the Group Areas Act reserving for whites land around the central business district, for coloureds and Indians land around the white areas, and placing Blacks on the urban periphery. Few areas for Black occupation exist at more central locations. In the DFR Blacks must reside mainly in nearby KwaZulu. Under present policy new Black housing development must, therefore, use land under tribal ownership.

Removal of the Group Areas Act will firstly allow central location immediately for anyone who can afford it. Secondly central land can be developed or redeveloped to high densities as part of housing policy for lower income families. There would be a (reduced) demand for serviced land on the periphery but by choice. Increased density near the central business district would also reduce commuter welfare losses.

Conversion from tribal land to urban development Land in KwaZulu is controlled by local chiefs who allocate use rights. As the urban boundary encroaches they allow development according to individual attitudes, often meaning high densities, with uncoordinated planning, and no services, amenities or formal local government. Urban authorities need to appropriate tribal land before development. Existing legislation provides for land acquisition for urbanization via a process of negotiation which is very slow and seems not to have overcome the reluctance of local chiefs. The reluctance of commercial farmers to give up land also affects attitudes. Changes in these procedures is critical for effective urban management.

Acquisition of commercial agricultural land Two large sugar companies own large areas of land on the DFR periphery. They are reluctant to sell even at prices relating to urban use (also pointing to tribal land as another option). Expropriation with compensation at existing-use value (or the introduction of a betterment tax) would reduce the cost of land giving the benefit of rezoning to society in general.

Regional land-use planning in the DFR In the DFR residential densities are generally highest on the periphery except near to the CBD, but average incomes decrease from the centre outwards. However, employment is mainly centrally located. This land-use pattern results from apartheid laws which have prevented low-income families from locating by choice near jobs. Normal patterns cannot be achieved quickly but rational land-use planning will remove some welfare losses.

An interventionist approach is needed. Deregulated activity corridors can reduce accessibility problems from the periphery and promote higher density living. High-density, infill residential development of vacant central land will also promote a more compact city. Industrial land must be developed accessible to peripheral settlements. Redevelopment of central areas to higher density can be encouraged by changes to planning regulations. All of these policies will reduce commuting times and costs.

Subsidization Subsidization of housing is a political issue, and post-apartheid governments will probably follow previous governments in helping low-income families afford an acceptable minimum standard. The effectiveness of this policy requires a target standard achievable within the government budget allocation to housing given the number needing assistance. Past policies in South Africa have generously assisted some but have been inadequate thus leaving a massive housing shortage (a common failing in housing programmes in the developing world). A sensitive problem for future government also arises from financial commitments under the existing system where the level of subsidization is very high.

The cost of a formal structure on a fully serviced site is unaffordable by about half the population. To bridge this affordability gap would be difficult given competing priorities (e.g. education and health services). Therefore, the method of subsidization should optimize use of funds. Existing formulae oversubsidize some but leave other occupants unable to afford the outlay. Criteria for a rational subsidy system are:

(a) earnings-related, decreasing in value up to a ceiling income;
(b) income of the whole family as the relevant measure (with regular verification);
(c) maintenance of the real value of subsidies and outlays by linking to the Consumer Price Index;
(d) ownership (rather than renting) with subsidized repayments to avoid

excessive maintenance and administrative costs (enormous burdens in existing schemes);

(e) financing of housing programmes from operating expenditure instead of loans to avoid stop-go policies when interest rates fluctuate.

Urban management/local government The institution best suited to manage urban development is legitimate local government with adequate fiscal capacity, authority and skills. Coordination is essential between central and local governments and in metropolitan areas between fragmented local authorities. Two-tier metropolitan government may be the best way to deal with the latter problem.

Measured against these criteria local government in South Africa (particularly in metropolitan areas) is inadequate. Firstly, local government fragmentation is substantial with no effective coordinating bodies. (In the DFR there are about 30 formal local authorities.) The recent introduction of Regional Services Councils (Joint Services Boards in Natal – see Ch. 16) hardly improves this. Secondly, local authorities are racially based and lack legitimacy (particularly those for Black people). Thirdly, except for white local authorities, funds are inadequate (rateable business property is located mainly in white areas). Finally, most informal settlements have no local government. The result is highly uneven provision of infrastructure with per capita expenditure on services far higher in wealthier (white) areas than in poorer (Black) areas (informal settlements have virtually none).

The debate on future local government has started but no clarity has emerged. The 'liberation' organizations have not yet officially published more than the statement in the ANC Constitutional Guidelines: 'Provision shall be made for the delegation of the powers of the central authority to subordinate administrative units for purposes of more efficient administration and democratic participation.' (The Freedom Charter does not deal with this subject.) The present government has published 'Proposed Uniform Legislation for Local Government in South Africa'. It contains nine points of departure from the present system, proposals for a uniform enabling act for local government, plus four constitutional and five institutional options. Suggested functions are also listed. These proposals after discussion are to be tabled as part of a constitutional package for negotiation.

Without appropriate local government structures housing policy will continue to be incomplete and less effective than desirable.

Amenities and services Housing policy is more than just provision of an affordable site and structure. Adequate services are essential for minimum standards of health, safety and convenience (e.g. water, electricity, refuse removal, sewerage and waste-water disposal, public transport, police, postal services, health clinics, fire and ambulance services, etc.) Minimum social requirements include access to educational, recreational and cultural facilities. Comprehensive planning would go further by encouraging shopping, entertainment and employment opportunities.

Existing townships for Indians or coloureds meet these standards fairly well but in formal Black townships many of the above are missing or inadequate. Informal settlements, for the most part, have none of them.

At present many authorities are responsible for provision of different services and amenities. Whoever is responsible for housing developments should either provide them or coordinate provision. Clear allocation of responsibility is essential for sound housing policy if other authorities continue actual provision. In particular adequate and timeous budget allocations must be negotiated, since settlements without electricity, schools, or post offices will not resolve the urban crisis in South Africa.

The housing process

The actual process of assisting low-income families to obtain acceptable housing requires consideration of several key issues. Many developing countries have failed to improve housing for the masses because one or more of these issues has not been successfully addressed.

The institutional framework Post-apartheid South Africa will certainly have a non-racial, uniform housing policy, but the correct new institutional framework is an important component. Placing housing as a direct responsibility of a central government department is one option. Setting up at arms length a national housing corporation, or using housing utility companies are others. Again if housing policy is implemented by a government department then the issue of either decentralization or devolution to lower tiers of government needs consideration. Whichever the framework central government is ultimately responsible.

Adopting an affordable housing plan Many families in South Africa need help to obtain adequate housing. A critical policy decision is the 'bottom line' in housing standards. This must in turn be compatible with the amount the government is willing and able to budget for its housing programme. The annual cost of the programme will be the difference between affordability and the 'bottom line' times the number of families newly requiring assistance plus planned reduction of the backlog. Incompatibility between the cost of the chosen minimum standard and the budget allocation will increase queues for assistance and perpetuate inadequate housing conditions.

A similar calculation is necessary for other expenditures in housing programmes. Other authorities (police, education, etc.) must budget accordingly to avoid unbalanced or incomplete urban development. A rule of thumb estimate is 4 other expenditure to 1 housing. Given the size of total expenditure on housing programmes suitable budgetary systems to facilitate coordination are necessary.

Assisted self-help programmes The sheer size of the backlog and growth gives rise to a huge demand for land, services and amenities. Finance for this from government borrowing is limited because of the cumulative impact of interest charges. The capacity of the operating budget to directly finance housing programmes is likewise limited. Provision of ready built formal houses affordable by all will therefore probably not be affordable. The optimum strategy then is to initiate many, carefully managed assisted self-help schemes. To be successful these must provide: permanent security of tenure; an acceptable level of services; technical assistance; access to building materials close at hand; available finance at affordable rates for both site and structure; community participation; opportunity for upgrading in future.

Rented housing for the very poor Self-help housing implies ownership. This is an appropriate housing delivery process for most families requiring subsidization. What policy, however, is suitable for those not wanting or unable to afford even self-help? Not everyone wishes to or can own a home.

Public-sector rental accommodation has proved expensive and socially problematic. Structurally it costs more to build than self-help. Maintenance is the government's (landlord's) responsibility which has proved very costly, particularly as structures age. It also creates an attitude of reliance on government rather than self. Finally, concentrating the poorest, often in multistorey flats, has created depressed social environments in South African cities and elsewhere.

An alternative for these families is to stimulate the private rental market. Assistance can be given through rent vouchers or suitable equivalent. Small-scale private investors will respond by producing and maintaining rental accommodation more cheaply.

There can be encouragement in self-help programmes as well as in central city areas in both existing or new high-rise developments. Town planning regulations can also be examined to allow additional development on suitable sites.

Range of options Within the 50 per cent of families needing assistance are relatively large differences in affordability and need. Income-related subsidization means that some families will live at the minimum acceptable standard while others can achieve better standards. These differences should be reflected in a range of options allowing natural stratification. Matching income (supplemented) to preferences will achieve the maximum welfare benefit from housing expenditure.

For similar reasons housing policy should facilitate upgrading. Loans, plans, and technical advice are necessary beyond the initial housing of a family. Creation of a cheap, simple second-hand housing market should be encouraged to enlarge options as family circumstances change. Existing methods of property sale and transfer are too costly for this segment of the market.

Loan finance The usual ratio of income to purchase price of a house necessitates loan finance to achieve optimum affordable housing standards. If housing must be provided from income alone then for many years a family has to rent or else buy lower quality housing than it is willing and able to afford. Private sector financial institutions provide for the upper part of the market. Until now government loans have assisted part of the middle to lower range. The balance have had no access to loan finance.

In assisted self-help schemes, loans for a site and structure are essential as are loans for improvements, or to erect rental accommodation. Allocating funds for subsidies is insufficient if appropriate capital is not freely available to the informal sector borrower. Either the state housing authority can provide loans or the private sector can be encouraged by ensuring profitability through guarantees. Most vital is that suitable channels must be created to get loan finance down to the lowest levels.

Recently a loan guarantee fund for amounts down to R12,500 has been set up to give access to private-sector housing finance. In addition a pilot scheme to extend unsecured loans of R1,000 to R5,000 to groups has been started in the Western Cape. These are at market interest rates now but if successful they could be later linked to subsidy schemes.

Rent control Starting with the post-war housing shortage for whites rent control has existed in South Africa since 1949, although in recent years controlled properties have become fewer as it is phased out. However, excess demand for certain categories of dwelling is reappearing in South African city centres as the Group Areas Act is largely ignored there. Cheaper flats, in particular, are now in short supply, rentals are rising, and increased occupational densities are noticeable. Now that the Group Areas Act is abolished these trends may be accelerated creating pressure to again broaden the scope of rent control.

Housing policy should not ideally trade apparent short-term gains for a long-term decline in the supply of cheap rental accommodation. Landlords will sell for owner-occupation, or will allow buildings to deteriorate. (This is already happening in parts of the Durban Central Business District.) Allocation will often not be to the needy, key money and other subterfuges will be used to circumvent the law, and profiteering will still occur. Stimulation of supply in the long run while subsidizing rentals of the poor in the short run is a more effective answer to housing problems in the city.

Conclusions

Housing and related policies of the South African state have, since the advent of rapid urbanization, paid at best limited attention to those most in need of intervention. Black urban residents and migrants, since their permanence has been accepted, have still received little material assistance to try to improve the quality of housing and urban life. More than a generation of neglect has created urban conditions worse than in many far poorer countries. The instability, violence and lawlessness of the late 1980s originate as much from these conditions as from political grievances. Half of all Black urban residents live in informal settlements where simplest tasks of obtaining water and fuel are costly and time-consuming. Even in formal Black townships electricity and other basic services are often absent. Where there are facilities such as schools they are too few and quality is very low. Local government, where it still functions at all in black townships, is a farce and a target for rent boycotts and even for assassination of elected representatives.

Post-apartheid governments will, therefore, inherit a tremendous backlog in housing and related fields. Housing, education, health services, and urban infrastructure will require huge expenditures to correct this, which will take at least a decade even to eliminate the worst conditions. In the meantime raised expectations will make a continuance of instability a risk when not satisfied. However, politicians would be well advised to adopt policies which are sustainable and which will reach everyone even if standards, particularly in housing, are less than ideal. Compromise will be essential to walk the tightrope between economic stagnation and political collapse.

22 *Contradictions in the transition from urban apartheid: barriers to gentrification in Johannesburg*

JONATHAN STEINBERG,
PAUL VAN ZYL & PATRICK BOND

Introduction

From the top of Africa's tallest building, the 50-floor Carlton Centre in Johannesburg, the panoramic view is disquieting. The surrounding skyline began changing rapidly in the 1980s, in a dangerously uneven way. Now, while the western side of the central business district (CBD) is graced by ever more sophisticated, shiny, postmodern architectural gems housing mainly financial institutions, the eastern end remains clogged with dozens of overcrowded 1930s low-rise flats. On the rooftops of these buildings Carlton Centre observers can see the cardboard shanties of poor and working families.

Hidden by the variegated surface-level mosaic of Johannesburg land-use patterns (Fig. 22.1), extremely deep-rooted political and economic dynamics have spawned classic competitive struggles over urban space. The most contradictory of these are the seemingly irreconcilable encounters involving capitalists as individuals versus broader capitalist class interests. The stakes are high, because half of all office space in South Africa is located in the Johannesburg CBD. (Threatening now to alter the balance of these forces, there is also to be found, competing for space on the east end of town, the militant tenant rights organization Actstop.)

From where did this struggle for the CBD of Johannesburg emerge, and what will result, especially in the portentous, so far relatively-untapped east end? We argue that the crusade for inner-city redevelopment has occurred in the context of a broader *economic crisis*; that it has been led by the structural *rise of finance* in the economy; and that it comes at a momentous time when urban desegregation permits the space of the city to play a greater role in the broader *formation of class alliances*.

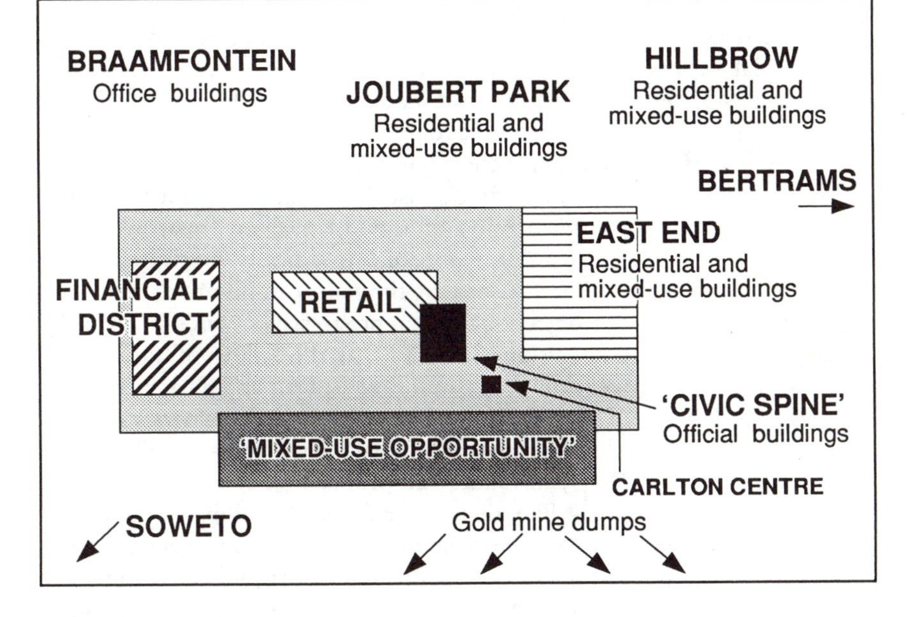

Figure 22.1 Simplified representation of land use in central Johannesburg.

Thus for the CBD's east end, the most coherent approach of the local state and capital is *gentrification*, but of a peculiar sort, aimed at an ascendant fraction of the black working class more than at traditional white yuppies. It is not, however, just the unusual target of this semi-gentrification project that creates the contradictions to which we refer. Instead the contradictions emanate from intercapitalist competition and the general tendency, deeply embedded in the South African economy, towards *overindebtedness* and *untrammelled financial speculation*.

It is in the struggle for this CBD space that the some of the most glaring contradictions of South African urbanization in the 1990s can be readily observed. This chapter attempts to construct a broader political-economic framework that can shed light on the *spatial fix* to economic crisis in other South African settings, and on the social and political contradictions of post-apartheid urbanization.

Overaccumulation crisis

But what do we mean by crisis? It is widely accepted that South Africa has experienced a structural slowdown in economic growth since around 1974, the exact causes of which are subject to debate. Business economists (e.g.

Dickman 1990) typically attribute the crisis to government policies ranging from apartheid to ineffectual monetary and fiscal policy, and seek remedies in free markets and a non-interventionist state. Progressive economists point to increased labour militancy and wage struggles beginning in the early 1970s (Nattrass 1990); the transmission of international crisis (Moll 1990); or some combination of the two which served to undermine a previously stable *racial Fordist* regime of capital accumulation (Gelb 1990b).

Meth (1990), however, is rightly opposed to arguments about the roots of crisis which 'lay a large portion of the "blame" on "apartheid" and much of the rest of it on the "unreasonable" (politically-motivated) wage claims of the workers' (p. 33). Many of the policy implications that stem from such analyses are objectionable to progressives, and in any case could not be expected to resolve the crisis, as the failure of anti-labour, anti-government intervention programmes of Reagan and Thatcher amply demonstrate (Clarke 1988). Yet, says Meth, even Gelb 'allows capital to slide too easily off the hook' of responsibility for the crisis (p. 32).

In contrast, Bond (1991) advances the proposition that South Africa's current economic crisis – like that, indeed, of the global system – is best described as one of *overaccumulation* of capital, a permanent tendency embedded in the normal operations of the system, at global, national and even regional levels (Harvey 1982; Clarke 1988). From this starting point, the rise of finance and its simultaneous search for investment outlets in the *built environment* follow logically. The conditions in Johannesburg are very much related to these broader processes.

In South Africa's case bottlenecks between investment for consumer goods and producer goods which emerged in the late 1960s were paramount to the particular form of the overaccumulation crisis (Clarke 1978). These were exacerbated by extensive multinational corporate direct investment in consumer goods industries; by aggressive import-substitution policies which gave enormous support to locally made consumer luxury goods; by enormously skewed distribution of income and wealth; and by South Africa's abundant mineral resource endowment, which promoted a *lotus-eating* effect – that is, it was easier to import producer goods than make them locally (Meth 1990).

The immediate results included a build-up of substantial retail and wholesale inventories in the late 1960s; a 'wild speculative boom' (Johannesburg Stock Exchange 1990) in the stock market from 1967 to 1969 (followed by collapse from mid 1969 to 1971); an intense but ruinous spurt of automation in private-sector manufacturing from 1970 to 1973; and the subsequent levelling off and decline of new private-sector manufacturing investment from 1973; all of which pointed to untenable structural bottlenecks, and presaged a long economic slump. The 1980s witnessed negative per capita growth; negative growth rates in productive investment; and an absolute decline in the value of capital stock, especially in manufacturing.

The economic crisis directly affected the development of South Africa's built environment, especially cities, in a manner reminiscent of the spatial fix that Harvey (1982) theorizes. For example, one early state reaction to the slump in private-sector investment was a huge increase in parastatal spending on transport, storage, communications, and the like. As the crisis set in, state construction of electricity grids and water lines also rose dramatically in comparison to the 1960s boom years.

These built environment investments mopped up some overaccumulated capital and helped overcome certain barriers to profitability that distance presented, but not enough to prepare the ground for renewed long-term private-sector investment and accumulation. And in any case the strategy was also built on an enormous foreign-debt build-up, which came home to roost with a vengeance from 1985. The debt facilitated intensive anti-apartheid financial sanctions pressure (a key reason for the political reforms described below); led to severe balance of payments constraints; forced the Reserve Bank to hike real interest rates; and caused a fiscal crisis from which the state attempted desperately and unsuccessfully to emerge through privatisation of some key parastatal corporations.

In the 1980s the built environment continued to play a role in the way the crisis unfolded. The state led the spatial fix into the bantustans (homelands), partly to support the pre-de Klerk political vision of apartheid. Direct state investment in township housing declined after the 1970s, but there was a huge increase in bantustan development grants. In addition, massive state subsidies were provided to the private sector, from the early 1980s, for decentralized production in bantustan border regions.

In spite of shifting some 15 per cent of manufacturing employment from urban areas to decentralization points during the mid 1980s, the strategy backfired when international anti-apartheid sanctions began to bite. Decentralized firms producing low-cost, low-technology, labour-intensive commodities had to turn from export to local markets in the late 1980s, thus squeezing metropolitan capital (Pickles 1991). Cities were by then already beginning to suffer from classic symptoms of deindustrialization, including vacant swaths of CBD manufacturing and warehousing space. The Urban Foundation, Development Bank of Southern Africa, South African Chamber of Business and Johannesburg City Council, among others, have since exerted pressure against Pretoria to cease subsidizing this temporary spatial resolution of overaccumulation, one beset, in any case, by rampant inefficiencies and corruption. (CBD warehousing districts also suffered due to the intensification of corporate *vertical integration* – hence less scope for merchant intermediaries – which accompanied South African monopoly capital's centralization and manufacturing diversification dating from the 1960s.)

This background is crucial to understanding the nature of the political-economic-spatial escape route favoured by the state and big business in the late 1980s, a route which will be followed through to a waystation in the Johannesburg CBD in the next sections. The crisis and its initial bantustan spatial fix helped hollow out the core productive space in the CBD, and the subsequent financial fix helped fill the spatial vacuum. More so than mainstream investment theory or French regulation theory favoured by COSATU's Economic Trends group (Gelb 1990b), *the theory of overaccumulation captures the frantic if uneven search for profits in different sectors and circuits, different regions, and over different time horizons in the space economy*.

To make that clear, it is the financial fix – the creation of unprecedented amounts of fictitious capital to forestall the devaluation of overaccumulated capital – that we turn to in more detail in view of its assertive role in restructuring the South African city during a period of crisis in the productive economy. Fictitious capital refers simply to the paper representations of value, whether securities, debt instruments, real estate titles, etc. The concept is important because it allows researchers to link certain forms of spatial development directly to powerful financial markets (Harvey 1982: chs 9–10).

Finance and the city

In South Africa, the decline in the productive economy from the mid-1970s and the inability of the state's spatial fix to fully mop up overaccumulation, as summarized above, were the precursors to the enthusiastic flow of capital into financial markets in the early 1980s. Not long after the 1981 collapse in the gold price, the Johannesburg Stock Exchange began an eight-year upward spiral that outpaced investment in fixed capital stock by a factor of eight. In 1989 alone, the market rose by more than 50 per cent in US dollar terms, the highest increase in the world, while real economic growth was limited to 2 per cent. During the first nine months of 1990, amidst a 20 per cent decline in average stock market values across the globe, the Johannesburg Stock Exchange industrial index ranked second best in the world, losing just 4 per cent.

'It is a damning indictment of the economy,' concedes Nedbank's chief economist, Edward Osborn,

> but we've also seen it in the States, in Japan, etc. This tremendous shift into simply pushing up values on the markets. So market values have risen extraordinarily, but the underpinning economic values have not been rising. So what we've really seen is financial inflation taking place at a horrendous rate. We haven't been immune from this international trend, in fact we've been one of the leading cases of this. (Interview, September 1990)

The financial rot has also extended deep into the credit system, not just the stock market. As a percentage of economic output, private-sector debt to banks doubled during the 1980s; personal and company foreclosures increased to record levels following business cycle downturn in the late 1980s; and the fallout included severe banking collapses and a merger wave that leaves South African financial institutions entering the 1990s with far too many paper assets relative to their underlying capital (Bond 1990a). Worst of all, the debt was not, by and large, pumped into new productive investments. 'A very large proportion of this has gone into bolstering the demand for financial assets,' Osborn acknowledges. 'So we've had the situation of a massive, exponential rise in financial flows, which are just swirling about, as it were, going into an inflation of values, financial assets, feeding a whole inflationary process in this country.'

In short, capitalists facing the overaccumulation crisis examined their balance sheets following the 1981 gold bust, withdrew funds that might have otherwise gone into production and, with the help of the big institutional investors and new borrowings, funnelled the money instead into non-productive outlets. In addition to the stock market, real estate also became a major speculative investment arena, in spite of enormous interest rate hikes in the late 1980s, with turnover up from around R12 billion per year in 1986 to R20 billion in 1989 (*South African Reserve Bank Quarterly*, December 1989).

'While there are anomalies resulting from South Africa's exclusion from world property, the similarities in development, investment and performance patterns are also remarkable,' notes the *Financial Mail* (27 April 1990). 'Cash-rich institutions in South Africa, freed from the constraints of prescribed assets and concerned about the short-term viability of equity, are queuing up to buy into hedges such as prime property investment.'

Consider the scope for *residential* investments, organized by financial and landed capital, in inner-city Johannesburg. The target is the black market, which, after all, accounts for 70 per cent of retail purchases in the CBD. White yuppies maintain stronger social connections to the northern suburbs, and don't appear to fit into the plans of developers at this juncture. Illustrating the potential market in the black urban professional class, a 1990 brochure for the Anglo American Corporation's Inner-City Property Development Fund promotes a R16 million investment in Hillbrow; a six-block site with two dozen residential buildings and 1,500 units. The site, now occupied largely illegally by working-class blacks, and

> undoubtedly one of the primary residential hubs of South Africa, provides the investor with excellent potential in terms of both rentals and sales. Yet it should also be said that by giving residents a desirable place in which to live – and therefore a stake in the area's future – it

> will help to stabilise and renew one of the CBD's critical buffer zones. This important project should be seen as a vital first step in the urban renewal process.

Similarly, the Urban Foundation has targetted nearby Joubert Park for residential redevelopment, and for manufacturing revitalization based on the small-scale furniture industry and potential linkages into the broader export-orientation that the state and capital now favour so strongly. Along with Hillbrow, the area has suffered severe devaluation, thanks in part to repossessions and redlining by financial institutions (Prinsloo 1988; *Weekly Mail*, 25 October 1990). Declines of up to a quarter in values of sectional title flats in repossession have been experienced by building societies. And the main drawback to the Urban Foundation's extensive redevelopment scheme, apparently, is that surveys conclude that current black residents are not interested in home ownership, even though their incomes could support bond repayments. Inculcating home ownership values thus remains a strong theme of the Urban Foundation (interviews with the Foundation's staff and city planners, August 1990).

Similar to township social restructuring which is now proceeding largely according to the dictates of housing finance (Bond, 1990a), the selected saturation of the inner city by capital is taking place within a broader context: the establishment of a post-apartheid social contract, of which top-level negotiations between Pretoria and the African National Congress are but a surface example.

A new political-economic dispensation?

As the 1990s begin, the broader avenue that the state and capital are paving to exit the economic crisis includes the following byways: export-oriented manufacturing production; capital-intensive production processes; greater use of foreign debt to import the means of production; continued liberalization of domestic financial markets; privatization of state assets, and deregulation of commerce; differentiation of black workers into a 30 per cent core of skilled workers and a 70 per cent periphery of unskilled, self-exploited, informal sector labour; for the core workers, vast increases in consumer indebtedness, supported through the collateral provided by newly-privatized housing; 'relatively unhindered labour mobility and a commitment to urbanization, but both based on individuals' capacity to purchase urban exchange-values (like houses) organized increasingly by established capitalists; for the poor, one-off site-and-service grants of R6,000 which will locate them in controlled peri-urban dumping grounds a great geographical distance from the formal economy; and, to ease the tensions inherent in the package, a social contract and unprecedented access to development finance intended to coopt leading sectors of the progressive movement, especially urban civic associations and the ANC.

While the seeds of this approach to solving dilemmas raised by over-accumulation have been sown for some years (Morris and Padayachee 1988), it is only recently that coherence has emerged from the diverse set of programmes within such powerful institutions as: the *Urban Foundation*, founded by Anglo American and other leading corporations to ameliorate social tensions and to carefully dismantle urban apartheid following the 1976 Soweto uprising; the R2.5 billion *Development Bank of Southern Africa*: and the *Independent Development Trust* – a R2 billion community development initiative announced by President de Klerk in February 1990, run by Judge Jan Steyn, long-time chair of the Urban Foundation.

These bodies have apparently assumed responsibility for guiding and grounding the overall political-economic strategy in the built environment, and for drawing into the package sufficient private capital and state resources, and the appropriate political actors. Swilling (1990) has labelled this aspect of the strategy – one constructed, he says, by the *econocrats* who replaced the P.W. Botha-era *securocrats* – as *deracialized urbanization*, although *deracialized apartheid* may be just as accurate to describe the likely spatial outcome of class differentiation, he suggests. 'Coordinated by the Urban Foundation,' the deracialized urbanization strategy 'is the first time all the major private sector organizations linked together to tackle the challenge of formulating new policy directions for the country,' the Urban Foundation itself testifies (*New African*, 10 September 1990).

Different corporate members of business organizations affiliated to the Urban Foundation have long tired of certain aspects of apartheid, including rigid labour markets, low cost and low quality consumption patterns, international isolation, and social and political instability. The constraints that apartheid urban land use impose on capital are damaging to capital's growth strategy for the 1990s, because, McCarthy and Smit (1987) argue, they even affect the capacity of manufacturers to expand their operations into international markets.

Transport bottlenecks that accompany apartheid geography are an especially important barrier to profitability, and this provides at least some of the logic behind the Urban Foundation's *compact city* philosophy. And in any event, the population density of inner-city Johannesburg is remarkably low in comparison to both First World and Third World cities around the world. Tomlinson (1987, 1990) argues that Johannesburg could comfortably accommodate a 150 per cent increase in population density, especially in such peripheral areas as the now sterile mining belt on the city's southern edge.

Aside from Hillbrow, Joubert Park and a few similar inner-city belt locations, the best indication of the success of the broader socio-spatial post-apartheid restructuring might be found in the east end of the Johannesburg CBD. And it is there that the contradictions in the whole process also become evident.

The east end of the Johannesburg central business district

Originally, the built environment in the east end of Johannesburg housed low-income whites as well as factory premises for small-scale manufacturing enterprises. In the early 1990s, the buildings are still predominantly owned by individuals and small firms, with financial and corporate capital only beginning to show noticeable interest at the margins of the district. (Nearby, in Bertrams, 78 per cent of property transactions involved investment companies, banks, building societies and property developers in 1989, up from 28 per cent in 1982 (Madisa *et al.* 1990: 49).) Yet in the 1980s, a number of processes radically altered the character and property values of the east end.

With the de facto erosion of the Group Areas Act, residential buildings became occupied entirely by illegal black residents. One result is a marked decline in the value of housing stock, as unscrupulous landlords milk the properties, leaving widespread squalor, despair and little legal recourse. However, due to the 330 per cent rise in South African CBD real estate values since the early 1980s (*Financial Mail*, 21 April 1989), land prices in the east end are disproportionately high relative to the value of the buildings.

It is this zone of land which has been earmarked by the Johannesburg City Council for an ambitious redevelopment project. Local government would embark upon a substantial infrastructural upgrading scheme to improve not only services, facilities and amenities, but also aesthetics. The Council then envisages a process whereby *large corporations purchase the buildings in this zone of the CBD, restore them, and then sell them to employees.*

The prime targets are 'thousands of middle and upper income earners . . . mostly employed in the high-rise banking complex taking shape in the west of the city,' according to one councillor (*The Star*, 22 April 1990). Even the upper strata of the working class might qualify for sectional title ownership of flats in slightly refurbished tenements. 'I'll be talking to the big boys (corporate capital),' says a leading city official. 'I'm sure they will jump at the opportunity. I can't even begin to describe what a problem it is for those chaps when their employees live in rented homes in Soweto.' (Interview with Planning Department official, August 1990).

The Council's scheme would, if successful, result in the movement of formally employed black workers into the inner city, but would also intensify the spatial marginalization of the seasonally employed and structurally unemployed. The employee home ownership feature establishes the conditions whereby the inner-city social structure would be regulated through a concrete relationship between monopoly capital and core workers. And the marginalization of excluded workers would occur in a *destatified* manner. 'When people get evicted no one can lump the blame on the Council's shoulders. Market forces will dictate where people can and cannot live,' says the city official, with some satisfaction. 'People complain that we are renovating theatres while people don't have roofs over their heads. But we cannot become a charity organization. It is not for us to mould this city's future. It is the role of the free market. We must leave the free market to its own devices.'

Thus the City Council's attempt to shape the east end of the CBD while remaining servile to the market would contribute, at a micro level, to a reorganization of the metropolitan region's class structure that corresponds to the larger political economic metamorphosis described in the previous section. But can the City Council's ambition be achieved?

At first blush, the eastern end of the CBD seems ideal for redevelopment based on classic gentrification processes (e.g. Smith 1979): disinvestment leading to the existence of a property (if not land) 'rent gap'; willingness by the local state to play a catalytic role; sufficient liquidity of property investment capital; and the social pressure and conditions in the spatially-organized labour market that ensure a client base for redevelopment. Yet what is striking about this area, at the beginning of the 1990s, is that the opportunity to carry out the Council project successfully may well already be lost.

Because of the high costs of underlying land, added to disproportionate new construction costs, investments in existing buildings are at optimum profitability in this area when directed to office, not residential, usage. The difference between building costs and yield for office space is now eight times higher than that for residential usage (interviews, August 1990). Furthermore, office blocks require less maintenance and do not carry with them the threat to the landlord of Group Areas Act violations and penalties. 'People think residential rentals have gone through the roof, but they are still not enough to attract developers,' says the financial director of Sable Holdings. 'The returns on residential property are way below those of office or industrial property . . . You have hundreds of tenants complaining about the plumbing and other stuff. Commercial property is a pleasure – you simply collect your cheque every month' (*Weekly Mail*, 18 October 1990).

Profitability calculations had become clear to the small property investors by mid 1990. At that time, in a 20-block area of the east end, no fewer than 33 buildings were, simultaneously, publicly advertised for sale, renovation, or offices to let. What was common to all the buildings was a unilateral move away from residential and industrial usage into office usage.

An increase in evictions of poor tenants is one predictable outcome of this process. In one typical case in mid 1990, rents were raised for run-down single rooms in Polly Lodge from R88 per month to R450 per month, leading Actstop members to construct tents outside for dislodged victims and to erect signs demanding, '*Nationalize the buildings!*' Adjacent to Polly Lodge stands Executive House, a one-way mirrored, upmarket office block boasting subdued hints of postmodern design. Home now to legal firms and government offices, Executive House was in 1985 an empty industrial building.

The trend emblematized by Polly Lodge and Executive House is clear. Lucrative small office development possibilities have fundamentally skewed the land use patterns in the east end of the Johannesburg CBD. Both the capital investment in (and structural transformation of) buildings in this area will make the return to marginally-profitable residential usage virtually impossible.

What does this mean in theoretical terms? During an overaccumulation crisis, the logic of capital accumulation – which in South Africa in 1990, is based increasingly on the political economic vision of high finance and its allies – confronts its own constraints simultaneously, as restructuring proceeds. The short-term investment strategies of individual capitalists, in the context of a CBD rife with land speculation, obstructs the realization of a new economic, social and spatial order that would serve broader capitalist class interests.

Conclusion

We have argued that flows of capital through the built environment are a crucial ingredient in the way the state and capitalist class have addressed South Africa's overaccumulation crisis. When combined with the rise of debt and financial speculation within the South African economy – which are themselves logical outcomes of overaccumulation – the impact on the form and functions of the Johannesburg CBD, as one example, are conspicuous. Add to this the conjunctures of racial-residential change and the search for a political settlement to South Africa's problems, and the CBD appears as a key node for the contradictions wrought by the crises of both apartheid and South African capitalism.

The contradictions boil down to time-honoured tensions and conflicts between individual capitalists and capitalist class interests. Because these do not appear anywhere near resolution at the urban scale, the attempt to redevelop the CBD through a limited gentrification project that simultaneously aims to support the broader restructuring of economy and society, is likely to be futile.

The broad arguments presented in this chapter are best considered *suggestions* – not irrevocable conclusions – about the dynamics now operating on South African cities. It is entirely possible that the dire warnings implied by the analysis will prove to be groundless, if, somehow, the South African economy does manage to recover from structural stagnation. For example, a chaotic and perhaps depression-ridden 1990s global economic environment could conceivably allow gold to again assert its traditional role as ultimate store of value, and, as in the 1930s, thereby help South Africa to play the role of '*prosperous undertaker during the plague.*' Based on our analysis of the depth of South Africa's economic crisis, this seems, sadly, to be the country's best hope.

Or as another example, the state and capital's optimal political solution – a social contract with leading segments of the progressive opposition, but enmeshed in a fully capitalist post-apartheid society – may well be achieved elsewhere, if not through a stillborn vision of gentrification in the Johannesburg CBD's east end. Joubert Park, the site of the Urban Foundation's ambitious redevelopment initiative, or Hillbrow, viewed with anticipation by Anglo American, are already undergoing a more thorough physical devaluation process that better sets the stage for the kind of financial and socio-economic restructuring which could, conceivably, overcome the contradictions faced in the east end of the CBD.

Only time and *praxis* will tell, as the inner-city poor, represented by Actstop, continue to show an extraordinary capacity to resist the mass evictions that will accompany either the gentrification or small corporate office strategy, not to mention the oppression they face in their current status (*The Star*, 26 August 1990). This sort of resistance should come as no surprise, for since the early 1970s, cities have been the sites of the main pressures for transformation of South African society. From early trade union organizing, to the 1976 Soweto uprising, to broad popular-front politics in the 1980s, South Africa's apartheid cities have hosted heroic resistance to both apartheid oppression and economic exploitation. And from rent strikes to bond boycotts to consumer boycotts to other forms of anti-apartheid and anti-capitalist organizing, this continues to be the case (Bond 1990b).

What is new and different about the recent rise of finance in the South African economy, though, is its totalizing capacity to both drive urban investment of a particular sort, and, because real-estate speculation has proceeded too far, to simultaneously block the urban political economic change necessary to stabilize a new social order. Hence the struggles for space in the CBD point to the acute need for further, deeper progressive analyses and political offensives against the powerful – yet potentially vulnerable – financial peaks at the South African economy's commanding heights.

References

Bond, P. 1990a. Township housing and South Africa's 'financial explosion': the theory and practice of financial capital in Alexandra. *Urban Forum* 1: 2. (forthcoming.)

Bond, P. 1990b. The struggle for the city is on. *Africa South*, 9 September/October: 11–12.

Bond, P. 1991. *Commanding Heights and Community Control: New Economics for a New South Africa*. Johannesburg: Ravan.

Clarke, S. 1978. Capital, fractions of capital and the state: 'neo-Marxist' analyses of the South African state. *Capital and Class* 5: 32–77.

Clarke, S. 1988. *Keynesianism, Monetarism and the Crisis of the State.* Aldershot: Edward Elgar.

Dickman, A. 1990. The economic debate. Speech to the South African Institute of Management, Johannesburg, 7 August.

Gelb, S. 1990a. Economic crisis and growth models for the future. Speech to the South African Institute of Race Relations, 22 May.

Gelb, S. (ed.) 1990b. *South Africa's Economic Crisis.* Cape Town: David Philip.

Harvey, D. 1982. *The Limits to Capital.* Oxford: Basil Blackwell.

Johannesburg Stock Exchange 1990. *South African Industrial Share Prices, 1920–89.* Johannesburg: Johannesburg Stock Exchange.

Madisa, E., Mason, P., Murphy, S. and Payne, B. 1990. Berea Street Blues: a study of tenure in relation to attitudes and experiences of living in Bertrams. Unpublished Human Geography research project. Johannesburg: University of the Witwatersrand.

McCarthy, J. and Smit, D. 1987. The internal structure of the South African city. Unpublished commissioned research paper. Johannesburg: The Urban Foundation.

Meth, C. 1990. Productivity and the economic crisis in South Africa. Research monograph. Durban: University of Natal Economic Research Unit.

Moll, T. 1990. From booster to brake? Apartheid and economic growth in comparative perspective. In N. Nattrass and E. Ardington (eds), *The Political Economy of South Africa*, pp. 73–87. Cape Town: Oxford University Press.

Morris, M. and Padayachee, V., 1988. State reform policy in South Africa. *Transformation* 7: 1–26.

Nattrass, N. 1990. Economic power and profits in post-war manufacturing. In N. Nattrass and E. Ardington (eds), *The Political Economy of South Africa*, pp. 107–28. Cape Town: Oxford University Press.

Pickles, J. 1991. Industrial restructuring, peripheral industrialization, and rural development in South Africa. Special Issue of *Antipode* 23 'Rural and Regional Restructuring in South Africa', pp. 68–91.

Prinsloo, D. 1988. Die Afsaking van die Groepsgebiedewet en die Invloed daarvan op die Woonomgewing: 'n Leningstrategie vir Finansiele Instellilngs. MBL dissertation. Pretoria: University of South Africa.

Smith, N. 1979. Gentrification and ideology in Society Hill. *Antipode* 11: 24–35.

Swilling, M. 1990. The money or the matchbox. *Work in Progress* 66 (May): 20–5.

Tomlinson, R. 1987. The year 2000: possible alternatives for the physical form and structure of the Johannesburg Metropolitan Region. Unpublished commissioned research paper. Johannesburg: The Urban Foundation.

Tomlinson, R. 1990. *Urbanization in Post-Apartheid South Africa.* London: Unwin Hyman.

Urban Foundation 1990. *The Urban Foundation Annual Review 1989.* Johannesburg: The Urban Foundation.

23 *'Turning grey': how Westville was won*

N A

KEYAN TOMASELLI &
RUTH TOMASELLI

Westville is a middle- to upper-class residential area bordering Durban (see Fig. 16.1). In 1987, the 6,000 households in Westville accommodated 20,000 whites. Practically all of the 4,000 Blacks legally living in the area were domestic workers, labourers and artisans. Only two black families occupied the main dwellings normally reserved for whites. The remainder lived in servants' quarters. The population of the adjacent, extremely wealthy, Indian Group Area, also part of Westville, was 1,800. These residents were represented by a Local Affairs Council, an all-Indian structure with very limited powers, linked to the Indian arm of the tricameral parliament, the House of Delegates. While property taxes from the Indian Group Area were paid to the Westville Council, only the white residents elected the Town Council, which was the chief executive body.

About 80 upper-middle-class Indian families had moved into the white area by June 1988. In the year that followed, another 120 Indian families moved in. By August 1989 the figure had risen to 374, and 400 by July 1990. This migration, in the face of legal, financial and racist impediments indicated severe land and housing shortages within Indian Group Areas and a desire to live in a stable middle- to upper-class community.

Our objective is to analyse the genesis and methods developed by a middle-class democratic movement which opposed racial discrimination and the Group Areas Act in Westville. We examine how this movement found a constituency, and discuss the class implications of its strategy.

Brief overview of the history of Westville Residents' Support Group

The Westville Residents' Support Group (WRSG) started as an informal response to the Mayor of Westville's May/June 1988 Newsletter, in which he referred to complaints about people of colour moving into Westville. Thirty white residents signed letters to the Mayor (dated June 16) suggesting that the Council take the lead in campaigning against the Group Areas Act, and that many residents would support it in this stand.

A manifesto drafted by the precursor to WRSG, the informal Westville Civic Action Group, received wide press coverage, particularly in the *Highway Mail* freesheet in early October 1988. The Group then changed its name to WRSG to clearly dissociate itself from the racist Durban-based Civic Action League.

Two inter-related circumstances provided the context in which WRSG coalesced. The first was the government's alienation of the Houses of Representatives and Delegates over its harsh amendments to the Act, which made provision for punitively high financial fines, and the confiscation of property in the event of a contravention of the Act, and the press publicity that the subsequent hung parliament engendered. This occurred during the run-up to the white municipal elections on 27 October. A pervasive perception amongst candidates and sitting town councillors was that the electorate was swinging to the right. Some candidates therefore took pro-Group Areas stances, while most hid behind suggestions of a referendum.

WRSG addressed a public meeting on 17 October 1988 at the Westville Civic Centre. The Group explained its stand and critiqued the racist myths that had grown up around the Act. One resident recalled aspects of the areas' multiracial history prior to the Act's implementation. Discussion from the floor was extremely heated. Supporters of the Act refused to accept arguments against, no matter how well researched. Of the 450 people present, however, less than 50 opposed a motion calling for the end of Group Areas in Westville.

At an inaugural meeting to formalize the WRSG on 8 November 1988, speakers drew attention to some of the results of the Group's intervention. The public meeting had altered the municipal agenda which previously took for granted that opponents of Group Areas were an easily ignored radical minority. The newly elected Council did not shift to the right, invalidating the fears of liberal candidates who questioned the timing of WRSG's intervention. More significantly, this meeting was the first in which both Indian and white residents debated a common problem at the same time at the same venue on the same agenda. Those Indians already living in the white area began to feel more secure and welcome.

Throughout 1989, WRSG organized public meetings on multiracial education, property values and the continuity of the home during times of change. Report-backs, legal advice, house meetings and social get-togethers were organized, and the monthly Council meetings monitored. In November, WRSG facilitated negotiations on merging the Westville Indian Ratepayers' Association with the Westville Ratepayers' Association. Both these bodies acted as pressure groups, and neither had anything more than advisory powers over the Town Council. The latter had an open constitution but had failed to involve Indian residents.

Initial opposition to Indians moving into a designated 'white area' came in the form of two petitions submitted to the Town Clerk and the (parliamentary) cabinet. The government ignored the petitions even though they were organized by the National Party official for the area. A right-wing counter group, the Westville Action Committee (WAC) emerged in May 1989, trading on racial hatred and white fear of 'non-whites' and the future. It tried to disrupt WRSG's campaign by bringing right-wing sympathizers from more conservative areas into Westville to disrupt public meetings called by WRSG. Other harassing incidents included the distribution in May 1988 of racist pamphlets calling for the eviction of 'people of colour' from Westville. A petition calling for the closure of a black soccer field leased from the borough was circulated soon after, while a census of houses in Westville occupied by Indian and black families was conducted. WAC's objective was to intimidate those 'illegally' living in these areas. At the same time, estate agents who were prepared to show or sell homes in the 'white area' to people 'of colour' were threatened with legal action. Some agencies complied with WAC's demands, which created a degree of anxiety amongst 'disqualified' residents. Further uncertainty was created by government prevarication on the Act. House-to-house pamphleting by WAC vilified the objectives and motivations of the WRSG, as well as anyone seen to be connected or supportive of the group, including sympathetic journalists and the liberal Democratic Party candidate in the general election of 6 September 1989. One of these pamphlets was considered by the Directorate of Publications as being injurious to harmonious race relations.

Though claiming 800 members, WAC was shown to be leaderless and short of finances and support. No committee was ever elected. The body used three tactics to try to evict Indian home-owners from the 'white' areas, all of which were acknowledged to be failures. Initially, they presented a delegation to the Attorney General to act against the 374 'disqualified' families living in 'white' Westville at that time. The Attorney General declined prosecution, although the government had turned down all 25 formal applications by Indian families for permission to live in a white Group Area. Next, WAC members proceeded to lay private criminal charges against 'offenders'. The police chose not to act on these charges. Finally, civil charges were considered. Costs for the complainant would be upwards of R30,000, but could not be considered in the absence of criminal convictions.

Although their strategies were largely unsuccessful, WAC complicated WRSG's campaign, particularly as it had the law (on Group Areas) on its side. However, its extreme racism and near-violent behaviour alienated moderate whites. It is significant that although WAC members supported the National Party, the party in turn abandoned them as too extremist.

The day before the general election on 6 September 1989, the *Natal Mercury* reported on an alleged pamphlet linking WRSG to the Mass Democratic

Movement (MDM). Government spokesmen and the parastatal South African Broadcasting Corporation had long demonized both the MDM and previous resistance organizations as 'ANC surrogates' and 'Communist dupes' fighting to make the country ungovernable. The bogus pamphlet's inflammatory language saluted 'new Indian residents to Westville . . . who have so courageously smashed the GROUP AREAS ACT by grabbing and moving into White houses . . .' It called on these COMRADES not to be intimidated by WAC (the pamphlet's probable producer) and demanded that white schools and the Town Council should be taken over and whites forced to leave the Borough. Replete with misspellings, half sentences and faulty grammar, the pamphlet ended with slogans distorted from the Democratic Party ('The tide is turning'), the Freedom Charter ('Forward to People's (sic) Power') and the MDM ('Forward to mass defiant action'). Both WRSG and MDM dissociated themselves from this disinformation.

WRSG called for WAC officials to make themselves known, and discredited it as an extremist outfit which resorted to clandestine methods and lies. Copies of WAC's racist lawyer's letters to the Council and estate agents were leaked to the press, as was a report on its secret meeting. WAC then threatened the *Highway Mail* which had taken the lead in exposing what one letter-writer called that WACky organization. Common to the pamphlets, anonymous telephone calls and threatening letters was a paranoia fuelled by an uncontrollable fear of Indians (and especially Blacks) as less than human, spuriously associated with violence, environmental degradation and filthy swamped living conditions. The images shown on South African television of Britain's Brixton, Toxteth and Birmingham riots were cited as 'proof' of what would happen in Westville. Only the third pamphlet distributed on 5 September 1989 showed a degree of restraint in its language, though continuing with fabricated interpretations of WRSG, and now, DP policies.

Winning support: general strategy

WAC's hysteria indicated WRSG's success. Where WRSG was initially perceived as the deviant minority, now the Group was able to project itself as responsible and constructive with both morality and history on its side. In the context of polarized South African politics, this image of reasonableness and middle ground was crucial to developing an alliance which crossed politics, culture and religion.

WRSG consisted of Jewish, gentile, Hindu and Moslem, and black people from a variety of European, African and Asian heritages. One of the experiences within the Group's committee was to overcome the religious and cultural divisions and suspicions created by the Act. Everyone had to resolve prejudices, preferences and attitudes which excluded or enforced sectarian cultural or religious practices. Initially, this proved difficult, but by uniting on the question of morality, members soon subordinated other agendas to the *single issue* of discrimination enforced by the Act.

Broadly, the aim of WRSG was to replace the fear that 'racial groups' had about living in proximity to and sharing facilities with each other, with the reassurance that comes from knowing people as neighbours, rather than as stereotypes.

WRSG strived to conduct itself as an open, knowable, consultative and democratic organization. A large accessible committee networked through religious and community organizations. Regular public meetings (attended by 200 to 450 people a time) injected new information into the debate on Group Areas. House meetings and intercultural activities went some way towards breaking racial misconceptions and stereotypes. Free legal services were offered to any 'disqualified' residents targeted by WAC.

Measuring support and perceptions

Initially, WRSG conducted its campaign by a 'seat of the pants' strategy, which was guided by a theoretical analysis of racism and residential land values and a projection of what was likely to happen when change occurred. Without a precise indication of the perceptions of the residents of Westville, or the numerical support base available for mobilization, WRSG relied on ad hoc attempts to identify the scale of support or opposition to the Act through a parish survey conducted by the Westville Catholic Church. This was cross-referenced with polls conducted on the Witwatersrand by Marketing and Media Research in October 1988 which found that almost half of white respondents in those areas wanted the Group Areas Act scrapped.

However, with the knowledge available to the Group, it was still not possible to specify the extent of its potential support, or the specific issues which alienated white residents, or at least left them disinterested in the possibility of integrated residential areas. By May 1989, it was clear that specific knowledge of the opinions of white Westvillians was necessary. Areas of concern were attitudes towards security, crime, schooling, class position, municipal by-laws and so on, correlated in terms of specific variables such as residential area, gender, age, income, language and political affiliation. Such data would help WRSG recognize white apprehensions and develop strategies to reassure this group about its future and security.

The Survey, undertaken during June 1989, was designed to establish how people felt about a variety of options. Structured interviews were administered, facilitating spontaneous open-ended responses. Interviews were conducted with 251 residing white home-owners, ensuring that age quotas and gender representations were properly spread. Ratepayers alone were chosen, on the assumption that they were more concerned about changes than people who were renting their accommodation. For the balance of the sample, four or five starting points were randomly selected on a grid map within each of Westville's nine wards. A maximum of five interviews were conducted from each starting point. One exception to an even geographical weighting was made with 98 interviewees chosen in two zones into which a disproportionately large number of Indian families had moved. These zones bordered on the 'Indian Group Areas' of Westville and Reservoir Hills. Fifty people contacted refused to be interviewed, 29 because they were not interested in taking part in a 'political' survey, and the other 21 because they were going out / did not have the time / were too busy. The refusal to answer questions on 'political' grounds may have biased the results slightly, a factor taken into account in the statistical reliability rating of 95 per cent.

The Survey results

Forty per cent of whites interviewed emphatically opposed the Act. Another 34 per cent indicated that they might be amenable to an open Westville if they could get more information about the positive consequences of getting rid of the Act. Only 28 per cent, mainly over 44 years of age, with a household income of less than R4,000 a month and who voted for the National Party, were adamantly opposed to the opening of Westville.

On the issue of opening Westville to other races, 73 per cent of the 'agrees' were between the ages of 30 and 34; 80 per cent were Democratic Party voters; and 73 per cent lived in three of the nine wards. Of those 'disagreeing', 49 per cent were over 45 years old; 54 per cent were Afrikaans-speaking; and 46 per cent lived in the ward which experienced a high crime rate.

Respondents who agreed with the opening of Westville were asked for their spontaneous, unprompted reasons. The most frequent included: 'If I can live in Westville others who can should be allowed to do so' (30 per cent); 'I don't believe in discrimination' (20 per cent); and 'It's a natural process and has to come' (15 per cent). Respondents disagreeing cited cultural, religious differences (24 per cent); the fear of over crowding, and of 'non-whites' taking over (24 per cent); that other 'groups' have their 'own' areas (19 per cent); and that they have different standards of living (19 per cent).

It was agreed by 71 per cent that children of other races should be allowed to attend local (white) schools. This figure dropped to 32 per cent for children of domestic workers. It seems that 39 per cent of the the total sample did not view domestic servants in the same light as they did middle- and upper-class residents of colour.

All respondents were questioned on whether they had sufficient information about the Group Areas Act, and its repercussions: 54 per cent considered themselves well informed, while 42 per cent felt they were not. Of the second group, the overwhelming majority (87 per cent) agreed that they wanted more information. In answer to a second follow-up question, 60 per cent of the 87 per cent felt that with more information, they could change their minds about the Act.

It was believed by 52 per cent of the sample that experience of integration in other countries would be useful in assessing what might happen in an open Westville. Of the 47 per cent positive responses, 28 per cent stated that Westville could learn from both the good and bad experiences of other countries; 15 per cent stated that integrated residential areas work worldwide and that few problems had occurred.

Of the 47 per cent negative responses to the question on the relevance of overseas experience, only 18 per cent said that mixed residential areas 'don't work' and 13 per cent said that whites would leave. The incidence of negative reasons was highest amongst respondents over 45 years of age (66 per cent), females (55 per cent) and National Party supporters (58 per cent).

In an open-ended question addressed to all respondents, spontaneous negative responses (86 of 251) against the opening of Westville were noted: people not wanting to live next door to 'non-whites'; whites would move out; whites would not get on with other races; there would be a decline in social or living standards; there would be an influx of Indians; Westville would degenerate into another Chatsworth (an 'Indian' satellite of Durban); property prices would drop.

Negative fears were offset by people agreeing to Westville being opened. In an open-ended question addressed to the 98 respondents (39 per cent of the total sample) in areas deemed to be close to 'Asian residents', the majority were *not* antagonistic to the removal Group Areas Act. Only 17 per cent of this group were unhappy about the situation. Of the remainder, most comments indicated that no problems existed between neighbours of different colours. In a series of overlapping responses, the most frequently cited remarks were: living close to Indian households did 'not worry them at all' (59 per cent); Indians 'never bother us' (21 per cent); 20 per cent commented on the 'quiet' nature of Indian households and that they 'keep their properties clean'; while *15 per cent did not even know that Indians lived near them.*

Other positive trends were identified, which included the perception of a harmonious society, in which all races would mix and understand one another, and in which the whole community would benefit. One result of this would be to show the world that South Africa was making efforts to integrate the country. Another theme was that only a few rich Asian families would move into Westville, rather than many poor Black families, and as a result, there would be no major changes in social standards.

Forty-one per cent initially reported that there should be no social qualifications before people of other races were allowed to live in Westville. *This percentage increased to 60 per cent for respondents living close to Asian homes.*

Amongst other respondents, income was the most frequently mentioned qualification (29 per cent), while some respondents mentioned size of the household, number of families and level of education.

Of all respondents, 39 per cent felt that the existing by-laws were inadequate to protect the existing character of an open Westville. In contrast, 46 per cent felt the by-laws to be adequate. Overall, 56 per cent of the sample were aware that the by-laws prohibited more than one family per household.

Strategic assessment

Though the Survey identified numerous internal contradictions held by respondents (for example, a minority of Democratic Party supporters wanting to retain the Act), certain clear trends were identified. Contradictions are important indicators of how common sense and prevailing myths overrule actual voting choices in everyday life but not at the polls.

WRSG had been partially successful in countering the myth that property values would drop, but the Group had not succeeded in allaying fears relating to cultural or religious differences, overcrowding, standards of living, or crime and security with regard to those opposing the abolition of the Act. This contradiction was not surprising as the government had replaced racist discourse with that of 'cultural difference'. This discursive shift underlay the government's new five-year plan which placed even more emphasis on the idea of 'groups' and 'non-group groups', the latter being permitted to live in Free Settlement Areas.

The parroting of state discourse by negative respondents – for example, 'each race has the right to preserve its own culture', 'friction between races' – encoded the assumption that the Group Areas Act was not itself discriminatory. In its pamphlets and at WRSG meetings people associated with WAC frequently mobilized the legal discourse of 'law and order' and used the official term 'disqualified' to describe Indians who now resided in white Westville. These interviewees did not see the inherent violence to families caused by this so-called 'right' and that it was a 'right' only claimed by fearful whites vulnerable to spurious information. In South Africa, legalism often obscures the real damage racism caused to society as a whole.

Class, race and prejudice: the geography of struggle

Timing, taking advantage of political and media moments, and unprecedented international pressure on the South African government created conditions which were seized by ordinary people working through organizations like WRSG. The Group's intervention was not decisive but worked towards creating the conditions for an earlier acceptance of a post-apartheid Westville. WRSG's campaign was broadly aimed at retaining the 40 per cent adamantly in favour of the abolition of the Act and winning over the floating 34 per cent.

Government disinformation about the need for Group Areas and 'group protection' inevitably rested on coercive and often brutal applications of the Act by the powerful over the unenfranchised. This viciousness was rationalized by whites supporting the Act by criminalizing ordinary people seeking to put a roof over their heads. Indian or Black movement into Westville was not a Communist plot to upset and wilfully destroy the character of the area as claimed by WAC.

After the mid 1980s, the UDF, and later the MDM evolved a seemingly paradoxical strategy of individuals mobilizing in their 'own' 'communities'. This strategy's apparent acceptance of Nationalist Party racial categories was philosophically problematic. However, it soon proved its worth in terms of broad-based trans-class mobilization. The realization was that whites who had the power to bring about change had been the one constituency ignored by the Movement. Previously, most whites in the MDM had concentrated on working outside their race and class positions, mainly with the black proletariat. As it was unlikely that conservative whites would be influenced by blacks, the MDM called on progressive whites to conscientize the areas in which they lived, and then link with the broader movement in whatever way they considered appropriate. In the case of WRSG, 'whites' initially mobilized whites; then a white–Indian alliance mobilized the white and Indian middle classes of Westville. Proletarian strategies, rhetoric and methods would only have served to alienate these more privileged classes. Instead, methods and language appropriate to each class or class alliance were developed in the context of the broader democratic movement. While only 2 per cent of whites polled in the Survey overtly supported the MDM, the majority were supportive of similar anti-apartheid goals if stated in less radical discourse.

The success of WRSG, however, was limited in class terms when assessed within the context of the struggle as a whole. Though the Survey identified a potential 62 per cent white support for the principles espoused by WRSG, the Group never came close to that figure in its actual membership. Middle-class white and Indian English-speakers do not emulate the mass-based mobilization found within the MDM. Both are generally more individualistic and less collectively minded than Blacks, and are uneasy with the idea of mass actions, pamphleteering and sustained interventions, as these were seen as the rabble-rousing tactics of the MDM. This was not understood by the hecklers and poster-carriers of the WAC and the Civic Action League who disrupted WRSG meetings. Ironically, the British immigrants who seemed to constitute the bulk of WAC imposed noisy tactics alien to a middle class, particularly English-speaking South African environment.

The white South African form of political response which leaves politicking to the politicians made it very difficult for WRSG to rally whites and even middle-class Indians in the way that the MDM had in its successful defiance campaigns against segregated beaches and hospitals during August 1989.

One result of this was a strategic choice for a public distance between the MDM and the WRSG. While some members of WRSG's committee participated in MDM campaigns and organizations with different agendas, the Group itself refrained from direct support of the MDM for fear of alienating WRSG's more conservative members. Nevertheless, the committee did organize meetings with representatives from other organizations at which the different strategies were discussed and explained. For example, WRSG encouraged whites to vote in the white municipal elections to ensure as progressive a council as possible, while the adjacent Indian Reservoir Hills Ratepayers' Association campaigned for a boycott against the Local Affairs Council elections, these bodies having little or no power.

The political ecology of South Africa is premised on both language and regional divisions. Nevertheless, it is criss-crossed with contradictions and anomalies. In order to make an impact, WRSG had to popularize amongst its members the idea that political action required more than passive voters exercising choices in periodic elections. A far greater commitment was needed to make a difference. WRSG had to borrow its theory of political action from the MDM but adapt the rhetoric to one more in keeping with liberal ideas.

Timing is another crucial element of political legitimacy in the strategic repertoire. When the Group started, it was necessary to concentrate only on the basic issues of residential integration. With the unbanning of previously banned organizations, and generally more open approach to the political agenda, it was possible for the WRSG to invite a senior ANC spokesman to address a public meeting in June 1990, an event impossible to contemplate just six months earlier.

By 1990, the majority of WRSG's members were Indian, which suggested its failure to sustain an increased overt white support under the conditions precipitated by the government's unanticipated unbanning of the ANC, UDF and other organizations in February 1990. While many English-speaking whites in Westville opposed the Act, their passive political culture precluded them from active participation, and many avoided WRSG public meetings.

In the course of the Survey, it became clear that most whites would tolerate Indians of a similar class moving into Westville, but explicit questions relating to a Black in-migration might have received very different answers. For example, the antagonism identified in the Survey of those not wanting an open Westville can be subsumed under two strands: the first of blatant racism (whites not wanting to live near other races, especially Blacks); and the second of a misunderstanding of the *class* of person who was likely to want, or afford, to move into Westville. This is particularly pertinent when it is recalled that Chesterville, a tightly-knit, extremely poor Black community, borders a section of Westville. The common-sense perception of conservative white (and Indian) residents of Westville saw Chesterville as a dangerous and volatile place, an ANC stronghold which needed to be contained by force.

The political violence which wracked Natal townships in 1985 turned Chesterville into a battle zone for the warring factions of the MDM and right-wing Black vigilantes. On-going internecine fighting between members of Inkatha and supporters of the UDF/ANC alliance created further instability in the area. The township also provided a conduit for criminals from further afield to move in and out of Westville, preying particularly on residential properties bordering the township, some of which were burgled repeatedly. As a result, the properties bordering Westville devalued greatly, and became a liability. Chesterville was seen by WRSG's opponents in the same light as was television reporting on the so-called race riots in Birmingham and Toxteth. The Westville Council's plan to establish a golf course as a buffer between the two towns was a cynical move notwithstanding its acknowledgment that Chesterville was not the criminal and terrorist cauldron alleged by WAC associates.

In the light of these circumstances, it was perhaps not surprising that the incorporation and integration of Chesterville into Westville, perhaps through the mechanism of a Regional Services Council (RSC), was never considered in early discussions about the future of Westville, though contacts were made with the UDF supporting Civic Association in that area. Yet, the only way to improve conditions in Chesterville would have been to incorporate it into Westville and shift resources into the area. Chesterville fell under the RSC that incorporated Durban, even though it borders Westville and is effectively isolated from Durban by a six lane highway.

What WRSG did was to spark moves towards a regional committee examining the opening of the entire Greater Durban region. It did this by initiating discussions with the Durban Central Residents' Association, the Chatsworth Housing Committee, the Urban Foundation and others. In this way, a cross-class, cross-racial and cross-cultural structure emerged which systematically placed pressure on *all* the municipalities in the region.

Important too, was the way that big capital threw in its lot with WRSG. The Survey was funded by Research International, and Shell SA was persuaded for the first time to place its human rights advertising in a mainly white-read liberal paper, other than the intellectual *Weekly Mail*. Shell funded extensive advertising in the *Highway Mail* on its anti-Group Areas position and on publicizing WRSG events. Whatever Shell's motives, what is significant is that its advertising now targeted *white* readers and that it took immediate action against one of its franchise holders in Westville who supported the closing of a Black soccer field.

WRSG's campaign during the study period should be seen as a first stage of resistance to apartheid and a means to exploit the government's vulnerability over its defensiveness about the Group Areas Act. WRSG tacitly acknowledged that on the issue of the Act it was useful to work alongside other groups, with which it disagreed in broader ideological terms. While WRSG was wary of the House of Delegates, the anti-Group Areas position taken by them (and the House of Representatives) provided the conditions to reassure relatively conservative whites and Indians. WRSG had to conduct itself in a way which did not catalyse a conservative backlash against either the Town Council, which included a number of members sympathetic to an integrated Westville, as well as a number who were not, or the Westville Ratepayers' Association, which also took an anti-Group Areas stance.

What of the future?

The Group Areas Act will be scrapped between the time of writing and the publication of this volume. The question then arises of whether a support group, such as the WRSG, will still have a meaningful role to play in assisting with the integration of all races into previous 'white' residential areas.

The answer is undoubtedly 'Yes'. Indeed, with the demise of the Group Areas Act, our true work will have begun. The real challenge which faces a middle-class area like Westville is the eventual integration of not only Indians and 'whites', but a breakdown of racial barriers across the board. The law is not the only hindrance to a fully integrated residential neighbourhood. While a start has been made among a small core of people, much work still needs to be done in bringing people of all races together on a social basis. The tragedy of apartheid is that it has worked so well: for most South Africans, of whatever 'colour', inter-racial mixing is still regarded with suspicion, mistrust and fear. It is this fear of the unknown which is at the heart of much prejudice and xenophobic feeling.

Acknowledgement

The Survey was conduced by Research International SA in conjunction with the Contemporary Cultural Studies Unit, University of Natal, Durban. Consultants were: John Butler-Adam, Director, Institute for Social and Economic Research, University of Durban-Westville, and Steve Piper,

Department of Surveying and Mapping, University of Natal. Thanks also to: Paul Mier, Louw-Hardt Stears and Westville Residents' Support Group for their advice, cooperation and help. See Tomaselli, K.G. and Tomaselli, R.E. 1990 Community in Transition, Westville and Group Areas. *Indicator South Africa* 7 (4): 59–62.

24 *Power, space and the city: historical reflections on apartheid and post-apartheid urban orders*

JENNIFER ROBINSON

Power, knowledge and spatial arrangements form a nexus of concerns in the works of Michel Foucault which have recently attracted the serious interest of geographers in general and which should receive similar consideration from those immersed in South African politics. As this chapter will demonstrate, such attention is appropriate both for writing a history of the apartheid city and for reflecting upon a possible post-apartheid urban order. In his genealogies Foucault sought to detail something of the power–knowledge connections which underpin modern society. His was a pessimistic view, one which saw systems of knowledge and various social practices intrinsically bound up with power and which interpreted supposed liberal reforms (for example, in prisons, or ultimately in the establishment of the welfare state) as elaborating upon and refining power techniques. As his commentators point out, for Foucault transformation and revolution would not necessarily bring freedom or liberation. Most likely, new forms of power would emerge, perhaps not disconnected with those of the past, which could very well survive a transformation focused upon the centre and not the extremities of power (Cocks 1989). As the new South Africa is being negotiated at the commanding heights of the power institutions of both the state and the opposition, it is appropriate therefore to consider the more routine power networks which have shaped and may in the future continue to shape South African cities.

It is unfortunately far too easy to draw some close parallels between Michel Foucault's (1977) carceral city of disciplinary institutions and the apartheid city. The scale at which order is imposed differs, though, and Rabinow's (1989) later study of urban planning and social order in modern metropolitan France and its colonies reflects a more relevant concern with the urban arena as a site of disciplinary and moral interventions. In South Africa, current events suggest that the construction of residential areas which resemble barracks, policing which makes these areas prison-like, and routine surveillance of township populations by army, police and state officials should soon be things of the past. In their place we look forward to cities made in the image of non-racial democracy. However, as other authors in this book suggest, many material and physical problems stand in the way of this goal. In addition to these, this chapter will suggest that the persistence of some of the more subtle aspects of power embodied in the urban form and in discourses concerning urban government will present obstacles to the creation of a democratic and equitable post-apartheid city.

In this chapter I outline something of the history of discourses and practices which have shaped South African urban areas. The initial focus is upon the formative period of apartheid, when the state was searching earnestly for a solution to a housing crisis of enormous proportions. Their efforts finally resulted in a coherent policy of racial segregation, closely linked to the creation of differential citizenship for each defined racial group. The contemporary housing crisis is posing similar challenges to the reforming apartheid state, now seeking to negotiate some form of settlement with the adversarial black population. A future government, presumably composed of some alliance of popular forces, will be faced with just the same housing crisis and, as a state, will search for appropriate political solutions around the question of urban form. In short, a new social contract, replacing the old differentiated notions of citizenship will have to be forged and, just as the old order shaped and was reinforced by a particular urban order, so we can expect that the new, emerging social order will have important urban consequences.

I argue, then, that the form of cities is closely connected with the nature of the political order and political power. In terms of electoral processes and government policies this has been well established in the literature. But in other ways, the links between urban space and power are less easy to uncover. Here, much more than simply the physical order invoked by Foucault's image of the carceral city must be considered – although this is in itself very important. Knowledge, professional discourses and the complex linkages which these forge between various aspects of spatiality and power also require our attention. This is substantiated below with respect to the formation of apartheid, after which we will reflect upon the significance of this theoretical claim for thinking about the post-apartheid city.

Discourses of urban administration in the formation of urban apartheid

Foucault's account encourages us to look for both continuities and discontinuities in the histories of discursive formations and practices. Elsewhere I have tried to demonstrate the persistence of a territorial mechanism for racial domination: the 'location strategy' in South African cities since the mid nineteenth century (Robinson 1990). The present period looks set to mark a key discontinuity in the deployment of this spatial technique. But at different times in the history of this strategy it has been incorporated and formulated within a variety of state and professional discourses which also require examination in terms of their continuity or otherwise. Amongst these we can include the imperatives of government, urban planning, public health and urban administration. All of these discursive networks have coalesced to promote urban segregation in the form of the location strategy for the better part of this century. All promoted concepts of physical order, social stability and visibility which were seen to be met by the enforced segregation of African people in formal, neatly laid out townships which were closely administered by a bureaucracy headed up by white officials.

Such was the model which the apartheid state promoted in its efforts to solve the housing crisis of the 1950s. This particular solution also mediated the political and social exclusion of African people upon which both national and urban political orders depended. But this political strategy employed a number of techniques of power, largely professionalised within local and central state apparatuses. Here we will consider some of the contributions which planners and administrators made to the apartheid urban solution.

Imagining geographies: order and the planning of townships

The creation of 'peaceful', 'orderly' urban environments for black people has constituted a major ambition of white authorities in South Africa throughout the twentieth century. Clearly the motivations underlying these ambitions varied in both time and space and of course depended upon who was conjuring up the vision. For example, in the early decades of the century a primary concern of many authorities was with the 'insanitary', 'overcrowded' and 'immoral' living conditions of many people in urban areas. Influenza epidemics (such as the one in 1918) and widespread tuberculosis encouraged a great deal of concern for the inadequately constructed and congested dwellings of the urban poor, including those of black people. Locally, Medical Officers of Health played a central role in encouraging authorities to engage in slum elimination and housing construction, and nationally, housing legislation and subsidies for municipal housing were introduced (Parnell 1987).

Slums have been universally portrayed by those in authority as dangerous, immoral and criminalizing places – although for those living there, they hold a much more subtle and lively meaning (Hart and Pirie 1984). But these descriptions of slums pervade the literature, as with this prominent Christian missionary and political observer who noted that

> Disreputable homes have a direct and traceable effect in creating disreputable people . . . slum yards are breweries, selling foul liquor. They are dens of immorality, filled with loose women. They

are the principle recreation grounds of the native people. (Phillips 1930: 111)

He claimed that here we can truly see the work of the Devil! And yet, the causes of concern to the authorities, such as the sale of liquor and the prevalence of prostitution, were central to the survival of women in urban areas. They provided sources of independence from men, whose absence and unreliability was not an incidental by-product of the migrant labour system and a sexually selective urbanization process.

A discourse concerned with intervention in the urban environment in pre-apartheid times was not confined to planners, then, but was bound up with a whole host of political, moral and social concerns. And however one finally answers the moral and historical question as to how closely planning as a profession was linked to the application of apartheid, the planning of South African cities both before and during apartheid was deeply infused with prevailing political and moral norms. In addition, planning as an independent discursive practice contributed to and enabled these norms, as did other such practices including what was then termed 'native administration'. In this section we will consider a number of schemes for African urban areas which were elaborated between the 1930s and 1950s – schemes which inhabited the intersection between planning and administration, incorporating the powers specific to these disciplines, but also quite obviously linked to wider social and political projects.

In one of the earliest of these schemes, F. Walton Jameson, an engineer and a member of the Central Housing Board as well as a Johannesburg City Councillor, took careful note of 'planning principles' in his formulations for 'native' housing. By virtue of his position on the Central Housing Board these ideas were of considerable influence in actual housing schemes. Together with other investigators in the 1930s and before, Walton Jameson found existing locations wanting in several respects. He singled out for attention the 'barrack-type' dwellings which up until that stage had been provided in municipal locations and he commented that: 'until we get out of the rut of thinking and building in terms of barrack type of locations, the problem of providing for the health and well-being of Natives will remain unsolved' (Walton Jameson 1937: 27). His praises were reserved for a number of 'model' schemes around the country – Orlando, Langa, Batho, Lamont – which he noted, 'in every way breathe an atmosphere of home life which is almost universally absent from other municipal schemes' (ibid: 28).

The promotion of a stable home life was one of the most persistent ambitions of planners of housing schemes – somehow 'loose women' and 'delinquent children' were to be eradicated from the landscape by the remaking of the environments in which they lived. Native Administration Departments attempted to foster this process by encouraging women in housewifely skills such as cooking, sewing, nutrition and health. Until the mass housing schemes of the 1950s the provision of a garden was seen as contributing to an 'atmosphere' of home life – in fact, Walton Jameson thought that an extra 25 feet of land on the usual 50 ft by 50 ft plots would transform the 'barrack-like dwelling' into a 'homestead' and emphasized the 'beneficient sociological influence on family life due to physical and mental occupation in the cultivation of the soil' (1937: 32).

The provision of family houses was also seen to have effects other than such 'sociological' benefits because 'neat little houses' as opposed to congested and overcrowded, inaccessible slum areas effected a division and a more controlled distribution of the population. Even when overcrowded such dwellings represented to Walton Jameson a 'lesser evil' than congested slum conditions (1937: 29). The relative absence of flats, or even of row houses in townships, together with the concerns which planners and administrators expressed for ensuring neatness, order and observability suggest that questions of 'control' are not inappropriate in this regard. Certainly in the planning of single-sex hostels, matters of control were explicitly considered. High surrounding walls were usually required and doors were restricted to the inward-facing walls. Only one entrance was allowed, in order that residents and visitors could more easily be supervised. The provision of cottage-style hostels in closed courts ensured that 'administrative control will be easier and order may more easily and quickly be restored should disturbances occur' (Reinecke 1959: 146). In an interesting corroboration of these intersections between order and physical planning, a contemporary manager of the single men's hostel in Port Elizabeth commented that it would be most useful if he could readily see all the rows of rooms easily at the same time, and then suggested that this would be most efficiently achieved if they were arranged as the spokes on a wheel, with his office at the centre – a peculiarly Benthamite proposition (Pick, interview, 1988). Clearly these issues of control have been important in the detailed construction of townships and specific buildings with the result that these physical arrangements have been a significant factor in determining the balance of power between those administering them and those living in these formal townships.

Planning's contribution to the urban order was not confined to this micro-scale organization, though. Practitioners also mediated and incorporated ideas and strategies which emanated from the state. An important case in point here is the organization of townships along ethnic grounds which reflected both the government's policy of Bantu education and their attempted resuscitation of tribal feelings as a strategy for urban social control. In Benoni an enthusiastic and efficient administrator of Native Affairs, Dr J. E. Matthewson, took up the 'rather interesting development' of ethnic grouping as part of the scheme for a new Location in which Dr Verwoerd as Minister of Native Affairs was apparently both interested and involved – 'Dr Verwoerd's personal interest in this scheme is no secret and he has given freely of his precious time and energy and has contributed tremendously towards its success' (Matthewson, in *Bantu*, May 1956: 39). In fact Verwoerd frequently referred to Benoni's township of Daveyton as a 'model', an example which other towns were exhorted to follow.

The grouping of 'tribes' around educational facilities not only permitted the coordination of government enforced mother-tongue instruction, but it was hoped that the reinforcement of tribal life would cultivate self-regulating 'discipline and order' and also allow tribally based headmen to operate within the administrative structures. Modern town planning principles, scientific surveys of housing needs (Matthewson, in *Bantu*, April 1956) and a comprehensive publicity campaign to combat opposition to the removals were the basic technologies of this township development.

Urban black locations, then, were not simply 'housing schemes' but places of manipulation, domination and control – and this was more generally the case than the above example of ethnic grouping suggests. The intersection of administrative imperatives and location planning was appreciated by both planners and administrators whose proposals for suitable schemes encompassed a number of alternatives. Amongst these, A.J. Cutten, a Johannesburg City Councillor, drew up a plan which reflected a dual concern for administrative and spatial order. He began by considering the South African housing problem within the same framework as nineteenth-century British housing, and also discussed plot size, street layout and service provision. But his Ideal Township employed the device of arranging the houses around a central point of interest (as was the case in the traditional African kraals, he noted) and this he suggested, 'provides an admirable basis not only for planning but also for guiding and controlling the lives of the individuals in the township' (Cutten 1951: 84).

The township was to be divided into precincts which were to be under the control of a headman who was responsible for law and order. These were then combined into blocks under 'blockmen' and further into superintendencies which were the responsibility of a white superintendent. As he notes, this administrative hierarchy was a usual practice in township administration but, he asks, 'how much more successful could it be in a township designed specifically with this end in view' (ibid.: 85). In addition each block would be centred around communal and social buildings which, in his words, 'radiating their influence around them, they become pivots on which the lives of the surrounding residents may be hinged, and by this means is reborn in the African the sense of social union that previously existed only in his native kraal' (ibid.: 87). This arrangement was one possible way of spatially interpreting the desire to reassert the values of the perceived stable and peaceful tribal life of the past as well as ensuring the implementation of an efficient administration which was responsible for overseeing and ordering the population. Other suggestions with similar ambitions were also put forward and implemented; such as a traditional 'kraal' scheme at

Vlakfontein, Pretoria. It was envisaged that this scheme would advance the restoration of the traditional system of Bantu rule and it was hoped that this would work to relieve the frustration which urban Bantu apparently felt for their lack of control over their existence, and also serve to promote stability and morality in Bantu life (Prinsloo 1950).

These and other contributions reveal an acute awareness of the value of the organization and delimitation of space for administration and social control. State bodies were similarly concerned with these matters – as the Benoni example illustrated – and where actual planning failed to produce spatially appropriate townships, the practice of administrators, their subdivisions and differentiations of residential areas, their 'local knowledge' and their regular movement into the environs in which the daily lives of inhabitants took place all represented a situated interaction between the human representative of state power and those gathered together for their more efficient incorporation into the purview of the state.

We can conclude two important points from this discussion. Firstly, planning discourse and the practical experiences of administrators were bound up with the political projects of the time, even as they made their own significant contribution to the elaboration of political power. Secondly, many of the concepts of planning itself, in terms of visual order, aesthetics and physical arrangements, fed into the particular nature of housing provision both during the earlier period of more 'liberal' politics and the later forging of apartheid urban policy. It is not the project of this chapter to extend the analysis of planning and administrative discourse into the present. Rather, the apparent significance of particular practices and discourses in shaping spatial organization will be considered within the context of the post-apartheid city.

Beyond apartheid? Political projects, discursive constructions and urban form

> When it comes to low-cost housing we work on urban planning principles, not group areas. (Dr Tertius Delport, Deputy Minister of Provincial Affairs, in *Sunday Tribune*, 16 September 1990)

To conclude I will consider some incipient discourses presently competing to define, shape and organize the South African urban arena. Perhaps all, one or none of these will prove persistent in the future power relations to be written into the apartheid city, but each bears consideration at the present moment and in the long run attention to language, texts and professional discourses will be at least one important requirement for making sense of the post-apartheid city.

Prior to the formalization of apartheid, much of the reorganization of

South African cities occured within the framework of slums legislation. Large-scale removals, frequently racially targeted, took place in all the major towns from the mid 1930s on the basis of the 1934 Slums Act (as case studies in this collection illustrate). Here, criteria regarding health, sanitation, fresh air, overcrowding, visibility and service provision were all meant to be employed in a technical fashion in an effort to improve and uplift the urban poor, and to alleviate a substantial housing crisis. In practice, slums were often identified on the basis of the 'race' of the inhabitants and slum clearances removed people to racially segregated public housing from what were frequently racially mixed neighbourhoods.

Slums legislation has remained in the statute books since this time with some procedural and technical amendments, but has played only a secondary role in shaping urban segregation since the advent of the Group Areas Act. In 1989, though, in the context of reorganizing urban policy in the wake of influx control legislation, slums legislation came to a brief, renewed prominence albeit with some significant proposed changes. Instead of the previous relatively rigid criteria for identifying a slum, the new legislation dwelt upon a much less clearly defined notion of a 'nuisance'. In tandem with Free Settlement Area legislation – which allowed a partial and limited relaxation of the Group Areas Act – the Slums Act was proposed as a means to reorder the disintegrating racial and land-use mosaic of the city.

Revisions of the bill produced a definition of a 'nuisance' as

> a condition which in the opinion of a local authority, within its area of jurisdiction, or . . . in the opinion of the Minister, within the area of jurisdiction of a local authority, constitutes a threat to the health or safety of the occupants of any premises, or the adjacent premises, or of any member of the public. (Slums Bill, B120–88(GA))

The technical criteria for slums embodied in earlier legislation were left unspecified. But in the Minister's view, any responsible local authority would continue to make decisions regarding what constitutes a 'nuisance' on the basis of these accepted technical definitions (Hansard 1988: Vol. 6, cols 16018–19). A local authority would 'use the expertise of its council, its town engineer, its health division and its town planner to take the correct decision in the end' (Hansard 1988: Vol. 6, col. 16005).

This would occur within a very different vision of the city from that which prevailed in either the 1930s or the 1950s, though. Then, in somewhat different ways during each of the two periods, a rigid, clinical urban order was envisaged in which visual disorder and areas which were administratively difficult to govern were to be replaced by regimented and thoroughly controlled townships. Now the government (and most actors in the urban housing arena) have accepted something of the visual and administrative disorder associated with informal settlements, but within this new approach the revised Slums Act was meant to provide the framework for a flexible method of permitting certain shack developments while simultaneously eradicating and controlling others. This would be possible through the discretion granted to different local authorities, although in terms of the Act such dicretion could be imposed or usurped by the Minister. This would allow, amongst other things, for aesthetically displeasing or politically troublesome settlements to be removed as well as those settlements occupying land designated for future, more profitable (for example commercial), or more politically desirable developments. Such an approach is certainly informing present state interventions in shack settlements, despite the withdrawal of the Slums Bill in the face of opposition.

Technical, professional practices and languages, then, will most likely persist with some role in shaping cities in South Africa for some time, although their position within a changing political and legislative context could vary. Nevertheless, for all states, stability, order and visibility ('surveillance') remain important ambitions and in these goals the planning and technical professions obviously play an important role. No doubt the political context within which stable, peaceful conditions are desired will be markedly different in a future South Africa, but the power relations embodied in their achievement in concrete urban form may yet be more familiar than some citizens of the post-apartheid state would wish. This would be at least partially a consequence of the strong relationship between disciplinary and governmental power and modernity and the strong prevalence of modernist projects amongst most players in the contemporary South African political scene. Certainly possibilities for more democratic planning procedures exist – and are evidenced by popular actions around land invasions and the creation of informal settlements. But technical constraints and the search for administrative order may prove to be significant obstacles along the path to the creation of cities of emancipation rather than domination.

In similar vein, a modernist fixation with economic growth has pervaded policy deliberations within the state, capital and the progressive forces. All seem to coalesce around the hope that 'revolution' might offer a solution to the current economic crisis and provide a stable path into the future, either to a free market, social democratic redistributive or some form of socialist society and economy. Such thinking has generated a language concerned with the efficiency of cities, their spatial form and functioning and their potential role in generating economic development. The Urban Foundation has given this most sustained consideration and has argued for 'compact' cities which are economically 'efficient' in many dimensions. A non-racial urban environment is seen to offer 'investment opportunities' which can be enhanced by urban government at a functional rather than administrative level. This would ensure the maximal use of available resources. Neighbourhood 'rights' and 'charters' are emphasized in the context of the visual appearance and quality of environments – an aesthetic concern central to planning discourse. In sum they advocate policy which would move towards the 'reintegration of South Africa's divided cities and towns where new development is channelled away from a dispersed and racially divided urban growth pattern, towards more compact, accessible, economically prudent and productive urban systems' (Urban Foundation 1990: 43).

If capitalism fails to write itself into the landscape through this modernist planning discourse, the accoutrements of postmodernism in the domain of consumption and architecture represent another modernist (i.e. economically progressive) language competing to structure the South African city. In addition, by including a new non-racial middle class in the spheres of both housing and consumption goods, the direction of change in the social structure of South African society and the fabric of its cities could well be managed by a new technocratic elite set upon recovering economic prowess in the face of international competition and internal political demands. In this scenario the fixity of categories such as 'Third World' sector (to describe black culture, the informal economy and non-formal living spaces – a particularly racist formulation – McCarthy 1990) will continue to be mirrored in the urban landscape, and will probably be increasingly accepted as South Africa discovers that, like most poor countries, the urban housing crisis cannot be solved within the disciplining constraints of contemporary capitalist economic growth and distribution. In this regard the planning principles which Dr Tertius (in the quotation at the beginning of this section) is quite happy to employ, very conveniently reinforce the effects of the Group Areas Act in keeping low-income developments out of the city centre on cheaper (or at least available) land. The physically peripheral underclass will not be abolished with these principles, even as the push for economic recovery will perpetuate this class division. The tangled net of the social and political order, urban practices, professional discourses and urban form may change shape and colour, but it will certainly remain of significance in explaining and describing South Africa's cities in the future.

Conclusion

And of liberation? Very helpfully for our purposes, Michael Walzer points to a major neglect in Foucault's work: a failure to consider politics in the form of constitutions, laws, forms of state and government. Most significantly, 'he has little to say about authoritarian or totalitarian politics, that is, about the forms of discipline that are most specific to his own lifetime' (Walzer 1986: 63). On this count we can surely look with satisfaction on any scale of human morality at the willingness of the present white minority government to negotiate itself out of power. However, and in the hopeful absence of serious obstacles to this course of action, this chapter has suggested that we could do well to be wary of the persistence of old forms of power and also to remain suspicious, or at least questioning, of new formulations for South African cities and society. For not only will the continuing dominance of global capitalism seriously threaten newly-won freedoms, but the durability of parts of the old order itself, especially that fixed in the built environment and embodied in professional knowledge and language, as well as the particular form which freedom itself might take, all suggest that the struggle for liberation is set to continue.

References

Cocks, J. 1989. *The Oppositional Imagination*. London and New York: Routledge.

Cutten, 1951. The planning of a native township. *Race Relations Journal* 18 (2): 74–95.

Foucault, M. 1977. *Discipline and Punish*. Harmondsworth: Penguin Books.

Hart, D. and Pirie, G. 1984. The sight and soul of Sophiatown. *Geographical Review* 74: 38–47.

McCarthy, C. 1990. Apartheid ideology and economic development policy. In N. Nattrass and E. Ardington (eds), *The Political Economy of South Africa*, pp. 43–54. Cape Town: Oxford University Press.

Parnell, S. 1987. Council housing provision for whites in Johannesburg: 1920–55. MA thesis, University of the Witwatersrand, Johannesburg.

Phillips, R. E. 1930. *The Bantu are Coming*. London: SCM Press.

Prinsloo, C. W. 1950. Bantoehuise vir die Bantoe. *Journal of Racial Affairs* 1 (3): 12–18.

Rabinow, P. 1989. *French Modern. Norms and Forms of the Social Environment*. Cambridge, Mass: MIT Press.

Reinecke, P. 1959. Contribution to discussion on Native Housing Policies and Problems. In Proceedings of the South African Institute of Town Planners Summer School, Johannesburg, 2–6 February 1959.

Robinson, J. 1990. 'A perfect system of control?' Territory and administration in early South African locations. *Society and Space* 8: 135–62.

Urban Foundation 1990. *Policy Overview: The Urban Challenge*. Johannesburg: The Urban Foundation.

Walton Jameson, F. 1937. The housing of Natives by public bodies. *Race Relations Journal* 4 (1): 27–34.

Walzer, M. 1986. The politics of Michel Foucault. In D. Couzens Hoy (ed.), *Foucault: A Critical Reader*, pp. 51–68. Oxford: Basil Blackwell.

25 *Lessons from the Harare, Zimbabwe, experience*

R. J. DAVIES

At the conclusion of the colonial era in Rhodesia in 1978 the city of Salisbury (now Harare) was the capital of a self-governing British colony that had unilaterally claimed independence in 1965. It was a modern city with a population of some 610,000 people of whom 78.6 per cent were African, 19.3 per cent white and 2.1 per cent Asian and coloured. Its structure embodied the economic, political, social and cultural asymmetry inherent in settler colonial society. The urban social formation was in large measure the outcome of processes which interrelated race and economic class in a colonial capitalist and mixed political economy. Productive forces were largely in white and foreign private ownership; Africans had experienced severe containment in their access to and control over the means of production and had been excluded from effective political influence and authority (Stoneman and Davies 1981).

The urban land system was one sphere in which asymmetrical relations of colonial society were strongly evident. Control over land ownership and occupation ensured control over production. It also provided a crucial device through which political and social control could be exercised over the subordinate indigenous population and which could subsequently be justified and reproduced. To 1979 land ownership and occupation in Salisbury had been strictly controlled by provisions of the Land Apportionment Acts of 1930 and 1941 and their amendments and later, from 1969, by the Land Tenure Act. Geographically embodying the structure of the colonial social formation, access to land was constrained by what may be called a *closed, orthogonal opportunity surface.* The city was segregated into two spatial components. One was for whites (and Asians and coloureds) for whom access to land was articulated by a freehold property market which reproduced the land use structure and socially differentiated residential patterns of the capitalist city in general. The other was for Africans, at that stage still dominantly subject to an allocation process in a controlled public land ownership system and rental housing. Whites occupied extensive, low-density and environmentally attractive residential areas in a broad sweep to the west, north, east and south-east of the central business district.

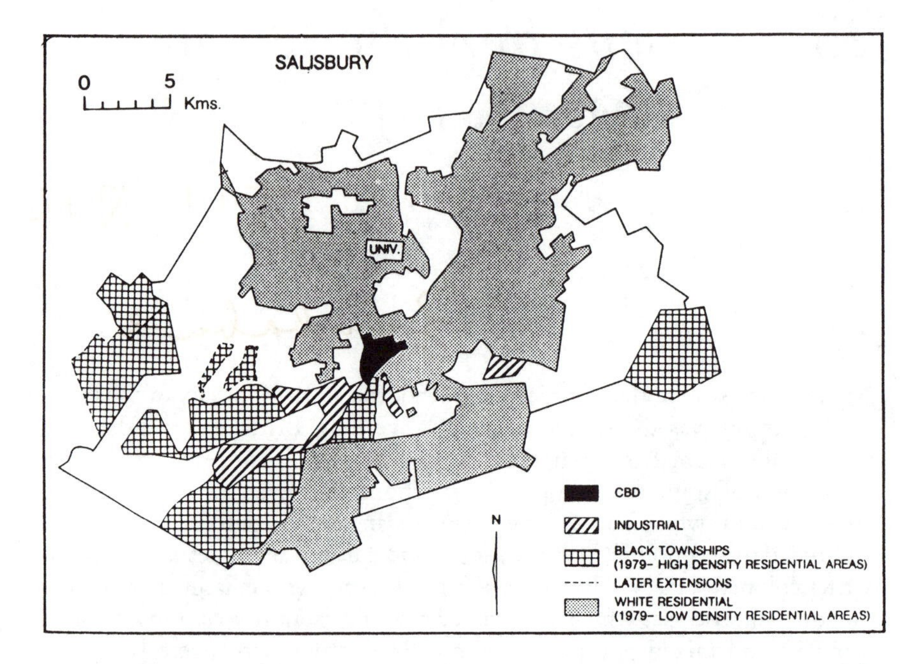

Figure 25.1 Structural outlines of Salisbury, 1978–9.

Non-residential land, not in the ownership of central or local government or their agencies, too, was controlled by white, Asian or coloured owners. Africans, other than those who occupied quarters on the properties of their employers in white residential and non-residential areas, were concentrated in high-density townships in the less desirable south-western sector of the city and on its north-western and western peripheries (Fig. 25.1). Though hemmed around by fewer legal constraints white–African space relations were as strongly structured as those of the apartheid city of neighbouring South Africa.

Ascendency of African political authority from 1978 first in the transitional Government of Zimbabwe–Rhodesia and from 1980, following full political independence under a government led by ZANU-PF (Zimbabwe African National Union-Patriotic Front), meant the immediate elimination of race as a primary reference in the social formation. The Land Tenure Act was repealed in 1979 and spatial constraint on the urban land system for ownership and occupation fell away. Overnight the closed opportunity surface of the city became an *open opportunity surface*. It is the African response to that phenomenon which is the subject of this chapter.

Research findings reported in the chapter concern the ethnic transformation of the former white residential areas of the city and lessons which might be drawn from the process for the future of the South African apartheid city.

Four issues are examined. They are black propensities to take up residence in former white residential areas, the body of potential black owners of residential property, the patterns of property acquisition over the first six years of political independence and the spatial dimensions of transformation. Attention is focused upon residential land ownership transfers using data drawn from the records of the Harare Deeds Registration Office.

The bases of transition

Though it had come to power with a goal to structurally transform Zimbabwean society on the basis of broad principles of Marxism–Leninism, the new government of Zimbabwe made no attempt to alter the essential structures of the existing urban land system. That, notwithstanding a deep-seated ideological commitment to an eventual form of public ownership and control of the means of production including land which a socialist order demands (ZANU Manifesto 1972; Nyangoni and Nyandoro 1978; Griffiths and Griffiths 1979; Smith *et al.* 1981; Zimbabwe 1982, 1984). Freehold tenure and the market process in the former white residential areas of Harare, now termed the Low Density Residential Areas (LDRA) remained fully operative. In the former segregated African townships, now renamed the High Density Residential Areas (HDRA), a drive to freehold tenure and commodification of existing rental housing, strongly initiated in the later colonial years, was accepted and extended. The processes of a capitalist urban property market were in effect to pervade the entire city save for a measure of control exercised over the HDRA to inhibit exploitive practices in those areas occupied by the urban poor.

Black propensities to take up ownership in the LDRA of Harare

Given the persistence of a capitalist mode of production operating over a now open opportunity residential surface, propensities for economically mobile blacks to enter the LDRA were high. Blacks had for long been exposed to the concept of individual tenure and were aware of its implications of security and market benefits. For those ready to exercise pent-up, latent or newly acquired economic mobility the former white residential areas provided an obvious and attractive alternative to the limited living environments of the townships. In contrast to the qualities of the HDRA the LDRA have historically projected positive images of desirability and diversity and were perceived as the traditional locus of the dominant, influential and economically successful. Failing the socialization of the residential system, to live in the LDRA became, for an emerging elite, a necessary precondition to the display of newly-won status. Political leaders quickly took up appropriate residential niches within the LDRA and coupled to the location selected by newly returning embassies, many from African countries, provided a powerful demonstration effect for others.

Potential black owners of residential property in the LDRA

The source of economic mobility which supports an African residential landowner class in the LDRA rested in the Africanization of formal employment. African employment in the public sector proceeded rapidly and by the mid 1980s only 6 per cent of public administration jobs in Zimbabwe as a whole were in white hands. Africans had penetrated all skill levels. In the private sector the process was slower, more particularly at higher skill levels from which potential entrants to the LDRA are drawn. Although the public service grew substantially from 1980, its absorptive capacity is not unconstrained and difficulties of entry to higher paid private-sector jobs since the mid 1980s is likely to have been more limited (Zimbabwe 1984).

Expanding African formal employment clearly generated a source of potential LDRA entrants, the scale of which is impossible to determine precisely. Employment and income surveys conducted in 1982 (Hoek-Smit 1983; World Bank 1983), however, suggested that the proportion of African households capable of affording the purchase of a market created home in Harare was less than 5 per cent of the total. In absolute terms that proportion amounted to some 8,000 households.

Data extracted in the present study show that Africans owned an estimated 7,033 dwellings in the LDRA of the city by 1985 representing some 4.5 per cent of the estimated total number of black households. In keeping with employment trends it is likely that the volume of future entrants will slow and be dependent upon further expansion of the public sector and growth and upward skill mobility in the private sector. Prospects for rapid movements in either sphere are not highly probably in the short to medium term.

The pattern of property acquisition by Africans over time in the LDRA

The rate of African residential property acquisitions in the LDRA was initially high. From 1981 to 1985, however, it slowed significantly though retaining an upward trend. By 1985 Africans owned approximately 23 per cent of dwellings in the LDRA of Harare.

The pattern is explained mainly by formal African employment trends in which the rapid Africanization of the public sector in particular initially generated a significant body of potential owners and a high demand for dwellings. In later years upward mobility of blacks has slowed in accord with growth trends in formal employment.

African demand for LDRA dwellings coincided with a high level of white emigration from Zimbabwe. Harare itself may have lost up to 25 per cent of its white population between 1979 and 1985. In these circumstances African demand for residential and economic property tended to be balanced by supply from emigrating whites. Though political uncertainties discouraged private investment in the immediate post-independence years, the balance, at least in part, explains the remarkably low rate at which new dwellings were added to the stock in the LDRA in the period.

The spatial dimensions of the transformation in the LDRA

As a starting point to the empirical analysis of the ethnic transformation of the former white residential areas it is instructive to examine aspects of the theory and experience of race–space relations in Western capitalist cities. This is not to suggest that this has direct applicability to the circumstances of the post-settler colonial city. Its utility lies in directing thought toward the formulation of hypotheses appropriate to those historically specific circumstances.

Several theoretical constructs of race–space relations in Western capitalist cities deserve attention. The first is the process of ethnic congregation or the tendency for distinctive racial or ethnic groups to cluster in particular parts of the city. The process, though not always strictly voluntary in operation, may be described as socially positive. It is one in which groups elect to reside together for cultural–social reasons and to attain a sense of social security in a strange environment. The specific locations of congregational clusters may be determined by historical or economic factors and they may persist over time. That structures of this kind are universal and not necessarily confined to Western cities is suggested by the work of Mehta (1969). Bearing in mind the strong racial and cultural differences between whites and Africans in a city like Harare, it might be thought that Africans filtering into former white residential areas may elect to form distinctive voluntary clusters within such areas.

The second construct concerns segregation and the ghetto formation process, ghetto growth and spatial expansion (Morrill 1968; Rose 1971). Certain superficial similarities do exist between the circumstances of black ghetto formation in the American city and the encapsulation of Africans in townships in Harare.

Although raw racism, prejudice and social practice are always factors in the segregation of ethnic groups, economic competitive race relations are frequently important and deep seated forces. Roof (1982) for example notes that (enforced) spatial concentration, by ensuring greater institutionalized separation and minority visibility, imposes barriers to effective economic competition and thereby aids in preserving dominant group privileges. That

Africans in Salisbury were not numerically a minority at the time is not at issue, their location in colonial social space is. In both the American black ghetto and the African townships of the colonial city the population was dominantly poor and economically immobile in a relative sense. A small, minority and economically mobile population in both cases experienced particularly severe frustrations of containment.

Forces containing the rights of black ghetto occupants from freedom of residential locational choice were outlawed in the USA through a Presidential executive order and by open occupancy legislation in the 1960s. In Zimbabwe the repeal of racially restrictive legislation in 1979 was designed to meet the same needs. In both cases the land ownership and occupation opportunity surface of the city was officially opened.

The form of the spatial response to these circumstances in the American city is a matter of record. It included a constrained, contagious diffusion process which led in some instances to a substantial outward expansion of the ghetto, but not to a general dispersal of blacks (Smith 1985). Observations of recent census data, furthermore, have shown that black areas are becoming more, rather than less, racially distinct, and that the general trend is towards polarization rather than dispersion of the non-white population (Clemence 1967; Smith 1985). Evidence suggests also that the distribution of welfare in the American city has become more unequal notwithstanding the economic advance of blacks in the past few decades (Smith 1985).

Expectations for the post-colonial city might be that the spatial diffusion of African owners into former white residential areas could take the same form. Examination of the operational and behavioural milieu of Africans in Harare, however, shows immediately that this is a false trail. Africans in Zimbabwe today enjoy dominant political authority directed at the ensurance of equal opportunity, at least, in those areas over which the state is able to exert direct control. Opportunity within the land system is included. A change of this type was anticipated in the shift to independence and there has been little overt evidence of white resistance to African acquisitions of residential property.

The manipulative role which the real estate and housing finance industry appears to play in the interests of whites in the American city, though undoubtedly present in Harare, particularly in the rental market (*Herald*, 14 July 1981) is unlikely to have been substantial. It would have been politically hazardous to attempt this and would, in any event, have operated against the economic interests of the industry and settlers in the circumstances of high white emigration.

The psychological and socio-economic barriers encountered by blacks in facing the hostility of the wider urban housing market in the American city have contributed to ghetto formation for the majority. In the Zimbabwean city, on the other hand, observations have shown that Africans have tended to adopt a positive, confident approach in exercising their new rights as prospective owners in the former white residential areas.

These and other factors suggest that the type of containment that has influenced black residential diffusion in the American (and for that matter, the British) city, is not likely as a spatial outcome of the process in Harare. Given new economic circumstances and the positive behavioural milieu of potential African owners, the open opportunity surface of the city has, to such individuals, meant just that. Expectations therefore are that the diffusion of Africans in the LDRA of Harare would have taken place on a city-wide basis. In contrast to the contained, contagious diffusion process that has characterized the expanding black ghetto of American and British cities, the process in Harare has been that of blanket relocation diffusion. These circumstances suggest, furthermore that the likelihood of African cluster and ethnic island formation in the LDRA is low. Variations in intensities of diffusion, however were likely to have been influenced by economic affordability.

It should be remembered that the source of containment that limits the scale of relocation diffusion is likely to remain for the majority of Africans in Harare. The economically immobile majority will remain contained within the HDRA of the city which will grow independently of the LDRA. The social implications of this structural division are significant.

Expectations for the transformation process are strongly borne out in the distribution of dwellings acquired by African owners in the LDRA between 1979 and 1985 (Fig. 25.2). Intensities of white to African dwelling transfers were highest in the LDRA peripheries and lowest towards the city centre. An overall index of concentration of 0.35, however, is modest and suggests that while spatial variations in the intensity of transfer was evident, the tendency for city-wide transformation to take place is the most important trend (Duncan and Duncan 1955; Davies 1964).

Transfer intensities in the peripheries varied and were highest in the south and north-west (Waterfalls, Hatfield, Queensdale, Cranborne, Mabelreign and Marlborough) where white to African transfers ranged from 54 per cent to 31 per cent. Moderately high transfer intensities, ranging from 16 to 23 per cent, occurred in the north, north-east and east peripheries (Mount Pleasant, Borrowdale and Greendale). The south and north-west peripheries showed proportional transfer intensities substantially above those that might have been expected had an even distribution of transfers taken place.

Inner peripheral zones, with moderately low transfer intensities ranging from 9 per cent to 16 per cent, occurred in the east and west of the LDRA in Highlands, Eastlea, Hillside and Braeside, on the one hand, and Belvedere and Ridgeview, on the other. Inner-city valuation areas had remarkably low transfer intensities ranging from 2.3 per cent to 7 per cent.

The equal opportunity surface of the LDRA created in 1979 has clearly been a conditional surface. In a city which remains dominated by a capitalist

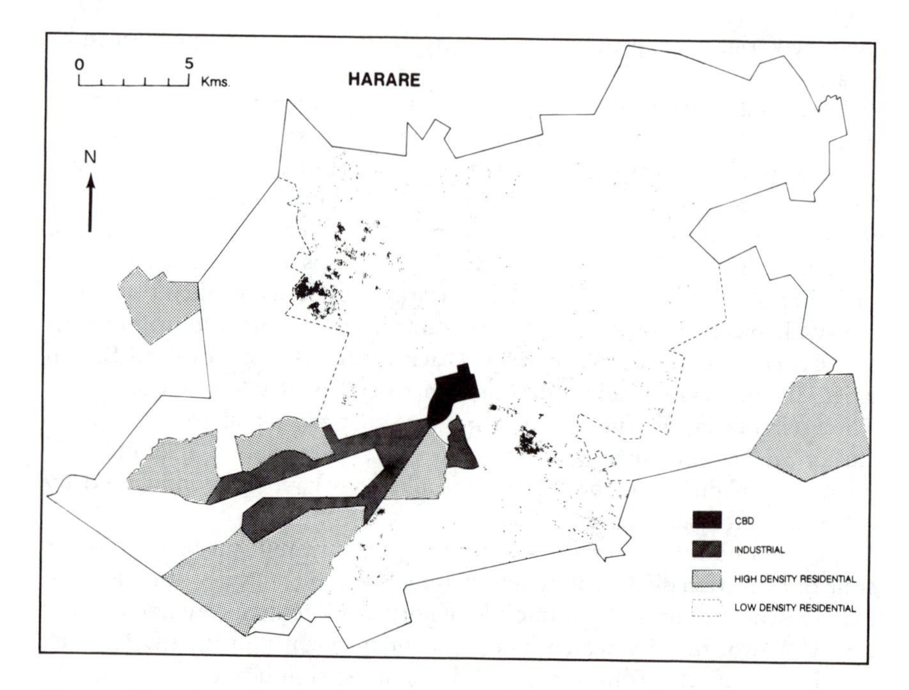

Figure 25.2 Transfer of dwellings from whites to blacks in Harare, 1979–85.

mode of production, economic forces of the urban land market that generate variations in land values and determine the spatial sorting of economic groups by ability to pay, are likely to be the major source of variation in African diffusion into the LDRA. A regression analysis of intensity of dwelling transfers to rateable values by valuation areas however, yields a correlation coefficient (*r*) of −0.2937 and a value of r^2 = 8.6 per cent. The correlation coefficient is low and other factors clearly enter into the explanation of spatial variation of dwelling transfers. Two factors suggest themselves.

The first is an ethnic related factor. Several statistical areas though low in dwelling land value experienced unexpectedly low levels of transfers to Africans (Arcadia, St Martins, Braeside and Ridgeview are examples). These areas contained a high proportion of coloured population. It is not suggested that Africans would avoid such areas for ethnic reasons, although that might be so. These areas were perceived as low-status areas in the colonial period. A more probable explanation is the residential stability of the resident population, a low propensity to emigrate and a consequent lack of vacancy opportunities. The removal of these areas from the regression analysis raises the correlation coefficient to −0.6568 and the value of r^2 to 42.94 per cent, a more satisfactory result. This result suggests that in general terms higher intensities of white to African transfers in the peripheries of the

LDRA and lower intensities in the inner city are explained by lower and higher property values respectively.

The second factor concerns the pattern of economic mobility attained by Africans in the post-independence period. Remaining constraints on African access to the occupation spectrum in the private sector tended to skew African economic mobility towards lower-echelon occupations. An outcome would be a concentrated demand for ownership in the lower valued areas of the LDRA. High intensities of white-African transfer in its south and north-west peripheral sectors may be explained by this factor.

Conclusion

Findings outlined in this chapter throw light on several important structural relations significant not only for the city of Harare but also for the state's goal of achieving a socialist transformation in Zimbabwe. They are significant also in providing empirical evidence that might inform understanding and social planning thought in anticipating the outcomes of structural change in similar circumstances such as in South Africa.

The retention of a mixed-capitalist mode of production has meant that an African, economically mobile, higher value property-owning class has been created. The class has been structurally incorporated and spatially segregated in what were formerly the purely white residential areas. Economic spatial asymmetries characteristic of the settler colonial city were structurally entrenched. In that sense it is clear that the state has not, in terms of its initial goals, achieved complete control over the forces of society.

Entry to the LDRA of the city was determined and limited by access to formal employment at appropriate income levels. It did not prove to be a simple linear process over time and it will in future be constrained by the rate at which higher levels of black economic mobility is attained. Growth prospects appear to be modest in the short to medium term.

An open residential opportunity surface coupled to a positive behavioural and operational milieu displayed by prospective African owners encouraged blanket relocation diffusion. The diffusing African property-owning class was economically sorted and spatially distributed in a pattern conforming to the established order of economic sub-classes of the earlier colonial city.

Variations in intensities of ownership transfers were strongly indicative of levels of economic mobility attained by Africans but other variables including ethnicity, demography and residential mobility of resident populations also possibly played a role. There was, however, no tendency for specific African ethnic clusters or ghetto-like formations to emerge.

The pattern of residential property ownership transfers in Harare produced a remarkable basin-shaped transformation in which the peripheries of the LDRA were most strongly affected. The pattern is possibly unique to the city and may not necessarily be reproduced in other cities of the same category including the South African city. It is nevertheless a characteristic which should be explored in a comparative context.

It should be recognized that the findings reported here only display mechanical elements of the transformation process. They do not penetrate the deeper question of whether spatial intermingling of white and African owners (and occupiers) also represents the development of effective functional and social interaction or integration. Effective neighbouring and friendship networks and, possibly, deeper relationships are examples. The most that can be said at present is that no significant evidence exists of inter-group conflict within the residential environment.

Results reported here concern the spatial transformation of only one facet of changing residential structure. Evidence from observation and telephone directories (however incomplete that record might be) suggest that owner-occupation is a dominant characteristic of African residential occupation in the LDRA of Harare. The rental market, which incorporates rental flats concentrated towards the city centre, adds a further transformation dimension. Such accommodation was rapidly taken up by Africans and creates an inner-city counterpoise to the pattern of dwelling ownership acquisitions.

The final issue is the crucial question of what the transformation process means for the state's aim of attaining a socialist transformation of society. If it is agreed that private ownership of property is a mechanism for fostering an individualistic and materialistic ideology based on commodity consumption (Marcuse, cited in Mabin 1983) then the transformation process in the LDRA of Harare is diametrically opposed to the socialist egalitarian view. In that sense the transformation process is not contributing structurally to the achievement of the socialist goals demanded by the Zimbabwean revolution.

References

Clemence, T. G. 1967. Residential segregation in the mid-sixties. *Demography* 4(2): 562–8.

Griffiths, J. and Griffiths, P. 1979. *Cuba : the Second Decade*. London: Writers' and Readers' Cooperative.

Hoek-Smit, M. C. 1982. Housing preferences and potential housing demand of low-income households in Harare, Zimbabwe. Unpublished report prepared for the Ministry of Housing, Government of Zimbabwe.

Horvath, R. J. 1972. A definition of colonialism. *Current Anthropology* 13:45–57.

Mabin, A. 1983. State houses for sale: re-examining home ownership. *Work in Progress* 26:4–7.

Mehta, S. K. 1969. Patterns of residence in Poona (India). In K.P. Schwirian (Ed.), *Comparative Urban Structure: Studies in the Ecology of Cities*, pp. 493–512. Lexington: Heath.

Morrill, R. L. 1965. The Negro ghetto: problems and alternatives. *Geographical Review* 55:339–61.
Nyangoni, C. and Nyandoro, G. 1978. *Zimbabwe Independence Movements: Selected Documents*. London: Rex Collings.
Roof, W. C. 1982. Residential segregation of Blacks and racial inequality in southern cities: towards a causal model. In H.P. Schwirian (ed.), *Comparative Urban Structure: Studies in the Ecology of Cities*, pp. 590–603 Lexington: Heath.
Rose, H. M. 1971. *The Black Ghetto: a Spatial Perspective*. New York, McGraw-Hill.
Smith, D., Simpson, C. and Davies, D. 1981. *Mugabe*. London: Sphere.
Smith, D. M. 1985. Inequality in Atlanta, Georgia, 1960–1980. Occasional Paper No.25, Department of Geography, Queen Mary College, University of London.
Stoneman, C. and Davies, R. 1981. The economy: an overview. In C. Stoneman (ed.), *Zimbabwe's Inheritance*. World Bank 1983. London: Macmillan.
Zimbabwe 1982. Ministry of Manpower, Planning and Development: National Manpower Survey, 1981. Harare: Government Printer.
Zimbabwe 1984. Annual Survey of Manpower. Harare: Government Printer.

Conclusion

DAVID M. SMITH

It would be tempting to conclude on an optimistic note, echoing the euphoria which accompanied the beginning of negotiations between the Nationalist government and the ANC in 1990 and subsequent indications of progress including the repeal of the Separate Amenities and Group Areas Acts. The days of apartheid, as generally understood, are clearly numbered. But what will really be the consequences for the South African city? It is proposed to round off this collection with a brief elaboration of what might, at first sight appear to be a rather rash if not unduly pessimistic proposition: *that the post-apartheid city is already here*, revealed by implication if not always directly in the contents of this book, if not in all its detail then at least in broad outlines. And this transition from the city of classical apartheid has arisen more from the internal dynamics and contradictions of the apartheid system itself than from any planned or negotiated process of reform.

Consider, first of all, the built form of the apartheid city. In certain respects it resembles that of the cities of the advanced (capitalist) industrial world, with a central business district surrounded by residential areas differentiated in quality according to the socio-economic status of their occupants. This is essentially the city reserved for the whites. Insofar as inner-city slums had been previously formed, by people other than whites, those who lived there had been relocated to the periphery during the era of massive township construction. Enclaves of better housing for the small Asian and coloured middle class came to relieve the otherwise uniform zones of basic dwellings built for those not classified as white. The urban landscape of group areas was thus not only segregated by race but also sharply differentiated according to housing and neighbourhood quality, availability of services and access to work.

This urban structure is still very much intact, and will inevitably survive the transition to a 'post-apartheid' society for some time. Unless there is a substantial redistribution of wealth, employment opportunities and the housing stock itself, the population distribution of the apartheid city cannot be expected to change very much, even with the repeal of the Group Areas Act. Few Africans will be able to move into affluent white areas, and even if there are more Asians and coloureds who can afford this, their incursion into white suburbs is likely to be carefully selective and, in places, resisted. Conservative cities and neighbourhoods will probably be able to perpetuate racial segregation informally, whatever the law may be. Strong local government autonomy may also assist white areas to preserve superior levels of services consistent with local revenue-generating capacity, even in the face of a central and perhaps regional government committed to more redistributive policies than in the past. The prospect of whites moving into Black townships, except as isolated individual demonstrations, is too ludicrous for serious consideration, and the same is probably true of most Asians and perhaps coloureds. It is also hard to see much movement of Africans into established coloured and Asian areas, except for those seeking better housing without challenging whites.

The changes most likely to take place in the classical apartheid city are already evident, and have been for some years. Racial integration is already taking place in the so-called 'grey areas', predominantly on the fringe of the central parts of Johannesburg, Cape Town and Durban. Access to jobs and services as well as the prevailing cosmopolitan atmosphere is likely to continue to make such locations more attractive to blacks than the more forbidding and higher priced white suburbs. The construction of up-market enclaves aimed specifically at Africans, coloureds or Asians is also likely to continue, though there could be more developments with a multiracial occupancy anticipated. Substantial changes in the townships where low incomes prevail are unlikely, except for a continuation of housing improvement at the hands of the occupants themselves and perhaps more of the upgrade and resale already taking place where the private sector sees profits in a form of gentrification.

It is the peri-urban shack accretions which represent the greatest change in the classical apartheid city. These have already transformed the major metropolitan regions as well as many smaller cities. And their continuing expansion is inevitable, given the inability of most Africans to afford anything else. More positive and larger-scale upgrade programmes might be anticipated on the part of a post-apartheid state and more progressive local government. And the spontaneous occupation of unused land in more central locations may be harder to resist. But otherwise, present trends towards more informal urbanization are unlikely to be greatly modified. And racial homogeneity will continue to characterize the shack cities, even if local upgrade generates some socio-economic diversity.

This vision of the post-apartheid city is, therefore, one of continuing segregation and separation, with little impact on the present highly unequal housing stock and associated local environment except probably to add much more to the bottom than the top end of the qualitative scale. Whites will have to tighten their belts, but few are likely to be forced down the housing market. The black political elite, bourgeoisie and administrative strata will grow, along with a skilled and well organized labour aristocracy, and they will again access to better residential areas. But for the mass of the Africans in the townships and shacklands, and for many coloureds and Asians, little if any improvement can be anticipated. This prospect is thus of class divisions steadily augmenting the racial separation inherited from the past, to produce a city characterized by some commentators as one of 'deracialized apartheid'.

What would it take substantially to change the present trajectory of change in the form of the South African city? The most dramatic shift of emphasis would be the introduction of a mass state housing programme for the poor, to reverse the tide of informal settlement. Advocacy of such a strategy invites reference to historical precedents in the township construction associated with a repressive state, and to the kind of urban development associated with what is held to be the failed socialist era in eastern Europe (Cuba is less frequently cited). There are also arguments concerning the cultural distaste of Africans for high-rise accommodation (happily occupied in places like Hillbrow, however), and to the positive attributes of individual initiative associated with self-help housing. But the most serious obstacle is cost. A substantial redistribution of national resources would be required to fund a mass formal housing programme, competing with other urgent needs such as improved education and health services for those hitherto discriminated against.

Another major departure from expectation would be substantial redistribution of the existing housing stock. As in eastern Europe and elsewhere with the advent of socialism, houses of the well-to-do (whites) could be expropriated, or even vacated voluntarily with incentives to move to smaller homes, and made available for multiple (black) family occupancy. The large gardens would make land available for extension and additional (shack) dwelling in convenient locations. More fancifully, whites might be moved to townships, or offered site-and-service plots on which to experience the benefits they often attribute to self-help. Redistribution is a practical means of matching resources with need, which can be implemented fairly quickly and justified by the blatant immorality of housing inequalities under apartheid. Yet merely to mention such proposals is sufficient to reveal their absurdity, as politically feasible solutions to housing the poor in anything short of a revolutionary socialist transformation of South African society.

This brings us to the crux of the matter – to the likely form of post-apartheid society and it institutions. This is, of course, the subject of intense current debate, with the growing realization that the negation of apartheid is only a beginning. While the Nationalist government, the white parliamentary opposition (insofar as it is still relevant) and business interests retain a staunch stand in favour of a 'free enterprise' (capitalist) economy and associated social relations, including the sanctity of private property, the ANC shows increasing signs of abandoning such socialist principles as nationalization (eg of banks and mining houses) to which it was wedded during the years of exile. At a practical level, it is hard to see the present government agreeing a new constitution without some protection for whites from what might otherwise be strong redistributive measures. And of course the internationalization of capital, and of the markets for the skills possessed by many whites, will constrain any post-apartheid government seeking to attract as well as retain investment and crucial personnel. Freedom of action is already qualified, if not severely compromised, as a succession of ex-colonial countries discovered on 'independence'.

The debate on future urban policy already risks closing around options which preclude major challenges to existing property ownership and to predominantly private sector solutions. Very simply, the powerful vested interests have the money to exercise a disproportionate influence, for example through the Urban Foundation and the Private Sector Council on Urbanization, the very name of which signals its partiality. By the time the ANC reveals its own urbanization policy, the room to manoeuvre may already have been narrowed. Yet if the cities of the new South Africa are truly to serve all the people, in an open, efficient and equitable way, all options must be given full and rigorous scrutiny, not merely those consistent with the ideology of those whose fortunes were built under apartheid. Otherwise, the post-apartheid city may really not be much different from that of today.

Index

AA0-6165